METHODS IN MOLECULAR BIOLOGY™

Series Editor
John M. Walker
School of Life Sciences
University of Hertfordshire
Hatfield, Hertfordshire, AL10 9AB, UK

For further volumes:
http://www.springer.com/series/7651

Intrinsically Disordered Protein Analysis

Volume 2, Methods and Experimental Tools

Edited by

Vladimir N. Uversky

*Department of Molecular Medicine, College of Medicine, University of South Florida,
Tampa, FL, USA
Institute for Biological Instrumentation, Russian Academy of Sciences,
Pushchino, Moscow Region, Russia*

A. Keith Dunker

*Department of Biochemistry and Molecular Biology, Center for Computational Biology
and Bioinformatics, Indiana University, Indianapolis, IN, USA*

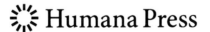 Humana Press

Editors
Vladimir N. Uversky, Ph.D.
Department of Molecular Medicine
College of Medicine
University of South Florida
Tampa, FL, USA

Institute for Biological Instrumentation
Russian Academy of Sciences
Pushchino, Moscow Region, Russia

A. Keith Dunker, Ph.D.
Department of Biochemistry
and Molecular Biology
Center for Computational Biology
and Bioinformatics
Indiana University
Indianapolis, IN, USA

ISSN 1064-3745   ISSN 1940-6029 (electronic)
ISBN 978-1-4614-3703-1   ISBN 978-1-4614-3704-8 (eBook)
DOI 10.1007/978-1-4614-3704-8
Springer New York Heidelberg Dordrecht London

Library of Congress Control Number: 2012939838

Humana Press is a brand of Springer
Springer is part of Springer Science+Business Media (www.springer.com)

Preface

Over the past decade, we have witnessed an explosive development of research dedicated to intrinsically disordered proteins (IDPs), which are also known as natively unfolded proteins among various other names. The existence of biologically active but extremely flexible proteins is challenging the century-old structure-to-function paradigm according to which a rigid well-folded 3D structure is required for protein function. Many structural biologists now recognize that the functional diversity provided by disordered regions complements the functional repertoire of ordered protein regions. The high abundance of IDPs in various organisms, their unique structural features, numerous functions, and crucial associations with different diseases show that there are enough grounds to conclude that these proteins should be considered as a unique entity, an unfoldome.

In comparison with "normal" globular proteins, IDPs possess increased amounts of disorder that can be detected by many physicochemical methods that were originally developed to characterize protein self-organization. On the other hand, due to the highly dynamic nature of IDPs, new and existing experimental methods need to be developed and extended, respectively, for the structural and functional analysis of these IDPs. These methods represent an instrumental foundation for experimental unfoldomics.

Information based on modern protocols is provided herein on virtually every experimental method used both to identify IDPs and to analyze their structural and functional properties. Hence, this book will be of interest to all scientists and students studying IDPs, whether the focus is on an IDP's (lack of) structure or on its function.

The general audience for this book includes scientists working in the fields of biochemistry, biophysics, molecular medicine, biotechnology, pharmacology and drug discovery, molecular and cellular biology; students of Medical Schools, departments of Biochemistry, Biophysics, Molecular Biology, Biotechnology, and Cell Biology, to name a few. We are aware that many scientists have encountered IDPs in their research, but have shied away from deeper studies due to the lack of knowledge of what to try next. By collecting the current methods for the analysis of IDPs in one place, our goal is to help such scientists further their investigations of these fascinating, dynamic molecules.

Tampa, FL, USA; Moscow Region, Russia   *Vladimir N. Uversky, Ph.D.*
Indianapolis, IN, USA   *A. Keith Dunker, Ph.D.*

Contents

Contributors

RINAT R. ABZALIMOV • *Department of Chemistry, University of Massachusetts-Amherst, Amherst, MA, USA*

JEFFREY N. AGAR • *Chemistry and Volen Center for Complex Systems, Brandeis University, Waltham, MA, USA*

TOSHIO ANDO • *Department of Physics and Bio-AFM Frontier Research Center, Kanazawa University, Kanazawa, Japan*

JARED R. AUCLAIR • *Department of Biochemistry and Rosenstiel Basic Medical Sciences Research Center, Brandeis University, Waltham, MA, USA; Department of Chemistry and Rosenstiel Basic Medical Sciences Research Center, Brandeis University, Waltham, MA, USA; Chemistry and Volen Center for Complex Systems, Brandeis University, Waltham, MA, USA*

PAU BERNADÓ • *Institute for Research in Biomedicine, Parc Científic de Barcelona, Baldiri Reixac, Spain; Centre de Biochimie Structurale. CNRS UMR-5048, Montpellier, France*

CEDRIC E. BOBST • *Department of Chemistry, University of Massachusetts-Amherst, Amherst, MA, USA*

SARAH E. BONDOS • *Department of Molecular and Cellular Medicine, Texas A&M Health Science Center, College Station, TX, USA*

MARK E. BOWEN • *Department of Physiology and Biophysics, Stony Brook University, Stony Brook, NY, USA*

MARCO BRUCALE • *Istituto per lo Studio dei Materiali Nanostrutturati (ISMN), Consiglio Nazionale delle Ricerche (CNR), Salaria, RM, Monterotondo*

LUIGI BUBACCO • *Department of Biochemistry, University of Bologna, Bologna, Italy*

MARIANO CARRIÓN-VÁZQUEZ • *Instituto Cajal, CSIC, Centro de Investigación Biomédica en Red sobre Enfermedades Neurodegenerativas (CIBERNED) and IMDEA Nanociencia, Madrid, Spain; Instituto Madrileño de Estudios Avanzados en Nanociencia (IMDEA-Nanociencia), Madrid, Spain*

ALEXANDRE CHENAL • *Unité de Biochimie des Interactions Macromoléculaires, CNRS UMR 3528, Institut Pasteur, Paris, France; Institut Pasteur, Unité de Biochimie des Interactions Macromoléculaires, Département de Biologie Structurale et Chimie, CNRS UMR 3528, Paris, France*

UCHEOR B. CHOI • *Department of Physiology and Biophysics, Stony Brook University, Stony Brook, NY, USA*

KELLY A. CHURION • *Department of Molecular and Cellular Medicine, Texas A&M Health Science Center, College Station, TX, USA*

STEFANO CIURLI • *Department of Agro-Environmental Science and Technology, University of Bologna, Bologna, Italy*

GUILLAUME COMMUNIE • *Institut de Biologie Structurale, CEA-CNRS-Université, Grenoble, France; Unit for Virus Host Cell Interactions, UJF-EMBL-CNRS, UMI 3265, Grenoble, France*

ASHOK A. DENIZ • *Department of Molecular Biology, The Scripps Research Institute, La Jolla, CA, USA*

PIOTR DOBRYSZYCKI • *Department of Biochemistry, Faculty of Chemistry, Wrocław University of Technology, Wrocław, Poland*

CHRISTINE EBEL • *Institut de Biologie Structurale, Université Grenoble, CNRS, CEA, Grenoble, France*

JAMES E. EVANS • *Department of Biochemistry and Molecular Pharmacology, University of Massachusetts Medical School, Worcester, MA, USA; Department of Proteomics and Mass Spectrometry, University of Massachusetts Medical School, Worcester, MA, USA*

ALLAN CHRIS M. FERREON • *Department of Molecular Biology, The Scripps Research Institute, La Jolla, CA, USA*

CHRISTIAN FIEDLER • *Institut für Biochemie und Biologie, Physikalische Biochemie, Universität Potsdam, Potsdam, Germany*

ANGELO FONTANA • *CRIBI Biotechnology Centre, University of Padua, Padua, Italy*

ERICA FRARE • *CRIBI Biotechnology Centre, University of Padua, Padua, Italy*

AGYA K. FRIMPONG • *Department of Chemistry, University of Massachusetts-Amherst, Amherst, MA, USA*

FRANK GABEL • *Institut de Biologie Structurale Jean-Pierre Ebel. UMR 5075 (CNRS, CEA, UJF), Grenoble, France*

KLAUS GAST • *Institut für Biochemie und Biologie, Physikalische Biochemie, Universität Potsdam, Potsdam, Germany*

KARIN M. GREEN • *Department of Biochemistry and Molecular Pharmacology, University of Massachusetts Medical School, Worcester, MA, USA; Department of Proteomics and Mass Spectrometry, University of Massachusetts Medical School, Worcester, MA, USA*

STEPHEN J. HAGEN • *Physics Department, University of Florida, Gainesville, FL, USA*

BASTIAN HENGERER • *Boehringer Ingelheim Pharma GmbH & Co. KG, Biberach, Germany*

RUBÉN HERVÁS • *Instituto Cajal, CSIC and Centro de Investigación Biomédica en Red sobre Enfermedades Neurodegenerativas (CIBERNED), Madrid, Spain; Instituto Madrileño de Estudios Avanzados en Nanociencia (IMDEA-Nanociencia), Madrid, Spain*

MARIUS IONUT IURASCU • *Laboratory of Analytical Chemistry and Biopolymer Structure Analysis, Department of Chemistry, University of Konstanz, Konstanz, Germany*

MASOUD JELOKHANI-NIARAKI • *Department of Chemistry, Wilfrid Laurier University, Waterloo, ON, Canada*

IGOR A. KALTASHOV • *Department of Chemistry, University of Massachusetts-Amherst, Amherst, MA, USA*

TOMASZ M. KAPŁON • *Department of Biochemistry, Faculty of Chemistry, Wrocław University of Technology, Wrocław, Poland*

JOHANNA C. KARST • *Unité de Biochimie des Interactions Macromoléculaires, CNRS UMR 3528, Institut Pasteur, Paris, France; Institut Pasteur, Unité de Biochimie des Interactions Macromoléculaires, Département de Biologie Structurale et Chimie, CNRS UMR 3528, Paris, France*

MAGNUS KJAERGAARD • *Structural Biology and NMR Laboratory, Department of Biology, University of Copenhagen, Copenhagen, Denmark*

NORIYUKI KODERA • *Bio-AFM Frontier Research Center, Kanazawa University, Kanazawa, Japan*

BIRTHE B. KRAGELUND • *Structural Biology and NMR Laboratory, Department of Biology, University of Copenhagen, Copenhagen, Denmark*

DANIEL LADANT • *Unité de Biochimie des Interactions Macromoléculaires, CNRS UMR 3528, Institut Pasteur, Paris, France; Institut Pasteur, Unité de Biochimie des Interactions Macromoléculaires, Département de Biologie Structurale et Chimie, CNRS UMR 3528, Paris, France*

PATRIZIA POLVERINO DE LAURETO • *CRIBI Biotechnology Centre, University of Padua, Padua, Italy*

PEDRO MADEIRA • *Laboratory of Separation and Reaction Engineering, Department de Engenharia Química, Faculdade de Engenharia da, Universidade do Porto, Porto, Portugal; Institute of Molecular Pathology and Immunology of the University of Porto, Porto, Portugal*

LARISSA MIKHEEVA • *Analiza, Cleveland, OH, USA*

SONJA MÜLLER-SPÄTH • *Department of Biochemistry, University of Zurich, Zurich, Switzerland*

DANIEL NETTELS • *Department of Biochemistry, University of Zurich, Zurich, Switzerland*

PAOLO NEYROZ • *Dipartimento di Biochimica "G. Moruzzi", Università di Bologna, Via San Donato, Bologna, Italy*

JAVIER OROZ • *Instituto Cajal, CSIC and Centro de Investigación Biomédica en Red sobre Enfermedades Neurodegenerativas (CIBERNED), Madrid, Spain; Instituto Madrileño de Estudios Avanzados en Nanociencia (IMDEA-Nanociencia), Madrid, Spain*

ANDRZEJ OŻYHAR • *Department of Biochemistry, Faculty of Chemistry, Wrocław University of Technology, Wrocław, Poland*

SERGEI E. PERMYAKOV • *Institute for Biological Instrumentation of the Russian Academy of Sciences, Pushchino, Moscow Region, Russia; Department of Biomedical Engineering, Pushchino State University, Pushchino, Moscow Region, Russia*

GREGORY A. PETSKO • *Department of Biochemistry and Rosenstiel Basic Medical Sciences Research Center, Brandeis University, Waltham, MA, USA; Department of Chemistry and Rosenstiel Basic Medical Sciences Research Center, Brandeis University, Waltham, MA, USA*

FLEMMING M. POULSEN • *Structural Biology and NMR Laboratory, Department of Biology, University of Copenhagen, Copenhagen, Denmark*

MICHAEL PRZYBYLSKI • *Laboratory of Analytical Chemistry and Biopolymer Structure Analysis, Department of Chemistry, University of Konstanz, Konstanz, Germany*

DAGMAR RINGE • *Department of Biochemistry and Rosenstiel Basic Medical Sciences Research Center, Brandeis University, Waltham, MA, USA; Departments of Chemistry and Rosenstiel Basic Medical Sciences Research Center, Brandeis University, Waltham, MA, USA*

ANDRÉS G. SALVAY • *Institut de Biologie Structurale, CEA-CNRS-Université, Grenoble, France; Instituto de Física de Líquidos y Sistemas Biológicos, Facultad de Ciencias Exactas, Universidad Nacional de La Plata, Argentina; Departamento de Ciencia y Tecnología, Universidad Nacional de Quilmes, Argentina*

BRUNO SAMORÌ • *Department of Biochemistry, University of Bologna, Bologna, Italy*

CELIA A. SCHIFFER • *Department of Biochemistry and Molecular Pharmacology, University of Massachusetts Medical School, Worcester, MA, USA*

BENJAMIN SCHULER • *Department of Biochemistry, University of Zurich, Zurich, Switzerland*

STEFAN SLAMNOIU • *Laboratory of Analytical Chemistry and Biopolymer Structure Analysis, Department of Chemistry, University of Konstanz, Konstanz, Germany*

MATTHEW D. SMITH • *Department of Biology, Wilfrid Laurier University, Waterloo, ON, Canada*

MOHAN SOMASUNDARAN • *Department of Pediatrics, University of Massachusetts Medical School, Worcester, MA, USA*

ANDREA SORANNO • *Department of Biochemistry, University of Zurich, Zurich, Switzerland*

ANA CRISTINA SOTOMAYOR-PÉREZ • *Unité de Biochimie des Interactions Macromoléculaires, CNRS UMR 3528, Institut Pasteur, Paris, France; Institut Pasteur, Unité de Biochimie des Interactions Macromoléculaires, Département de Biologie Structurale et Chimie, CNRS UMR 3528, Paris, France*

BARBARA SPOLAORE • *CRIBI Biotechnology Centre, University of Padua, Padua, Italy*

DMITRI I. SVERGUN • *European Molecular Biology Laboratory, Hamburg Outstation, Hamburg, Germany*

TOSHIYUKI TANAKA • *Graduate School of Life and Environmental Sciences, University of Tsukuba, Tsukuba, Japan*

AGNES TANTOS • *Institute of Enzymology, Biological Research Center, Hungarian Academy of Sciences, Budapest, Hungary*

ISABELLA TESSARI • *Department of Biochemistry, University of Bologna, Bologna, Italy*

PETER TOMPA • *Institute of Enzymology, Biological Research Center, Hungarian Academy of Sciences, Budapest, Hungary*

KIT I. TONG • *Department of Medical Biochemistry, Tohoku University Graduate School of Medicine, Sendai, Tohoku, Japan*

VLADIMIR N. UVERSKY • *Department of Molecular Medicine, College of Medicine, University of South Florida, Tampa, FL, USA; Institute for Biological Instrumentation, Russian Academy of Sciences, Pushchino, Moscow Region, Russia*

ALEJANDRO VALBUENA • *Instituto Cajal, CSIC and Centro de Investigación Biomédica en Red sobre Enfermedades Neurodegenerativas (CIBERNED), Madrid, Spain; Instituto Madrileño de Estudios Avanzados en Nanociencia (IMDEA-Nanociencia), Madrid, Spain*

CAMELIA VLAD • *Laboratory of Analytical Chemistry and Biopolymer Structure Analysis, Department of Chemistry, University of Konstanz, Konstanz, Germany*

KEITH R. WENINGER • *Department of Physics, North Carolina State University, Raleigh, NC, USA*

MAGDALENA WOJTAS • *Department of Biochemistry, Faculty of Chemistry, Wrocław University of Technology, Wrocław, Poland*

MASAYUKI YAMAMOTO • *Department of Medical Biochemistry, Tohoku University Graduate School of Medicine, Sendai, Tohoku, Japan*

BORIS ZASLAVSKY • *Analiza, Cleveland, OH, USA*

Part I

Single Molecule Techniques

Chapter 1

Immobilization of Proteins for Single-Molecule Fluorescence Resonance Energy Transfer Measurements of Conformation and Dynamics

Ucheor B. Choi, Keith R. Weninger, and Mark E. Bowen

Abstract

Fluorescence resonance energy transfer provides information about protein structure and dynamics. Single-molecule analysis can capture the information normally lost through ensemble averaging of heterogeneous and dynamic samples. Immobilization of single molecules, under conditions that retain their biological activity, allows for extended observation of the same molecule for tens of seconds. This can capture slow conformational transitions or protein binding and unbinding cycles. Using an open geometry for immobilization allows for direct observation of the response to changing solution conditions or adding ligands. Here we provide detailed methods for immobilization and observation of fluorescently labeled single proteins using total internal reflection microscopy that are widely applicable to the study of intrinsically disordered proteins.

Key words: Intrinsically disordered proteins, Single-molecule fluorescence, FRET, Vesicle, Encapsulation, Reconstitution

1. Introduction

Although depleted in hydrophobic amino acids, which cause chain collapse in folded proteins, intrinsically disordered proteins (IDPs) are prone to intramolecular interactions that affect chain dynamics (1). The balance between net charge and hydropathy has been shown to determine the compaction of the polypeptide (2, 3) ranging from extended random coils to disordered globules that are closer in volume to folded proteins (4). Some IDPs can be induced to fold in the presence of ligands or protein-binding partners while others are perpetually disordered. Under native buffer conditions, an ensemble IDP sample contains a dynamic mixture of structures. The distribution of conformations under native conditions and

Vladimir N. Uversky and A. Keith Dunker (eds.), *Intrinsically Disordered Protein Analysis: Volume 2, Methods and Experimental Tools*, Methods in Molecular Biology, vol. 896, DOI 10.1007/978-1-4614-3704-8_1, © Springer Science+Business Media New York 2012

their importance is an open question. Conformational selection of rarely populated states may contribute to the biological function of IDPs (5).

The distance dependence of fluorescence resonance energy transfer (FRET) has been used for decades to characterize polypeptide dynamics (6, 7). The fluorescent donor and acceptor are attached to the polypeptide, through synthesis or mutagenesis, with a defined separation in the primary sequence. In a dynamic polypeptide chain, FRET efficiency can be related to the root mean squared (rms) displacement in any direction, $R_{rms} = \sqrt{\langle R^2 \rangle}$, or absolute distance in stable structures. Such methods have been used to examine polymer models of the denatured state (8) and IDPs under native conditions (9, 10). Ensemble FRET can provide valuable insights into IDPs, but single-molecule FRET (smFRET) can resolve sample heterogeneity and to some extent probe dynamics.

FRET can measure distances and conformational fluctuations on the scale of 2–8 nm (7). For a disordered random coil, this corresponds to fluorophore separations of 50–175 residues in the primary sequence. Conformational dynamics occur on a variety of timescales. Dynamics of extended random coils are faster than the time resolution of current Electron Multiplied Charge Coupled Device (EMCCD) cameras, which leads to a narrow distribution of FRET values. Thus even at the single-molecule level, the details of molecular behavior are time-averaged and are equivalent to ensemble measurement of R_{rms}. Dynamics approaching the exposure time lead to peak broadening in the FRET distribution (11). EMCCD detection is best at resolving subpopulations, capturing dynamics >10/s, and direct observation of the response to non-equilibrium conditions (e.g., ligand binding). With immobilization, an individual molecule can be recorded for tens of seconds using Total Internal Reflection Fluorescence (TIRF) microscopy with EMCCD camera detection (12). TIRF illumination decays within 100–200 nm of the surface allowing immobilized molecules to be selectively excited. Unfavorable interactions with a surface can destroy a protein sample and must be prevented. In many cases it is possible to retain the biological activity of immobilized proteins. Familiar examples in biochemistry include the interactions between glutathione-S-transferase (GST) and Glutathione Sepharose resin and the biotin–streptavidin system. Retention of activity must be validated by comparison to non-immobilized samples. Here we describe methodology for the immobilization of proteins for TIRF microscopy while retaining biological activity. We have used smFRET to characterize IDPs involved in synaptic transmission, which serve to illustrate our methods for smFRET measurements of immobilized IDPs. We will demonstrate three experimental preparations that we have used for smFRET measurements of IDPs (Fig. 2). First, immobilizing the protein via a biotin–streptavidin interaction on a "passivated" surface provides an open geometry that allows manipulation of buffer conditions or addition of ligands or binding partners. Second, encapsulation confines the protein inside a phospholipid vesicle where it is protected from aggregation and adsorption. Co-encapsulation of multiple proteins within one liposome can be used to study weak protein–protein interactions since the liposome volume is approx. femtoliter. Finally, for membrane proteins, reconstitution into a phospholipid vesicles or deposited phospholipid bilayers can approximate the physiological membrane environment and probe the influence of lipids on IDP conformation.

2. Materials

2.1. Reagents and Supplies

1. 1 in. × 3 in. Quartz microscope slide (G. Finkenbeiner, Inc., Waltham, MA).

2. Diamond grinding bit, 0.029 in. tip diameter, #115005 (Starlite Industries Inc., Rosemont, PA).

3. 400 XPR Rotary Tool with Model 220 stand (Dremel, Racine, WI).

4. 24 × 30 mm #1.5 micro cover glass (VWR, Radnor, PA).

5. Double and single-sided tape (Scotch/3M, St. Paul, MN).

6. 5 min epoxy (ITW Devcon, Danvers, MA).

7. Optical adhesive 63 (Norland Products, Cranbury, NJ).

8. Electro-lite CS-410 UV Light Curing System (Thorlabs Inc., Newton, NJ).

9. Benchtop UV Transilluminator (UVP, Upland, CA).

10. Egg PC and 18:1 Biotinyl Cap PE in chloroform (Avanti Polar Lipids, Alabaster, AL).

11. Hand-held Mini-Extruder Kit (Avanti Polar Lipids, Alabaster, AL).

12. TBS buffer (20 mM Tris, 150 mM NaCl, pH 7.5).

13. Sepharose CL-4B packed in a NAP-5 column (GE Healthcare, Piscataway, NJ).

14. Biotinylated bovine serum albumin (bBSA) (Sigma–Aldrich, St. Louis, MO).

15. *Escherichia coli* strain AVB101 (Avidity LLC, Aurora, CO).

16. Streptavidin (Invitrogen, Carlsbad, CA).

17. Beveled pipette tips (VWR #53503-566).

18. MATLAB Software (The Mathworks, Nattick, MA).

2.2. Observation Chambers for TIRF Microscopy

Here we provide a procedure to prepare flow channels for single-molecule observations from quartz microscope slides, which assumes the use of prism-based TIRF excitation. The use of quartz or fused silica is paramount to avoid fluorescent impurities found in glass (see Note 10).

2.2.1. Preparation and Cleaning

1. Submerge the quartz slide in water. Drill holes using the diamond bit with the Rotary tool set to 35,000 rpm. To avoid breaking the slides, replace worn out bits often. Up to five flow chambers can be made on a slide as long as their total size does not exceed the width of the cover glass (see Note 1).

2. Before use, slides are carefully cleaned by sequential sonication in acetone, ethanol, and 1 M KOH. Place slides in a standard staining jar and cover with solvent. Bath sonicate for 30 min. Use a separate container for each solvent and dedicated tweezers to move slides. Finally, slides are extensively rinsed with deionized water.

2.2.2. Construction of Flow Channels

1. Make sure work area is clean. Briefly pass the slide through the flame of a propane torch until dry. Allow the slide to cool by placing it on a metal ring stand with the slide interior surface facing down (see Note 1). We reuse the ~2 in. metal lids from chemical and restriction enzyme containers for this purpose.

2. Cut tape to form flow channel walls. Use double-sided tape unless lipid bilayers are being used. Attach a ~2 in. piece of tape lengthwise on a clean glass microscope slide. Cut with a razor blade into 1–2 mm wide strips. Trim to ~1.5 in. in length.

3. Immediately before assembling the chamber, turn the slide face up to expose the interior surface (see Note 1). Place strips of tape on the slide as spacers between the drilled holes to form the flow chambers (Fig. 1). Do not allow the tape to touch the channel interior. Avoid air bubbles beneath the tape as this will result in variations in the channel thickness. Work out any bubbles using pressure from the tweezers or a similar tool.

4. To remove antistatic coating and dust, quickly pass both sides of the cover glass through a propane flame (<1 s). Cover glasses can easily warp or crack if held in the flame too long and should be discarded if this happens. Allow to air cool for 30 s while holding with tweezers.

5. Align the long edge of the cover glass over the slide to cover all the chambers. Lower the cover glass onto the slide in a continuous motion. Once stuck to the tape, the cover glasses are difficult to remove. Use pressure from the tweezers or a similar tool to work out any bubbles trapped between the glass and the tape. Cut off any overhanging tape with a new razor blade.

Tape Strips with Standing Beads of Optical Adhesive

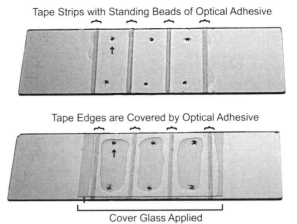

Tape Edges are Covered by Optical Adhesive

Cover Glass Applied

Fig. 1. Assembly of multichannel flow chambers using UV adhesive. (*Top*) A quartz slide is shown with six holes drilled (e.g., *arrow*) to create three flow chambers. Strips of single-sided tape (*curved brackets*) demark the walls of the flow chambers. A thin bead of UV adhesive sits on the center of each strip of tape but does not exceed the width of the tape. (*Bottom*) The same slide is shown after the cover glass has been applied (*straight bracket* denotes edges of the cover glass). The adhesive has flowed to cover the tape edges and now constitutes the walls of the flow chamber. The flow chambers are then cured with UV light.

6. Combine the 5 min epoxy and allow 1–2 min for partial curing. If the epoxy is not sufficiently viscous, capillary action will draw epoxy into the channel, which can possibly cover the drilled holes. Using a pipette tip, apply a thin bead of epoxy to both ends of the flow chamber. Avoid covering holes as this will render the chamber unusable. To avoid leakage, ensure that each chamber has a continuous seal. Allow to cure completely.

7. If using lipid bilayers, single-sided tape is used in conjunction with UV curable optical adhesive to form the channel. Lipid bilayers extract fluorescent impurities from double-sided tape.

8. Prepare the optical adhesive ahead of time. Remove the plunger and fill a 1-ml syringe with adhesive. Attach at 30G1/2 needle and insert plunger driving glue to tip.

9. All procedures up to step 4 of Subheading 2.2.2 are the same only using single-sided tape. Apply a thin bead of adhesive along the center of each strip of single-sided tape (Fig. 1). There should be just enough adhesive to cover the tape *after* compressing the cover glass to the slide. Too much adhesive will fill the chamber and may cover the holes while too little will expose the sample chamber to the edge of the tape.

10. Apply the cover glass as described in Subheading 2.2.2, step 4. Take care as the cover glass can only be applied once. Mild pressure can be used to direct adhesive flow and completely cover the tape edges. Use judicial application of the UV light

curing system to harden the adhesive in each chamber for 30 s. The goal is not to fully cure the adhesive but only to fix the cover glass in place and insure optimal coverage of the tape.

11. Finally seal the ends of the flow chamber with adhesive. Avoid covering the holes with excess adhesive by capillary action. Place the slide cover glass up on a UV light box. Use ~2 mm wooden dowels to elevate the slide off the surface. To insure a uniform chamber thickness, a glass microscope slide is placed atop the cover glass with ~70 g of weight applied to the center (e.g., a 3 in. optical post). Expose to UV for 30 min to fully cure the adhesive.

2.3. Preparation of Phospholipid Vesicles

Phospholipids can be used to encapsulate soluble proteins in vesicles, reconstitute membrane proteins or to form a deposited bilayer on the microscope slide surface. A wide variety of phospholipids are available to mimic biological membranes, prevent nonspecific interactions or add functional groups or affinity tags including biotin and Ni-NTA. We typically use phosphatidylcholine extracted from chicken egg (egg PC), which shows low levels of fluorescent impurities and minimal nonspecific binding of proteins.

2.3.1. Formation of Unilamellar Vesicles by Extrusion

1. Place 30 mg of lipid in a chloroform solution in a 13×100-mm glass tube. Inside a chemical fume hood, dry the lipid under a gentle stream of argon or dry nitrogen gas while rotating. Form a thin but compact film as this is easier to recover during resuspension. Once lipids are dried, place the tube under vacuum for 30–60 min to remove residual chloroform. A pump is recommended as house vacuum is often insufficient.

2. Add 1 ml of TBS to make a 30 mg/ml lipid solution. The high concentration of lipids facilitates encapsulation and reconstitution. Vortex for 30 s or until all dry lipid is suspended.

3. Use 4–5 freeze–thaw cycles to aid in the solubilization of the lipids. The goal is to take the sample above and below the lipid phase transition temperature, which depends on the lipids in use. For each cycle, flash freeze the lipid suspension in liquid nitrogen and thaw in a 37°C water bath followed by 30 s vortex.

4. Assemble the extruder according to the manufacturer's instructions. The membrane filter pore size depends on the experiment. A 50 nm diameter is used for deposited bilayers while 100–200 nm is used for encapsulation. Test everything for leakage with TBS buffer before extruding.

5. Draw lipids into the first syringe. Go slowly for the first few passes to avoid rupturing the membrane. Extrude with 20–30 passes back and forth through the extruder. The opacity will decrease noticeably depending on the vesicle size.

6. Collect liposomes from the second syringe (i.e., NOT the first syringe used to initially draw up the lipids). This insures that the collected material has passed through the extruder. Spin the sample at $14,000 \times g$ in a microcentrifuge and collect the supernatant containing vesicles.

2.3.2. Encapsulation of Soluble Proteins in Vesicles

1. For immobilized vesicles, prepare a stock of egg PC with 0.1 mol% of 18:1 Biotinyl Cap PE in chloroform. Otherwise, egg PC in chloroform is sufficient. Dry lipids as described in Subheading 2.3.1, step 1.

2. For encapsulation, labeled protein(s) can be added to the lipid samples at Subheading 2.3.1, step 1 or after step 3 (see Note 2). Freeze–thaw cycles can facilitate encapsulation efficiency but are not tolerated by all proteins. Aggregation of proteins during repeated freezing should be tested.

3. Remove unencapsulated proteins by desalting on Sepharose CL-4B. Vesicles elute at the void volume while free proteins are retained. Columns are created by replacing the Sephadex G-25 in commercial NAP-5 columns with Sepharose CL-4B to a bed height of 25–30 mm and replacing the frit. Measure the void volume with extruded vesicles. The presence of lipids can be confirmed with a spectrophotometer by measuring scattering at 400 nm.

4. Equilibrate a CL-4B desalting column in TBS. Load 200 µl of sample. Allow all liquid to drain before adding additional buffer. Use two applications of 100 µl TBS to rinse the frit. Add additional TBS to bring the total added volume (including sample) within 100 µl of the measured void volume of the column. Collect 200-µl fractions. Sample is typically recovered in 400–600 µl and spun at $14,000 \times g$ in a microcentrifuge. Light scattering by vesicles makes measurement of protein concentration difficult.

2.3.3. Detergent-Assisted Insertion to Reconstitute Membrane Proteins in Vesicles

1. Membrane proteins are exchanged into TBS containing 100 mM β-D-octyl glucoside (βOG) by affinity chromatography. Mix the protein sample at a 1:4 ratio with 30 mg/ml extruded 50 nm liposomes in TBS. Incubate on ice for 30 min.

2. Dilute sample 1:1 with detergent-free TBS. Incubate at room temperature for 5 min.

3. Remove detergent and unincorporated protein by desalting with Sepharose CL-4B as described in Subheading 2.3.2, step 4. This method can lead to uniform orientation of membrane proteins with their cytoplasmic domain facing outwards. Orientation can be tested by proteolysis in the presence and absence of 1% Triton X-100.

3. Methods

To monitor protein conformational changes or protein interactions over time, molecules must be immobilized so they will not diffuse out of the observation volume. Here we describe three different approaches to immobilize single molecules. First, soluble proteins are biotinylated and immobilized directly to the surface using a biotin–streptavidin linkage (Fig. 2a (1, 2)). Before immobilization, the surface must be "passivated" through the use of blocking agents to prevent adsorption and loss of activity (13). Second, vesicles containing encapsulated soluble proteins are immobilized to the surface using a biotin–streptavidin linkage (Fig. 2a (3)). Most proteins are an order of magnitude smaller than the vesicle diameter, so they undergo free diffusion within the vesicle interior. The vesicle protects the encapsulated protein from interactions with the surface and other proteins. Third, proteins containing a

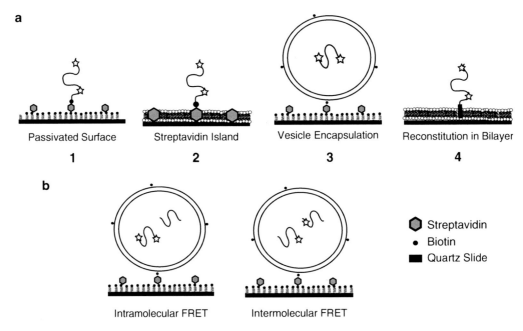

Fig. 2. Experimental approaches to immobilized single protein molecules. (**a**) From *left* to *right*, (1) shows a schematic of a biotinylated single molecule directly immobilized on a passivated surface via streptavidin. Dyes are indicated by *stars*. The passivation can use bBSA or PEG (Subheadings 3.1.1 and 3.1.3). (2) Shows a protein immobilized on a streptavidin island surrounded by a deposited lipid bilayer (Subheading 3.1.2). (3) Shows a protein encapsulated within a lipid vesicle (Subheading 2.3.2), which is immobilized to a passivated surface (Subheading 3.2). (4) Shows a membrane protein reconstituted via the transmembrane domain into lipid bilayer (Subheading 2.3.3). A deposited bilayer is formed from vesicles containing reconstituted membrane proteins (Subheading 3.3). (**b**) Schematic showing co-encapsulation experiments to study protein binding. Intramolecular FRET (*left*) can monitor changes in a protein induced by binding to an unlabeled ligand protein. Intermolecular FRET (*right*) between two proteins, each singly labeled with donor or acceptor, can directly report on protein binding events.

transmembrane domain are reconstituted into vesicles, which are then used to form a planar lipid bilayer on the slide surface (Fig. 2a (4)). Although imperfect, deposited bilayers can be used to examine the effect of lipids on conformation or binding studies. We describe methods where rinsing and sample application is implemented using hand-activated pipettes although automated buffer exchange schemes are also possible (see Note 10).

3.1. Direct Immobilization of Proteins to a Passivated Surface

Proteins can be biotinylated enzymatically or during recombinant expression using commercial reagents and the AVB101 bacterial strain. The effectiveness of a surface passivation strategy must be tested for each protein. First, the level of nonspecific binding is tested by omitting streptavidin or forgoing biotinylation (Fig. 3). Ultimately the results of studies using the immobilized protein should be compared to encapsulated or freely diffusing controls.

3.1.1. Biotinylated Bovine Serum Albumin Surface (14)

1. Use the flow cell shown in Subheading 2.2.2. Using a beveled pipette tip to form a tight seal, wash the channel with 4×200 µl of deionized water followed by TBS (see Note 5).

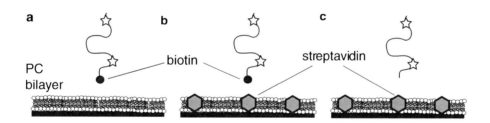

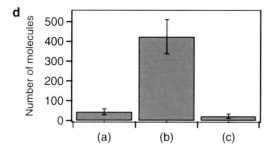

Fig. 3. Determining the extent of nonspecific binding to a passivated surface. Data is shown for streptavidin islands (Subheading 3.1.2) but the same method can be used to test any surface preparation. (**a**) As a control, biotinylated, fluorescently labeled protein was applied to a deposited lipid bilayer containing no streptavidin. After 5 min incubation, the proteins were rinsed out and the extent of binding was assessed. (**b**) The same protein solution in (**a**) applied to a streptavidin island surface and rinsed. (**c**) As an additional control, the same protein was expressed, purified, and labeled without biotinylation and applied to the streptavidin island surface. (**d**) To assess the degree of binding, the number of molecules retained on the surface was counted. Shown are the average number of molecules per field of view for the experiments shown in (**a**)–(**c**), respectively. The degree of nonspecific binding can be assessed by comparing retention of the protein to the streptavidin islands to the surface without streptavidin and/or protein lacking biotin (see Note 9).

2. Add 100 μl of 0.2-μm filtered 1 mg/ml bBSA to the channel (see Note 4). After the incubation, rinse the channel with 4 × 200 μl TBS to remove unbound bBSA.

3. Nonspecific binding to bBSA can be further reduced by adding 100 μl egg PC vesicles (50 nm at 15 mg/ml). BSA is a fatty acid-binding protein so phospholipids can block exposed hydrophobic sites. Incubate and rinse as described above.

4. Add 100 μl 0.2 μm filtered 0.1 mg/ml streptavidin (SA). Incubate and rinse as described above.

5. The fluorescently labeled, biotinylated protein sample should be diluted in TBS to ~pM (see Note 4). Incubate 5 min and rinse with TBS (see Note 5). The sample is now immobilized and can now be imaged or subjected to manipulations of the external solution (Fig. 2a (1)).

3.1.2. Streptavidin Islands in a Deposited Lipid Bilayer (15)

1. Use the optical adhesive flow chamber described in Subheading 2.2.2, step 7. Equilibrate the flow channel with TBS as described in Subheading 3.1.1, step 1. Inject 100 μl of 1 nM SA. Incubate 5 min to allow the sparse deposition of individual SA molecules, which retain specific binding activity when adsorbed. Rinse with TBS.

2. Inject 100 μl 50 nm egg PC vesicles at 3 mg/ml in TBS. Incubate 1 h in a humid environment to allow the spontaneous condensation of a lipid bilayer around the adsorbed SA molecules (see Note 3). Rinse slowly with TBS (see Note 5).

3. Nonspecific binding can be reduced by further incubation with 1 mg/ml BSA and/or 30 mg/ml egg PC liposomes.

4. The fluorescently labeled, biotinylated protein sample should be diluted in TBS to ~pM (see Note 4). Incubate 5 min and rinse with TBS. The sample is now immobilized and can now be imaged or subjected to manipulations of the external solution (Fig. 2a (2)).

3.1.3. Biotinylated Poly (ethylene glycol) Surface

Poly(ethylene glycol) (PEG) coated surface is an effective and popular way of preventing nonspecific binding. A detailed protocol is described elsewhere (16).

1. A major difference from the protocols above is that PEG is applied to both the quartz slide and cover glass *before* the observation chamber is constructed. The interior surfaces of the quartz slide and the cover glass are amine functionalized by incubation in 1% Vectabond in acetone for 5 min. Rinse exhaustively with deionized water. Dry with dry nitrogen or under vacuum.

2. Assemble the optical adhesive flow chamber as described in Subheading 2.2.2, step 7. Take care to protect the interior surfaces (see Note 1).

3. Dissolve 100 mg of 99 mol% m-PEG-SPA: 1 mol% biotin-PEG-NHS in 1 ml 100 mM sodium bicarbonate. Quickly inject 100 μl directly into each of the flow channels. Incubate for 1–2 h in a humidity-controlled environment (see Note 3). Wash exhaustively with deionized water to remove unbound PEG. PEG coated slides can be stored after drying in air for a few days.

4. When ready to use, exchange channel into TBS. Add 100 μl 0.1 mg/ml SA and incubate for 5 min. Wash with TBS. Now the surface is ready to immobilize biotinylated samples. Non-specific binding can be further reduced by incubation with 1 mg/ml BSA prior to the streptavidin (17).

5. The fluorescently labeled, biotinylated protein sample should be diluted in TBS to ~pM (see Note 4). The sample is now immobilized and can now be imaged or subjected to manipulations of the external solution (Fig. 2a (1)).

3.2. Immobilization of Vesicles to a bBSA Surface (12)

1. Use the epoxy flow cell shown in Subheading 2.2.2. Prepare the bBSA and SA surface as described in Subheading 3.1.1.

2. Fluorescently labeled protein(s) should be vesicle encapsulated as described in Subheading 2.3.2. To study the effects of intramolecular interactions on IDP conformation, an unlabeled ligand protein can be encapsulated with the labeled sample using the procedure described in Subheading 2.3.2, step 2. (Fig. 2b, left). The concentrations of the unlabeled ligand can vary depending on the binding affinities to monitor a partial or a full effect on the conformational changes of the labeled samples. To monitor intermolecular interactions, two singly labeled proteins (one labeled with donor and the other with acceptor) are co-encapsulated at a 1:1 ratio using the procedure described in Subheading 2.3.2, step 2 (Fig. 2b, right).

3. Dilute the encapsulated sample ~1:10^4 in TBS (see Note 2). Inject 100 μl into the flow channel and incubate for 5 min (see Note 3). Remove free vesicles by gently rinsing with TBS (see Note 5). The sample is now immobilized and ready for measurement (Fig. 2a (3)).

3.3. Formation of a Deposited Bilayer Containing Membrane Proteins (see Note 8)

1. Use the optical adhesive flow chamber described in Subheading 2.2.2, step 7. Equilibrate the flow channel with TBS.

2. Membrane proteins are reconstituted as described in Subheading 2.3.3. Empirical dilution with unlabeled 50 nm egg PC liposomes may be required to achieve optical resolution (see Note 4). Inject 100 μl of protein containing vesicles. Incubate 15 min to allow the spontaneous condensation of a lipid bilayer at the surface of the slide.

3. Rinse the channel gently with TBS (see Note 5). An additional incubation with 15 mg/ml protein-free 50 nm egg PC liposomes for 1 h can block defects remaining after step 2.

4. For membrane proteins, the integrity of the bilayer is critical. Diffusion of fluorescently labeled lipids can be tested by Fluorescence Recovery After Photobleaching (FRAP) or single particle tracking. Similarly, defects can be identified by incubation of the bilayer with fluorescently labeled proteins or liposomes and assessing the degree of retention. Protein orientation, which may differ from vesicle samples, can be checked with proteolysis (Fig. 2a (4)).

3.4. Data Collection and Analysis

The assembly of prism TIRF microscopy instrumentation is beyond the scope of this chapter and has been thoroughly described elsewhere (16). A prism-based TIR instrument can be constructed from a standard inverted microscope and commercial optomechanical components with a small amount of custom milling (Fig. 4). Briefly, excitation of the Alexa555 (or Cy3) donor and measurement of FRET efficiency uses a circularly polarized 532-nm laser, while direct excitation of the Alexa647 (or Cy5) acceptor uses a 635-nm laser. The lasers are brought to the microscope stage by means of standard geometrical optics and introduced to the sample through a quartz prism (e.g., PLBC 5.0-79.5-SS CVI Laser) coupled to the quartz slide with an index matching oil (e.g., Cargille Type FF). Immediately before imaging, samples are exchanged into imaging buffer (see Note 6) to stabilize fluorescence emission. The microscope image is passed through a series of optical elements to separate the donor and acceptor emission (Fig. 4). This is achievable using commercial image splitters (e.g., Optosplit II; Cairn Research Ltd., Kent, UK), which results in separate donor and acceptor images of the sample. Fluorescent beads (e.g., Fluospheres; Invitrogen, Carlsbad, CA) are used to determine the pixel registry between the donor and acceptor images, which allows identification of corresponding donor and acceptor fluorophores attached to a single molecule. Since the protein is randomly labeled with acceptor (A) and donor (D) dyes, all combinations are possible [AA, DA, AD, DD]. Also, more than one molecule can be encapsulated inside a single vesicle. To distinguish these behaviors, we used alternating laser illumination (18) to determine the number of dyes in diffraction-limited spot (see Note 7).

3.4.1. Extraction of smFRET Efficiency from the Microscope Image

1. Each EMCCD camera exposure can be treated as a numerical array and is readily processed using MATLAB. Single molecules appear as diffraction-limited spots, which are easily detected as pixels of maximum intensity separated by five or more pixels from any neighboring maxima. A ten-frame average taken under 635 nm illumination is used to locate active acceptor fluorophores. Once identified, these pixels are reexamined at each frame of the movie.

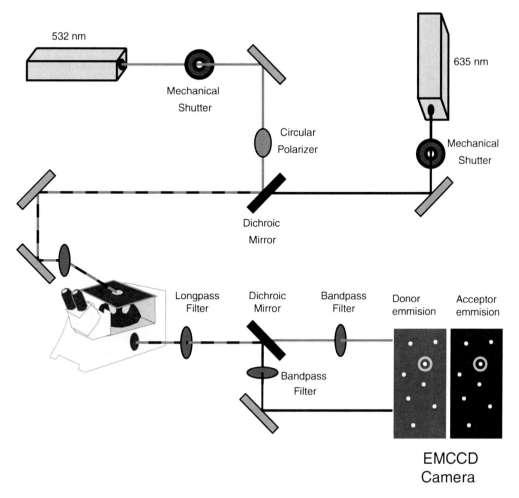

Fig. 4. Schematic of a prism-based total internal reflection microscope for single-molecule spectroscopy. A circularly polarized 532-nm laser and a linear polarized 635-nm laser are combined using a dichroic mirror and routed to a quartz prism on the microscope stage. Mechanical shutters control the excitation wavelength. The fluorescence emission is split by color into two images using a dichroic mirror, which are passed through additional optical elements to isolate donor and acceptor signals. The two replicate images are collected by an EMCCD camera and are processed using MATLAB to identify single molecules containing a donor and acceptor fluorophore.

2. Using a 60×, 1.2 NA water immersion objective with a 512 × 512 pixel EMCCD camera, a single molecule is contained within a 3 × 3 pixel array about the most intense pixel. Fluorescence intensity for a single molecule can be determined simply by taking the sum of the 3 × 3 matrix. In practice, most of the single-molecule intensity falls on the top four of the pixels so tracking this smaller set of pixels is sufficient. Such undersampling decreases the width of the single-molecule intensity distribution (unpublished data) reflecting lower noise.

3. For each molecule, the donor and acceptor intensity in each frame of the movie is plotted as a function of the frame number

to examine the time dependence of fluorescence emission (Fig. 5). These time traces are first analyzed to confirm that the diffraction-limited spot contains a single molecule labeled with one donor and one acceptor. Single fluorophores are confirmed by the observation of a single step in the photobleaching decay to baseline (Fig. 5). If labeling efficiency is high and molecules are optically well resolved, two single fluorophores in a diffraction-limited spot can be attributed to a doubly labeled protein. To characterize a sample, observations of hundreds to thousands of single molecules are collected.

4. The hallmark of smFRET is anticorrelated donor intensity during acceptor photobleaching (Fig. 5). A critical normalization factor, γ, is taken from the magnitude of this change and used to adjust for differences in quantum yield and detection efficiency, which can vary between samples and even between molecules (19). To calculate smFRET efficiency, E, requires correction for the leakage of donor emission into the acceptor channel, β. This value can be obtained from control experiments and are constant as long as the dyes and instrument are unchanged. Depending on the experimental setup, correction for direct excitation of the acceptor may also be required but is not significant at the laser powers used in our system.

$$E = \frac{I_A - \beta I_D}{((I_A - \beta I_D) + \gamma I_D)}.$$

3.4.2. Analysis of smFRET Efficiency Time Traces

1. The most straightforward analysis is to compile all the observed smFRET efficiency values into a histogram that represents the probability distribution. For a random coil, rapid conformational dynamics lead to the appearance of a single peak that is well described by a single Gaussian function using nonlinear least squares fitting. The mean FRET efficiency and the distribution width both provide information about IDP structure and dynamics. For compact globular IDPs, more complicated smFRET distributions are observed that require multiple functions to describe (Fig. 5).

2. The averaging of rapid intramolecular dynamics in random coil-like IDPs gives rise to steady FRET efficiency levels over time (Fig. 5a–d). The resultant FRET distributions typically show a single Gaussian peak of shot noise-limited width. Because of the time-averaging, this single-molecule data can be related only to ensemble properties using polymer models.

3. More complex, stochastic conformational transitions have been observed in globular IDPs or protein complexes involving IDPs. The transitions are slow; occurring within a single frame (Fig. 5e–h). Data of this type can be analyzed with Hidden Markov models (20) or Edge detection strategies (21).

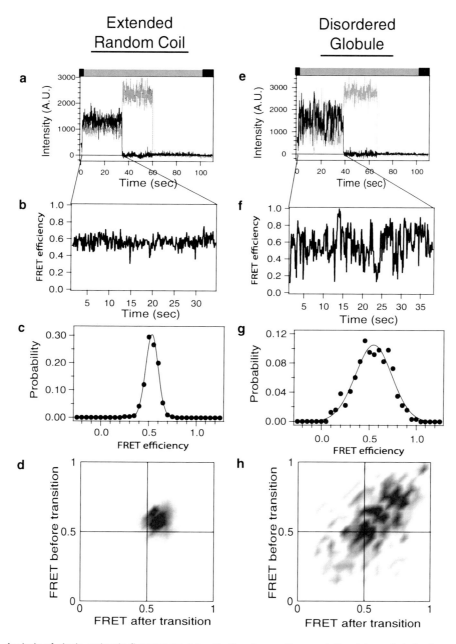

Fig. 5. Analysis of single-molecule fluorescence intensity time traces. Representative data analysis is shown for two classes of IDP that have been labeled with donor and acceptor. *Left*: Extended random coil IDP. *Right*: Disordered globular IDP. (**a** and **e**) Background corrected fluorescence intensity of a single molecule. Donor emission is colored *gray* while acceptor is colored *black*. Note the single-step photobleaching of each dye to baseline, which confirms a single molecule. *Colored bar* above the panel describes the alternating laser excitation (see Note 7). (**b** and **f**) Intensities above have been converted to FRET efficiencies for the period before the acceptor dye molecule photobleached. Note the steady FRET of the random coil compared to the stochastic switching in a disordered globule. (**c** and **g**) FRET efficiency at each time point is compiled into histograms to examine the probability distribution. These histograms contain data from only one molecule, while a typical experiment would compile hundreds to thousands of molecules. Random coils give a single peak of shot noise-limited width. (**d** and **h**) The time series of changes in FRET are examined using transition density plots, which plot the FRET state before a transition (*y*-axis) against the FRET state after that transition (*x*-axis) for molecules (**a**) and (**e**), respectively. The *darker areas* indicate more favored transitions between the indicated FRET states. The random coils show only one peak while the globular IDP makes transition to many different FRET states.

From such analyses one obtains the distribution of FRET states (Fig. 5c, g), the dwell times for each state (i.e., state lifetime) and the sequence of state transitions. Each of these parameters can be analyzed to extract information about the sequence and nature of conformational fluctuations as well as the stability of intermediate states.

4. For both types of IDP, the use of immobilized molecules with EMCCD detection allows one to capture individual binding events and the instantaneous response to changes in solution conditions.

4. Notes

1. A three channel slide will need three holes at the top and bottom (Fig. 1). When drilling slides, the exit hole can be slightly irregular, which prevents the formation of a good seal with the pipette during rinsing. This side of the slide is used as the channel interior. As such, it should be protected from contacting anything during assembly.

2. Encapsulation efficiency depends on the absolute concentration of protein and lipids. Efficiency is highly sample dependent and typically <10%. Micromolar protein concentrations at the lipid concentration described results in most liposomes being empty or containing a single molecule. Membrane protein reconstitution is similarly variable.

3. Longer incubations are carried out in a humidified, vapor-saturated chamber, such as an empty pipette tip box with deionized water in the bottom. This prevents samples from drying in the chamber. A drop of TBS is placed over each hole of the chamber to further reduce evaporation. Should this dry, apply a drop of TBS to one hole to force air out the other before rinsing.

4. The goal is to empirically achieve spacing between immobilized molecules larger than the optical resolution by adjusting the sample concentration (typically <250 molecule/field). Because reconstitution and encapsulation efficiency can vary, the dilutions need to achieve optical resolution of samples may vary.

5. It is critical to remove any trapped air bubbles, which present an air–water interface, and prevent the introduction of bubbles at all future steps. Similarly, rinsing should be carried out slowly to avoid excessive shear forces at the surface. Once active enzymes or deposited bilayers are present, either bubbles or shear can destroy the sample.

6. Imaging buffer is TBS with 1% glucose and includes oxygen scavengers (20 U/ml glucose oxidase and 1,000 U/ml catalase), which extend fluorophore lifetime and 100 μM cyclooctate-traene, which prevents fluorophore blinking.

7. First we illuminate with 635 nm (0–1 s) to excite acceptor-labeled molecules; next 532 nm illumination is used (1.5–100 s) to observe FRET events. Finally, 635 nm (100.5–110 s) is used to verify whether the acceptor dye molecules have photobleached.

8. This technique is best for type 1 or 2 (single spanning) membrane proteins that only protrude from one side of the membrane. In these cases, free diffusion within the plane of the bilayer is observed. There is only ~2 nm between the deposited bilayer and the slide surface. Polypeptide protruding beneath the bilayer adsorbs to the quartz and becomes immobilized or denatured.

9. With multichannel slides, the channels are used sequentially to prevent cross contamination. The slide is tilted vertically along its long axis for rinsing and the bottom channel is used first so outflow does not cross the other channels.

10. The presence of fluorescent contamination can be checked at any point using the microscope. Contamination can be introduced at any step and is checked by testing all reagents and processes involved.

References

1. Uversky VN, Dunker AK (2010) Understanding protein non-folding. Biochim Biophys Acta 1804(6):1231–1264

2. Mao AH et al (2010) Net charge per residue modulates conformational ensembles of intrinsically disordered proteins. Proc Natl Acad Sci U S A 107(18):8183–8188

3. Muller-Spath S et al (2010) Charge interactions can dominate the dimensions of intrinsically disordered proteins. Proc Natl Acad Sci U S A 107(33):14609–14614

4. Uversky VN (2002) What does it mean to be natively unfolded? Eur J Biochem 269(1):2–12

5. Boehr DD, Nussinov R, Wright PE (2009) The role of dynamic conformational ensembles in biomolecular recognition. Nat Chem Biol 5 (11):789–796

6. Flory JP (1969) Statistical mechanics of chain molecules. Interscience, New York, NY

7. Stryer L, Haugland RP (1967) Energy transfer: a spectroscopic ruler. Proc Natl Acad Sci U S A 58(2):719–726

8. Chen H, Rhoades E (2008) Fluorescence characterization of denatured proteins. Curr Opin Struct Biol 18(4):516–524

9. Ferreon AC et al (2009) Interplay of alpha-synuclein binding and conformational switching probed by single-molecule fluorescence. Proc Natl Acad Sci U S A 106(14):5645–5650

10. Weninger K et al (2008) Accessory proteins stabilize the acceptor complex for synaptobrevin, the 1:1 syntaxin/SNAP-25 complex. Structure 16(2):308–320

11. Kalinin S et al (2010) On the origin of broadening of single-molecule FRET efficiency distributions beyond shot noise limits. J Phys Chem B 114(18):6197–6206

12. Boukobza E, Sonnenfeld A, Haran G (2001) Immobilization in surface-tethered lipid vesicles as a new tool for single biomolecule spectroscopy. J Phys Chem B 105(48):12165–12170

13. Rasnik I, McKinney SA, Ha T (2005) Surfaces and orientations: much to FRET about? Acc Chem Res 38(7):542–548

14. Ha T et al (1999) Ligand-induced conformational changes observed in single RNA molecules. Proc Natl Acad Sci U S A 96(16):9077–9082

15. Graneli A et al (2006) Organized arrays of individual DNA molecules tethered to supported lipid bilayers. Langmuir 22(1):292–299

16. Roy R, Hohng S, Ha T (2008) A practical guide to single-molecule FRET. Nat Methods 5(6):507–516

17. Pertsinidis A, Zhang Y, Chu S (2010) Subnanometre single-molecule localization, registration and distance measurements. Nature 466 (7306):647–651

18. Lee NK et al (2005) Accurate FRET measurements within single diffusing biomolecules using alternating-laser excitation. Biophys J 88(4):2939–2953

19. McCann JJ et al (2010) Optimizing methods to recover absolute FRET efficiency from immobilized single molecules. Biophys J 99 (3):961–970

20. McKinney SA, Joo C, Ha T (2006) Analysis of single-molecule FRET trajectories using hidden Markov modeling. Biophys J 91 (5):1941–1951

21. Sass LE et al (2010) Single-molecule FRET TACKLE reveals highly dynamic mismatched DNA-MutS complexes. Biochemistry 49 (14):3174–3190

Chapter 2

Application of Confocal Single-Molecule FRET to Intrinsically Disordered Proteins

Benjamin Schuler, Sonja Müller-Späth, Andrea Soranno, and Daniel Nettels

Abstract

Intrinsically disordered proteins (IDPs) are characterized by a large degree of conformational heterogeneity. In such cases, classical experimental methods often yield only mean values, averaged over the entire ensemble of molecules. The microscopic distributions of conformations, trajectories, or sequences of events often remain unknown, and with them the underlying molecular mechanisms. Signal averaging can be avoided by observing individual molecules. A particularly versatile method is highly sensitive fluorescence detection. In combination with Förster resonance energy transfer (FRET), distances and conformational dynamics can be investigated in single molecules. This chapter introduces the practical aspects of applying confocal single-molecule FRET experiments to the study of IDPs.

Key words: Intrinsically disordered proteins, Fluorescence spectroscopy, Single-molecule detection, Förster resonance energy transfer, FRET, Diffusion, Correlation spectroscopy, FCS, Confocal detection, Photon statistics

1. Introduction

Single-molecule Förster resonance energy transfer (FRET) has developed into a mature technique that is used to probe a wide variety of biomolecular processes, including protein folding and dynamics. Like other single-molecule techniques, single-molecule FRET offers a fundamental advantage: it can resolve and quantify the properties of individual molecules or subpopulations inaccessible in classical ensemble experiments, where the signal is averaged over many particles. Fluorescence spectroscopy is a particularly appealing technique, owing to its extreme sensitivity and versatility. In combination with FRET (1, 2), it enables us to investigate intramolecular distance distributions and conformational dynamics of single proteins.

Vladimir N. Uversky and A. Keith Dunker (eds.), *Intrinsically Disordered Protein Analysis:*
Volume 2, Methods and Experimental Tools, Methods in Molecular Biology, vol. 896,
DOI 10.1007/978-1-4614-3704-8_2, © Springer Science+Business Media New York 2012

Time-resolved ensemble FRET can also be used to separate subpopulations and to obtain information on distance distributions (3), but data interpretation is typically less model independent (4). For kinetic ensemble studies, the reactions need to be synchronized, which is often difficult.

Already the first experiments using single-molecule FRET in the context of protein folding (5–8) demonstrated the potential of the method for separating subpopulations and for probing conformational dynamics in proteins. Since then, single-molecule FRET has been used to address a wide range of questions in protein folding and dynamics (9, 10). Because of the possibility to resolve conformational heterogeneity, the method proved to be particularly helpful for probing the properties of proteins in their denatured and other nonnative states (9). In the last few years, the application of single-molecule FRET to IDPs has increased strongly, and the progress in the field has been reviewed recently (11). Here, we will present some of the basic aspects that need to be taken into account for single-molecule FRET experiments on IDPs. We will focus on confocal experiments, mostly on freely diffusing molecules. For experiments on immobilized molecules and their detection via total internal reflection fluorescence (TIRF), see Chapter 1.

The basic idea of a single-molecule experiment using FRET is very simple: a donor dye and an acceptor dye are attached to specific residues of a protein. If a protein molecule resides in the volume illuminated by the focused laser beam, excitation of the donor dye can result in energy transfer to the acceptor. The efficiency of energy transfer, which can be determined from the rates of detected donor and acceptor photons, depends on the distance between the fluorophores, and can thus be related to the separation of the dyes. The changes in fluorescence intensity from donor and acceptor can thus be used to distinguish between different conformational states of a protein and to determine the dynamics of their interconversion.

1.1. Förster Resonance Energy Transfer

The quantitative relationship between the probability of transfer—the transfer efficiency—and the inter-dye distance is given by the theory developed by Förster in the 1940s (1). Accordingly, the transfer efficiency E for the dipole–dipole coupling between a donor and an acceptor chromophore depends on the inverse sixth power of the inter-dye distance r:

$$E = \frac{R_0^6}{R_0^6 + r^6},$$ (1)

where R_0 is the Förster radius, the characteristic distance that results in a transfer efficiency of 50%. Due to the strong distance dependence of the efficiency, FRET can be used as a "spectroscopic ruler" on molecular length scales, typically between 2 and 10 nm. R_0 is calculated as

$$R_0^6 = \frac{9,000 \ln 10 \kappa^2 Q_D J}{128 \pi^5 n^4 N_A},\tag{2a}$$

where J is the overlap integral (defined below), Q_D is the donor's fluorescence quantum yield, n the refractive index of the medium between the dyes, and N_A is Avogadro's number (1, 12). The orientational factor is defined as $\kappa^2 = (\cos \Theta_T - 3 \cos \Theta_D \cos \Theta_A)^2$, where Θ_T is the angle between the donor emission transition dipole moment and the acceptor absorption transition dipole moment, Θ_D and Θ_A are the angles between the donor–acceptor connection line and the donor emission and the acceptor absorption transition moments, respectively. κ^2 varies between 0 and 4, but complete averaging of the relative orientation of the chromophores during the excited state lifetime of the donor results in a value of $2/3$, the value most frequently used in practice (however, see Subheading 3.3). Q_D and n need to be measured (see Note 1), and J is calculated from the normalized donor emission spectrum $f_D(\lambda)$ and the molar extinction coefficient of the acceptor $\varepsilon_A(\lambda)$ according to

$$J = \int_0^\infty f_D(\lambda) \varepsilon_A(\lambda) \lambda^4 d\lambda.\tag{2b}$$

The accuracy of the determination is mainly limited by the accuracy of n, which is often nonuniform and difficult to estimate for a protein (but probably very close to n of the solvent for an unfolded protein), and ε_A, which cannot easily be determined independently and is provided by the dye manufacturer with an uncertainty of at least a few percent. Fortunately, the influence of such uncertainties is moderated by the fact that R_0 depends on these quantities only in a sub-linear fashion [Eq. (2a)].

Experimentally, transfer efficiencies can be determined in a variety of ways (12), but for single-molecule FRET, two approaches have proven particularly useful. One is the measurement of the number of photons (13) emitted from the donor and the acceptor chromophores, n_D and n_A, respectively, and the calculation of the transfer efficiency according to

$$E = \frac{n_A}{n_A + n_D},\tag{3}$$

where the numbers of photons are corrected for the quantum yields of the dyes, direct excitation of the acceptor, the detection efficiencies of the optical system in the corresponding wavelength ranges, and the crosstalk between the detection channels (see Note 2). A second approach to measure E, which can be combined with the first (14), is the determination of the fluorescence lifetime of the donor in the presence (τ_{DA}) and absence (τ_D) of the acceptor, yielding the transfer efficiency as

$$E = 1 - \frac{\tau_{DA}}{\tau_D}. \qquad (4)$$

Frequently, we have to consider a distance distribution instead of a single distance, especially in unfolded proteins. If information about the distance distribution is available from simulations or additional experiments, it can be included in the analysis (15). In general, it is important to be aware of the different averaging regimes, because the timescales of both, conformational dynamics of the protein and reorientational dynamics of the dyes, influence the way the resulting transfer efficiency has to be calculated (15).

1.2. Outline of the Procedures

Performing a single-molecule FRET experiment on IDPs requires several steps. First, protein samples suitable for labeling have to be prepared, either by chemical synthesis or by recombinant expression in combination with site-directed mutagenesis. After identifying a suitable dye pair with the Förster radius in the desired range, the fluorophores need to be attached to the protein as specifically as possible to avoid chemical heterogeneity. Control experiments may need to be performed to ensure that the conformational properties of the proteins are not altered by the labels. After customizing the instrument for the sample, data can be taken either on freely diffusing molecules or on immobilized molecules, depending on the observation times required. Finally, the data are processed and analyzed, taking into account the specific dynamic and structural properties of the system under study.

2. Materials

2.1. Instrumentation

The details of confocal single-molecule instrumentation can be found in recent reviews (16, 17). An important development for the wide application of confocal single-molecule methods to the study of biomolecules is the recent availability of comprehensive commercial instrumentation (18).

Experimental setups for confocal single-molecule FRET typically involve excitation with pulsed or continuous wave (cw) lasers and single photon counting with avalanche photodiodes (APDs) or similar detectors. Particular advantages of such instruments compared to camera-based imaging systems (cf. Chapter 1) are the more detailed and quantitative spectroscopic information available and the much better time resolution. Imaging of immobilized samples can be performed by scanning the sample. Advantages of camera-based imaging systems are the possibility to monitor many molecules in parallel, and that the instruments are significantly cheaper and easier to set up.

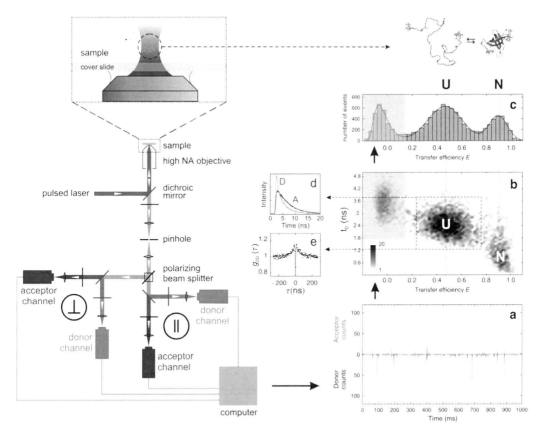

Fig. 1. Overview of instrumentation and data reduction in confocal single-molecule spectroscopy. The scheme on the *left* illustrates the main components of a four-channel confocal single-molecule instrument that collects fluorescence photons separated by polarization and wavelength and records their individual arrival times. (**a**) Sample of a trajectory of detected photons recorded from molecules freely diffusing in solution (in this example Csp*Tm* in 1.5 M GdmCl), where every burst corresponds to an individual molecule traversing the diffraction-limited confocal volume (see *upper left* of the scheme). (**b**) 2-Dimensional histogram of donor fluorescence lifetime τ_D versus transfer efficiency E calculated from individual bursts, resulting in subpopulations that can be assigned to the folded and unfolded protein, and molecules without active acceptor at $E \approx 0$ (shaded in *grey*). (**c**) Projection of the histogram onto the E axis. (**d**) Subpopulation-specific time-correlated single photon counting histograms from donor and acceptor photons from all bursts assigned to unfolded molecules that can be used to extract distance distributions (19). (**e**) Subpopulation-specific donor intensity correlation function, in this case reporting on the nanosecond reconfiguration dynamics of the unfolded protein (20).

Figure 1 shows a schematic with the main optical elements for confocal epifluorescence detection. A laser beam is focused with a high numerical aperture objective to a diffraction-limited focal spot that serves to excite the labeled molecules. In the simplest experiment, the sample molecules are freely diffusing in solution at very low concentration, ensuring that the probability of two molecules residing in the confocal volume at the same time is negligible. When a molecule diffuses through the laser beam, the donor dye is excited; fluorescence from donor and acceptor is collected through the objective and gets focused onto the pinhole, a small aperture serving as a spatial filter. The photons are then separated by polarization and/or wavelength by polarizing beam splitters and dichroic mirrors, and directed to the corresponding detectors. State-of-the-art counting electronics record the arrival time of every photon with picosecond time resolution (21–23).[1] From the resulting photon record, a range of quantities can be derived, e.g., donor and acceptor fluorescence intensities, transfer efficiencies, fluorescence lifetimes, anisotropies, burst durations, photon counting histograms, intensity correlation functions, etc. (24).

Intensity correlation functions are particularly useful for monitoring conformational dynamics. For completely unfolded proteins, diffusive chain dynamics of segments in the range of 50–100 amino acids typically occur on timescales of 10–100 ns (20, 25–27). For IDPs with significant amounts of secondary or tertiary structure, internal dynamics may occur on significantly longer timescales. For freely diffusing molecules, dynamics from the nanosecond (25) to the millisecond range are accessible. Correlation functions of data from immobilized molecules (cf. Chapter 1) can be used to extend this timescale further.

An advantage of observing freely diffusing molecules is that perturbations from surface interactions can largely be excluded, but the observation time is limited by the diffusion times of the molecules through the confocal volume. Typically, every molecule is observed for no more than a few milliseconds in the case of proteins, even though this timescale can be extended to ~100 ms by taking into account the recurrence of individual molecules to the confocal volume (28). Alternatively, molecules can be immobilized on the surface and then observed for a more extended period of time, typically a few seconds to minutes, until one of the chromophores undergoes photodestruction or bleaching. For this purpose, the instrument can, e.g., be coupled to a piezo *xyz* flexure stage for sample scanning, which allows the acquisition of fluorescence images and the reproducible positioning of the laser beam on individual molecules. Potential complications in this case are interactions with the surface, which may perturb the sensitive equilibria and dynamics of IDPs.

For sample design, especially for choosing the chromophores, it is important to be aware of the characteristics and limitations of the instrument, such as the laser lines available for excitation or the time resolution and signal-to-noise ratio achievable with the detectors. The lasers typically used range from simple cw systems with a single fixed wavelength to tunable, pulsed lasers, or supercontinuum sources (29) that make a broad range of wavelengths accessible. With a pulsed source, fluorescence lifetimes become available in addition to intensities, which can provide additional information and improved separation of subpopulations (14). Temperature control of the sample can be implemented, but since the objectives

---

[1]The limiting factor for the time resolution currently is the jitter of APDs in the range of 100 ps.

should not be heated to high temperatures, a temperature gradient across the sample may need to be taken into account by a suitable calibration (30).

2.2. Chemicals

Obviously, single-molecule fluorescence experiments make great demands on buffer preparation and sample purity, since we rarely have the means to distinguish sample molecules unequivocally from contaminants. Some solutes, e.g., denaturants or osmolytes, may need to be used at molar concentrations. Highest purity buffer substances are therefore strictly required. Common buffers such as phosphate salts and Tris can be obtained in excellent purity from most major suppliers as spectrophotometric grade chemicals, other substances only from more specialized sources (e.g., GdmCl and Tween 20 from Pierce Biotechnology, IL). As a general rule, all solutions have to be tested in the single-molecule instrument for fluorescent impurities prior to use. Quartz-bidistilled water is recommended; water from ion exchanger water purification systems (MilliQ) is usually suitable, but needs to be monitored more regularly for contaminations. Fluorophores for protein labeling can be obtained with a variety of reactive groups from several manufacturers, such as Molecular Probes/Invitrogen (Alexa Fluors), Amersham Biosciences (cyanine dyes), Atto-Tec (Atto dyes).

2.3. Chromatography

Purification of labeled proteins proceeds very much the same way as any other protein purification, but again, buffers should have very low fluorescence background, especially in the final steps. HPLC or FPLC systems with fluorescence and diode array absorption detectors can greatly simplify the identification of correctly labeled species.

3. Methods

3.1. Choosing the Fluorophores

Several criteria must be met by chromophores for single-molecule FRET:

1. They must have suitable photophysical and photochemical properties, especially a large extinction coefficient ($\sim 10^5$/M/cm or greater), a quantum yield close to 1, high photostability, a low triplet state yield, and little intensity fluctuations (blinking) from the broad range of photochemistry that can affect FRET measurements.

2. The absorption maximum of the donor chromophore must be close to a laser line available for excitation.

3. Good spectral separation of donor and acceptor emission is necessary to minimize direct excitation of the acceptor and to reduce crosstalk between the detection channels.

4. Acceptor absorption and donor emission spectra must give an overlap integral that results in a suitable Förster radius [calculated from Eqs. (2a) and (2b)]. The best sensitivity for distance changes can be obtained for distances close to R_0.

5. The dyes must be available with suitable functional groups for specific protein labeling (typically succinimidyl esters for amino groups or maleimides for sulfhydryl groups).

6. The dyes must be sufficiently soluble in aqueous buffers, otherwise they may induce protein aggregation, a problem that has been minimized by the introduction of charged groups in many of the popular dyes.

Note that some of the fluorophores' properties may change upon attachment to the protein. In many cases, it is thus advisable to screen a series of dye pairs. The most commonly used dyes are organic fluorophores developed specifically for sensitive fluorescence detection. Examples of common dye pairs are Cy3/Cy5 and Alexa 488/Alexa 594. Semiconductor quantum dots (31) are promising candidates due to their extreme photostability, but they are not yet available with single functional groups, so far they can be used only as donors because of their broad absorption spectra, and they are themselves of the size of a small protein, which increases the risk of interference with the conformational dynamics.

3.2. Protein Labeling

Specific placement of fluorophores on the protein ideally requires groups with orthogonal chemistry. For simple systems, such as short peptides, sequences can be designed to introduce only single copies of residues with suitable reactive side chains (8, 15), e.g., one thiol and one amino group. In solid phase peptide synthesis, protection groups and the incorporation of nonnatural amino acids can be used to increase specificity, but for longer chains, chemical synthesis becomes inefficient and shorter chains must be ligated (32) to obtain the desired product.

The production of proteins of virtually any size and sequence by heterologous recombinant protein expression is the method of choice to obtain very pure material in sufficiently large amounts for preparative purposes. But the number of functional groups that can be used for specific labeling is then very limited. Sufficiently specific reactivity in natural amino acids is provided only by the sulfhydryl groups of cysteine residues, the ε-amino groups of lysine side chains and the free α-amino group of the N-terminal amino acid. However, except for small peptides, the statistical and therefore often multiple occurrence of cysteine and especially lysine residues in one polypeptide prevents the specific attachment of labels. Increased specificity can be achieved by removing unwanted natural cysteines by site-directed mutagenesis or introducing cysteines with different reactivity caused by different molecular environments within the protein (33). Labeling is usually combined with multiple chromatography steps to purify the desired adducts. Alternative methods (34) are native chemical ligation of recombinantly expressed and individually labeled protein fragments or intein-mediated protein splicing (35), the specific reaction with thioester derivatives of dyes (36), puromycin-based labeling using in vitro translation (37), or introduction of nonnatural amino acids (38). Most of the latter methods are not yet used routinely, are not openly available, or must be considered under development.

Currently, the most common approach is still to rely on cysteine derivatization. Here we present a short outline for labeling an IDP with a FRET pair:

1. The labeling positions should be chosen such that the process of interest can be monitored optimally. If you plan to monitor the conformational distribution or dynamics in the completely disordered or unfolded state, choose the labeling positions such that the sequence separation results in an average transfer efficiency near 0.5. Depending on the compactness of the chain, this will typically be in the range of 50–100 amino acids.[2] If the conformational transition to a folded state in the presence of a ligand or binding partner is at the focus of interest, a sufficient difference between the transfer efficiencies in the disordered and ordered states will be required. All solvent-accessible cysteine residues that might interfere with the labeling reaction need to be removed by site-directed mutagenesis, and two surface-exposed cysteines are introduced.

2. Express the protein, purify it under reducing conditions and concentrate it to at least 200 μM.

3. Remove the reducing agent and adjust the pH by passing the protein over a desalting column equilibrated with 50 mM sodium phosphate buffer pH 7.0. Ensure that the resulting protein concentration is at least 100 μM.

4. React the protein with the first chromophore (see Note 3) by adding the maleimide derivative of the dye at approximately a 1:1 molar ratio, incubate 1 h at room temperature or at 4°C overnight.[3]

---

[2]The mean squared distance between the attachment points can be estimated from $\langle r^2 \rangle = 2l_p l_c$, where l_p is the persistence length of the IDP, typically in the range between 0.2 and 0.4 nm (19, 39) in the absence of strong charge repulsion (40), and l_c is the contour length of the segment, which can be calculated as $N \cdot 0.38$ nm, where N corresponds to the number of peptide bonds in the segment (19). The observed transfer efficiency can then be estimated from Eq. (5).

[3]For preliminary screens or in case the sequential labeling procedure is not feasible, both dyes can be added simultaneously. In many cases, the fraction of molecules that contain only donor chromophores can be reduced by empirically varying the ratio of donor and acceptor dye in the reaction. Even relatively large proportions of "donor only" molecules can be tolerated in single molecule experiments because of the separation of subpopulations, but a high-quality sample preparation will simplify data acquisition, analysis, and interpretation greatly.

5. Separate unlabeled, singly labeled, and doubly labeled proteins by chromatography, e.g., by ion exchange chromatography, taking advantage of the net charge on many common chromophores. In favorable cases, this method even allows the separation of labeling permutants. Including low concentrations (0.001–0.01%) of detergents such as Tween 20 can reduce protein losses due to nonspecific adsorption to the column material.

6. Concentrate the singly labeled protein to at least 100 μM and react it with the second chromophore as in steps 3 and 4. Make sure that the pH is adjusted properly.

7. Separate singly and doubly labeled protein as in step 5. Frequently, size exclusion chromatography will suffice.

3.3. Controls

Interactions of the dyes with the protein surface or the polypeptide chain can reduce the mobility of the chromophores, reduce their quantum yield by quenching, and affect the stability of the protein. This needs to be taken into account both for the design of the labeled variants and the control experiments. Due to the substantial size of the fluorophores, they can usually only be positioned on the solvent-exposed surface of the protein if the folded structure is to be conserved. Even then, the use of hydrophobic dyes can lead to aggregation of the protein. Positions in proximity to aromatic amino acids, especially tryptophan, should be avoided to minimize the risk of fluorescence quenching (41). Important controls are equilibrium or, better, time-resolved fluorescence anisotropy measurements on the ensemble or single-molecule level (15, 24), which are sensitive to the rotational flexibility of the dyes and can therefore provide indications for undesirable interactions with the protein surface. In the disordered state of IDPs, the rotational mobility of the dyes is often affected very little compared to folded proteins. It is also essential to ensure by direct comparison with unmodified protein that labeling has not substantially altered the properties of the IDP.

Several factors can complicate the extraction of quantitative distance information from single-molecule fluorescence experiments. Most common effects are photobleaching, possible interactions of the chromophore with the polypeptide (resulting in a reduction of quantum yields or lack of fast orientational averaging of the dyes) or a change of solvent conditions, which can affect the refractive index and the photophysics and photochemistry of the dyes. A suitably labeled control molecule that essentially provides a rigid spacer between the dyes, and whose conformation does not change under denaturing conditions can thus be valuable for avoiding a misinterpretation of the results. Two suitable types of molecules are double-stranded DNA (2) and polyproline peptides (15). Since the type of attachment chemistry and the characteristics of the immediate molecular environment can influence the photophysical properties of the fluorophores (42), it is desirable to use a polypeptide-based reference molecule.

In general, it is always essential to compare the results from single-molecule experiments quantitatively with ensemble data (see Note 4). Even though it may be tempting to analyze only the results from a few selected molecules, the overall result must agree with the ensemble measurement, and the criteria for singling out molecules for analysis have to be as objective and clearly defined as possible.

3.4. Other Technical Aspects

3.4.1. Optical Elements

Customizing the single-molecule instrument will involve the installation of suitable spectral filters specific for the dye pair used. A compromise between maximum collection efficiency, minimal background from scattering, and cross-talk between the channels (especially donor emission leakage into the acceptor detector) has to be established. At least two dichroic mirrors are required, but the signal-to-noise ratio can usually be improved by additional filters, e.g., long-pass filters to reject scattered laser light, a laser line filter, and band-pass or long-pass filters for the individual detection channels. A broad range of filters and dichroic mirrors is available from companies such as Chroma, Omega Optical, or Semrock. Other variables involve the choice of excitation light intensity and laser pulse frequency (which should be optimized carefully by systematic variation), the objective used (e.g., water immersion vs. oil immersion (see Note 5)), and the size of the pinhole in a confocal setup.

Cells for single-molecule measurements are usually assembled using glass cover slides with a thickness corresponding to the optical correction of the objective. Generally, fused silica results in lower background due to the high purity of the material. Impurities on the surface of cover slides can give rise to background, especially in experiments on immobilized molecules close to the surface of the glass. It can thus be crucial to clean them carefully. In experiments on freely diffusing molecules with water immersion objectives, the focus can often be placed sufficiently far away from the cover slide to avoid such complications. In this case, commercially available chambered cover glasses can be used conveniently.

3.4.2. Additives

Several substances have been identified as additives in single-molecule fluorescence experiments that reduce the complications from photobleaching and blinking of the fluorophores. Photobleaching is assumed to be dominated by excited state (probably triplet state) reactions with highly reactive molecules in solution, such as singlet oxygen. At the same time, however, oxygen is an efficient quencher of long-lived triplet states, whose population at excitation close to optical saturation can decrease the overall fluorescence intensity. The causes of blinking, i.e., the population of transient dark or dim states, are even less well understood. Processes that are known to contribute are, e.g., radical formation by photo-induced electron transfer (43) and related redox reactions (44), cis–trans isomerization as in cyanine dyes (45), and spectral shifts of unknown chemical origin (46). Photobleaching can be reduced by removing oxygen (47), but often at the cost of increased triplet state lifetimes and a resulting decrease in emission rates. Oxygen removal is thus frequently combined with triplet state quenchers such as trolox (48). 2-Mercaptoethanol and cysteamine have also been found to increase photostability and emission rates (30, 48). For several dyes, a detailed photochemical characterization and sophisticated strategies to reduce blinking and bleaching are now available (44, 49).

3.5. Free Diffusion Experiments

The observation of fluorescence from freely diffusing molecules at a concentration of about 10–100 pM is a very simple and robust way of performing single-molecule FRET experiments. In this concentration range, the probability of two or more protein molecules residing in the femtoliter confocal volume at the same time is very small, and the signal bursts observed (Fig. 1) arise from individual molecules, provided aggregation can be excluded. Since the molecules are only observed for about a millisecond each, bursts from hundreds to thousands of individual molecules are typically collected in several minutes to hours, depending on the protein concentration and the statistics required. The simplest way of analyzing the data is by binning them in intervals approximately equal to the average burst duration, typically about 1 ms, and identifying photon bursts by a simple threshold criterion (13). An approach that reduces the contribution of background and optimizes the accurate identification of the beginning and end of each burst is to use the corresponding increase and drop in the photon arrival frequency (14). The photon counts of a single burst are integrated, and a threshold for the total number of photons is used to discriminate signal from noise and to select the largest bursts (13, 14). Typically, burst sizes between 50 and 500 counts are reached, depending on the brightness and the translational diffusion coefficient of the molecules.

For further analysis of the identified bursts, several corrections need to be taken into account: background, differences in quantum yields and detection efficiencies of donor and acceptor, cross-talk between the channels, and direct excitation of the acceptor. A general analysis procedure that can be used to obtain bursts corrected for these effects and that takes into account that the corrections are interdependent is given in Note 2. The corrected bursts can then be analyzed in terms of distributions of transfer efficiencies, fluorescence lifetimes, anisotropies, burst size distributions, correlation functions, or other derived parameters (Fig. 1). Alternatively, the corrections can be included directly to calculate the expected distributions of the observables based on a model for

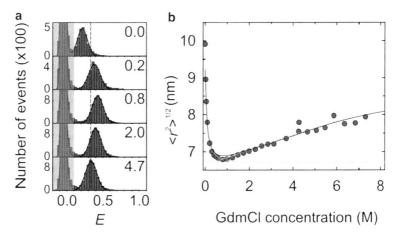

Fig. 2. Single molecule FRET efficiency (*E*) histograms can be used to monitor unfolded state dimensions. (**a**) *E* histograms of FRET-labeled prothymosin α (ProTα) at different guanidinium chloride concentrations (molar concentrations given in the *upper right corners*). The peaks at higher transfer efficiencies correspond to the doubly labeled protein, the peak at $E \approx 0$ corresponds to molecules lacking an active acceptor chromophore. (**b**) Conversion of the mean transfer efficiencies from (**a**) to root mean squared distances shows the pronounced collapse of ProTα upon charge shielding at low, and the re-expansion at very high denaturant concentrations (40).

the underlying distance distribution and the measured burst size distribution, which can then be compared to experiment (cf. Subheading 3.7.1). The most common parameter used for data analysis is the transfer efficiency, which reports on distance distributions or average distances in the molecules under study. As an example, Fig. 2 shows the pronounced change in the transfer efficiency of labeled prothymosin α with guanidinium chloride concentration and the resulting changes in chain dimensions calculated by numerically inverting Eq. (5) and using distance distributions from simple polymer physical models (40).

In many cases, it is helpful to investigate the relation between several of these parameters, for instance in two-dimensional histograms of fluorescence lifetime versus the transfer efficiency calculated from Eq. (3) (24). This multi-dimensional analysis can be used to improve the separation of subpopulations, to recognize characteristic signatures of certain conformational states, and to identify possible complications from quenching or a lack of orientational averaging of the dyes (24). The separation of subpopulations is one of the biggest strengths of single-molecule experiments, since it allows us to assess the heterogeneity of the sample very directly, such as the presence of slowly interconverting conformational states. Even in a simple case, the presence of the folded and unfolded forms of an IDP and the donor-only signal (Fig. 3) would render a quantitative analysis with ensemble fluorescence methods virtually impossible. Additionally, in single-molecule experiments, the

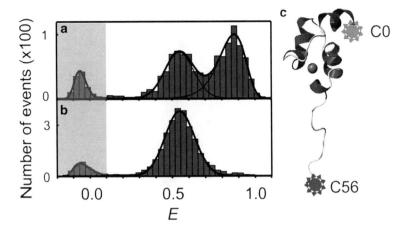

Fig. 3. Single molecule FRET efficiency histograms can be used to assess the conformational heterogeneity of IDPs. (**a**) The N-terminal domain of HIV integrase in the presence of 100 μM Zn^{2+} shows an unfolded subpopulation ($E \approx 0.55$) and a folded subpopulation ($E \approx 0.9$). (**b**) Upon addition of EDTA, the Zn^{2+} is removed from the folded state and all molecules are unfolded. (**c**) Schematic structure of the folded integrase domain.

photon records from individual subpopulations can be combined to obtain subpopulation-specific information with high resolution, e.g., on fluorescence lifetimes (19, 50) or correlation functions (20). A very helpful way of detecting the presence of protein aggregates that are commonly observed for IDPs and other marginally stable proteins are subpopulation-specific burst size distributions (51) (Fig. 4). Fluorescence intensity correlation functions calculated from the entire measurement or subpopulations (52) can also provide a good indication for the presence of aggregates because of their slow translational diffusion (51, 53).

The analysis of subpopulations can additionally be aided by alternating excitation of donor and acceptor (54, 55), which allows the presence of an active acceptor dye to be probed. This can be particularly important if states with very-long-range inter-dye distances are populated, which can lead to an overlap with the donor-only population.

3.6. Experiments on Immobilized Proteins

An approach to extend the observation time of individual proteins beyond their millisecond diffusion time through the confocal volume is their immobilization on a surface. However, nonspecific interactions with the surface can disturb the equilibrium and dynamics of the conformations (6). Strategies for minimizing such interactions include the optimization of surface functionalization (56, 57) or the encapsulation of individual protein molecules in surface-tethered lipid vesicles (58). The absence of binding to the surface can be tested by single-molecule polarization measurements or by quantitative comparison with experiments on freely diffusing molecules.

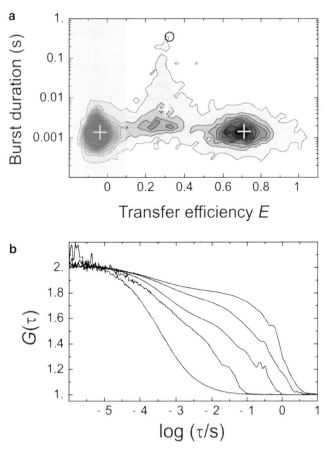

Fig. 4. Identification of protein aggregation in single-molecule FRET experiments of rhodanese (51). (**a**) Contour plot of a two-dimensional histogram calculated from burst duration versus transfer efficiency. (**b**) Fluorescence correlation spectroscopy (FCS) analysis of data from samples containing aggregates of rhodanese prepared under different conditions, resulting in increasing degrees of aggregation (from *left* to *right*). Correlation functions were calculated from the donor fluorescence signal in cross correlation mode to avoid the influence of detector afterpulsing, and the amplitude was normalized to 1 at $\tau = 10$ μs.

Immobilized molecules prepared in this or other ways can be observed either by wide field imaging with evanescent wave excitation (cf. Chapter 1), or with a confocal system in combination with sample scanning or laser scanning. In the case of wide field imaging, the information of all molecules in the field of view is recorded simultaneously, albeit with lower time resolution than in confocal measurements and less spectroscopic detail available. In confocal measurements, the surface is first scanned to identify individual molecules, which are then targeted by the laser individually and observed sequentially. In both cases, trajectories of fluorescence intensities are obtained, terminated by photobleaching of a chromophore, which typically occurs after several milliseconds to minutes, depending on the exciting laser intensity. Transfer efficiencies or other parameters are calculated from the trajectories and corrected in a similar way as for the bursts from freely diffusing molecules (cf. Subheading 3.5). An important criterion for identifying transitions between states or conformational is the anticorrelated signal change between donor and acceptor channels that is expected for a change in distance between the chromophores. The resulting trajectories of transfer efficiencies or other derived parameters can be analyzed by a wide range of methods similar to those pioneered in the field of single-channel recording (59).

3.7. Data Analysis

3.7.1. Photon Statistics

A large number of different and often complementary types of analysis and theoretical approaches have been applied to single-molecule measurements, which in many cases has been crucial for extracting new information from the experimental data. Examples include the use of hidden Markov models (60) and other methods based on information theory (61), generating functions (62, 63), and the concepts of network theory (64), to name but a few. Whereas initial developments in the field originated from seemingly very different concepts, we begin to see some convergence in the methods used. Here, we can just point out some of the most important developments, without describing them in detail.

Frequently, a simple analysis of transfer efficiency histograms does not allow for more than determining the mean transfer efficiencies from the peak positions. The widths and the shapes of the peaks are often ignored although they may contain useful information about structural heterogeneities and dynamics present within and between the subpopulations (65). A main prerequisite is to separate the intrinsic width caused by heterogeneities from shot noise broadening (13). The latter is due to the small and fluctuating number of photons collected and depends on experimental details such as the diffusion time through the confocal volume, the intensity distribution of the laser focus, and the brightness of the fluorophores. Early theoretical contributions (63, 65) culminated in the rigorous treatment of photon statistics in single-molecule experiments by Gopich and Szabo (66–68). They showed how to calculate the joint probability $P(n_D, n_A)$ of finding n_D and n_A donor and acceptor photons in a burst and how to obtain from it theoretical transfer efficiency histograms, which can then be compared with the measured ones. Nir et al. (69) and Antonik et al. (70) simplified FRET efficiency histogram analysis for rigid molecules by factorizing $P(n_D, n_A)$ into two components, one of them being the distribution of the sum of donor and acceptor photons, which can be taken directly from the measured data and includes the influence of translational diffusion, such that detailed knowledge of the shape of the confocal volume is no longer needed. This type of approach can also be used to analyze single-molecule anisotropy histograms (71) and heterogeneous mixtures (72). The method can be used to fit

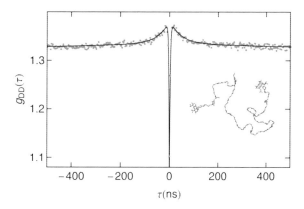

Fig. 5. Measurement of rapid chain dynamics in unfolded proteins with nanosecond fluorescence correlation spectroscopy. The donor intensity autocorrelation function from a Hanbury Brown and Twiss experiment on unfolded Csp*Tm* labeled with a donor and acceptor dye shows the global unfolded chain dynamics in the tens of nanosecond range (20). The fast component in the range of a few nanoseconds is caused by photon antibunching (77).

the measured transfer efficiency histograms directly assuming an underlying, shot noise-free, transfer efficiency distribution. Alternatively, this underlying distribution can be obtained from a more general and rigorous maximum likelihood method (68, 73, 74). In the absence of a suitable model, e.g., because of a combination of orientational and distance distributions, singular value decomposition has proven useful for a quantitative analysis (75).

Recent work shows that conformational dynamics such as the kinetics of protein folding can be obtained from single-molecule FRET data even when the dynamics occur on the same timescale as translational diffusion. Again, modeling the diffusion through the observation volume is not needed. With time-binned raw data, kinetic rate constants can be obtained by analyzing transfer efficiency histograms (68). If the data are available, it is preferable to apply a rigorous maximum likelihood approach to the arrival times and colors of each individual photon (74). The elegant theoretical framework developed by Gopich and Szabo also allows any conformational dynamics or photophysical effects other than FRET to be taken into account (67). It can, e.g., be employed to model rapid conformational dynamics and obtain intramolecular diffusion coefficients of unfolded proteins from fluorescence intensity correlation functions (20, 26, 76) (Fig. 5). Rigorous theoretical tools of this type and the other methods mentioned above will clearly be an important step towards extracting all the information contained in single-molecule measurements.

3.7.2. Timescales and Distance Distributions

An important issue to consider for the analysis of single-molecule FRET experiments is the dynamics of the molecular system. The relative magnitude of the timescales of at least four different processes will have an influence on the position and the width of the FRET efficiency histogram: (a) the rotational correlation time of the chromophores, (b) the fluorescence lifetime of the donor, (c) the intramolecular dynamics of the chain connecting the fluorophores, and (d) the observation timescale.

The rotational correlation time of the chromophores influences the value of the orientation factor κ^2 (Eq. 2a): if dye reorientation is sufficiently fast such that the relative orientation of the donor and acceptor dipoles average out while the donor is in the excited state, κ^2 can be assumed to equal $2/3$. If, in the other extreme, the donor fluorescence decay is much faster than dye reorientation, a static distribution of relative dye orientations can be assumed. Intermediate cases are difficult to treat analytically (12), and simulations become the method of choice (78). $\kappa^2 = 2/3$ is often a good approximation for IDPs, because the rotational correlation times of dyes attached to unstructured polypeptides are typically in the range of a few hundred picoseconds, while their fluorescence lifetimes are in the nanosecond range.

For unfolded proteins, the characteristic times of the fluorescence decay and of long-range inter-dye distance changes are often well separated, i.e., the chain is essentially static on the timescale of the fluorescence decay. The distribution of transfer rates resulting from the distance distribution will thus give rise to highly non-exponential fluorescence decays, which can be used to obtain information about the shape of the underlying distance distribution (3, 19, 50).

Finally, the timescale at which the inter-dye distance distribution is sampled, compared to the observation timescale (more accurately, the inter-photon times (65)), will affect the width of the measured transfer efficiency distributions. As shown by Gopich and Szabo (65, 66), the observation time must be approximately an order of magnitude smaller than the relaxation time of the donor–acceptor distance to obtain physically meaningful distance distributions or corresponding potentials of mean force. Otherwise, only the mean value of the transfer efficiency of the respective subpopulation can be used to extract information about the distance distribution, and an independent model for the shape of the distance distribution is needed. In practice, this means that distance distributions can be determined from free diffusion experiments on proteins if the underlying dynamics are on a timescale greater than about 1 ms, assuming photon count rates of $\sim 10^5/s$ typically achieved during fluorescence bursts (65). A noticeable influence of dynamics on the width, however, is already expected for fluctuations in the 10–100 μs timescale (8, 65). Recently, methods to obtain reaction dynamics from the shape of transfer efficiency histograms have been developed (79) and applied to protein folding dynamics (80).

The three physically most plausible limits for the averaging regimes and the resulting mean transfer efficiencies $\langle E \rangle$ given a normalized distance distribution $P(r)$ are (15):

1. If the rotational correlation time τ_c of the chromophores is small relative to the fluorescence lifetime τ_f of the donor (i.e., $\kappa^2 = 2/3$), and the dynamics of the peptide chain (with relaxation time τ_p) are slow relative to τ_f,

$$\langle E \rangle = \int_a^{l_c} E(r)P(r)\mathrm{d}r \quad \text{with} \quad E(r) = \left(1 + \left(\frac{r}{R_0}\right)^6\right)^{-1}, \quad (5)$$

where $P(r)$ is the normalized[4] inter-dye distance distribution, a is the distance of closest approach of the dyes, and l_c is the contour length of the peptide. This is the most commonly used approximation for unfolded proteins.

2. If $\tau_c \ll \tau_f$ and $\tau_p \ll \tau_f$,[5]

$$\langle E \rangle = \frac{\int_a^{l_c} (R_0/r)^6 P(r)\mathrm{d}r}{1 + \int_a^{l_c} (R_0/r)^6 P(r)\mathrm{d}r}. \quad (6)$$

3. If $\tau_c \gg \tau_f$ and $\tau_p \gg \tau_f$,

$$\langle E \rangle = \int_0^4 \int_a^{l_c} E(r, \kappa^2)P(r)p(\kappa^2)\mathrm{d}r\,\mathrm{d}\kappa^2 \quad \text{with}$$

$$E(r, \kappa^2) = \left(1 + \frac{2}{3\kappa^2}\left(\frac{r}{R_0}\right)^6\right)^{-1}. \quad (7)$$

The theoretical isotropic probability density $p(\kappa^2)$ for the case in which all orientations of the donor and acceptor transition dipoles are equally probable (12) is

$$p(\kappa^2) = \begin{cases} \dfrac{1}{2\sqrt{3}\kappa^2} \ln(2 + \sqrt{3}) & 0 \leq \kappa^2 \leq 1 \\[3mm] \dfrac{1}{2\sqrt{3}\kappa^2} \ln\left(\dfrac{2 + \sqrt{3}}{\sqrt{\kappa^2} + \sqrt{\kappa^2 - 1}}\right) & 1 \leq \kappa^2 \leq 4 \end{cases} \quad (8)$$

---

[4] $\int_a^{l_c} P(r)\mathrm{d}r = 1$.

[5] Note that the averaging has to be done over the transfer rate constant k_t, i.e. $\langle E \rangle = 1/(1 + k_D/\int_a^{l_c} k_t(r)P(r)\mathrm{d}r)$, where $k_t(r) = k_D(R_0/r)^6$, and k_D is the fluorescence decay rate constant of the donor in the absence of the acceptor.

It is important to recognize that even for a molecule with a single fixed distance or very rapid conformational averaging, the resulting FRET efficiency histograms are broad because of shot noise, the variation in count rates about fixed means due to the discrete nature of the signal (only small numbers of photons observed from an individual molecule), but in practice broader histograms than expected from shot noise alone are frequently observed. The origin of this excess width is often unclear (8, 81), but there are factors other than slow distance fluctuations that can contribute, e.g., fluctuating fluorescence quantum efficiencies due to quenching, differences in quantum yields and thus transfer efficiencies of labeling permutants, slow reorientation of the dyes (78), or other optical effects such as a mismatch of the confocal donor and acceptor observation volumes. Consequently, without a suitable reference, it is difficult to assign a width in excess of shot-noise to slow conformational dynamics or distance distributions.

An important alternative approach for obtaining dynamic information is the analysis of correlation functions (82). The details are beyond the scope of this chapter, but the basic idea of these experiments is very simple: fluctuations in the distance between donor and acceptor will lead to fluctuations in transfer efficiency, and thus in the intensities of donor and acceptor. Such intensity fluctuations can be monitored most effectively by fluorescence correlation spectroscopy, and recent developments now allow the investigation of dynamics in unfolded proteins and polypeptides down to the nanosecond range (20, 25) (Fig. 5). The experiments are often done at slightly higher concentrations (~1 nM) to optimize the signal, but the same type of analysis can also be applied to experiments at picomolar concentrations.

3.8. Current Developments

Both the instrumentation and data analysis methods for single-molecule FRET are still developing rapidly. Two important developments that begin to be applicable on a routine basis are the use of microfluidic devices and the extension to more than two colors.

The combination of confocal detection with microfluidic mixing devices (83) is a very useful extension of steady state experiments on freely diffusing molecules and has been evolving rapidly in the past years. This approach allows nonequilibrium processes to be monitored with single-molecule resolution (75, 83–86) and sub-millisecond dead times (86). The application of microfluidic mixers to IDPs is particularly promising for the investigation of coupled folding and binding processes (86).

Similarly promising for the molecular interactions and dynamics of IDPs is the extension of the FRET system to three or four dyes (87). While the demands on sample preparation, instrumentation, and data analysis increase substantially, this approach allows multiple distances to be monitored simultaneously. Especially the site-specific labeling of proteins with more than two fluorophores is biochemically still very demanding. Reported experiments on proteins have thus remained limited to intermolecular FRET, but applications to coupled folding and binding reactions are expected to benefit greatly from this approach.

4. Notes

1. The quantum yield of a chromophore can change upon attachment and should therefore ideally be determined from a protein sample labeled with only one dye. Note, however, that static quenching of the donor does not influence the Förster radius and the observed transfer efficiency.

2. The relation between the raw photon counts $n_{A,0}$ and $n_{D,0}$, as measured in the two detection channels for acceptor and donor emission, respectively, and the corrected values n'_A and n_D can be expressed by the matrix equation

$$\begin{pmatrix} n_{A,0} \\ n_{D,0} \end{pmatrix} = \begin{pmatrix} a_{11} & a_{12} \\ a_{21} & a_{22} \end{pmatrix} \begin{pmatrix} n'_A \\ n_D \end{pmatrix} + \begin{pmatrix} b_A \\ b_D \end{pmatrix},$$

where the matrix a_{ij} describes the cumulative effect of the differences in quantum yields, the different collection efficiencies of the detection channels, and cross-talk (bleed-through), i.e., acceptor emission detected in the donor channel and donor emission detected in the acceptor channel. b_A and b_D are the background count rates in the acceptor and the donor channel, which can be estimated from a measurement on blank buffer solutions.

The elements of matrix a_{ij} can be determined for a specific single-molecule instrument (except for a scaling factor α) from a measurement of two samples containing donor and acceptor dye, respectively, with a concentration ratio equal to the ratio of their extinction coefficients at the excitation wavelength (ensuring that, at identical laser power, the same mean number of excitation events take place per unit time in both samples). By inverting the resulting matrix, the correction matrix $c_{ij} = a_{ij}^{-1}$ is obtained, which transforms the background-corrected raw counts $n_{A,0} - b_A$ and $n_{D,0} - b_D$ into the corrected values n'_A and n_D. Note that the factor α remains unknown, but cancels if intensity ratios are computed, as in the case of the transfer efficiency. Also note that this correction procedure can easily be extended to more than two channels by using a matrix of higher rank. Finally, n'_A has to be corrected for direct excitation of the acceptor according to $n_A = n'_A - (n'_A + n_D)/(1 + \varepsilon_D/\varepsilon_A)$, where ε_D and ε_A are the extinction coefficients of donor and acceptor, respectively, at the excitation wavelength. Ideally, these corrections should already be taken into account for burst identification.

3. Adhere to the labeling instructions given by the manufacturer. Especially the notes provided by Molecular Probes/Invitrogen are very detailed and helpful.

4. These experiments will take up the great majority of the labeled protein samples, which should be taken into account for the preparation scale.

5. With oil immersion objectives, the focal volume must be positioned very close to the cover slide surface to minimize chromatic aberration. In this case it is particularly important to use fused silica cover slides to reduce background from glass luminescence.

6. The signal from molecules with a transfer efficiency of zero is a notorious phenomenon in free diffusion experiments, which can of course be due to incomplete labeling or impurities, but may also be caused by light-induced inactivation of the acceptor. There is no problem if the transfer efficiency of the intact molecules under study is sufficiently different from zero, or if the intact molecules can be separated from the zero transfer events in combination with other observables, such as lifetime or polarization. An elegant general solution is the alternating excitation of donor and acceptor (88, 89), which independently probes the acceptor chromophore and allows all molecules with an inactive acceptor to be excluded from the analysis. Alternatively, the "donor only" peak can be included explicitly in the analysis as a separate population (24).

Acknowledgments

This work has been supported by the Swiss National Science Foundation, the Swiss National Center of Competence in Research for Structural Biology, and a Starting Researcher Grant by the European Research Council.

References

1. Förster T (1948) Zwischenmolekulare Energiewanderung und Fluoreszenz. Annalen der Physik 6:55–75

2. Ha T, Enderle T, Ogletree DF et al (1996) Probing the interaction between two single molecules: fluorescence resonance energy transfer between a single donor and a single acceptor. Proc Natl Acad Sci U S A 93:6264–6268

3. Haas E, Katchalskikatzir E, Steinberg IZ (1978) Brownian-motion of ends of oligopeptide chains in solution as estimated by energy-transfer between chain ends. Biopolymers 17:11–31

4. Vix A, Lami H (1995) Protein fluorescence decay—discrete components or distribution of lifetimes—really no way out of the dilemma. Biophys J 68:1145–1151

5. Jia YW, Talaga DS, Lau WL et al (1999) Folding dynamics of single GCN4 peptides by fluorescence resonant energy transfer confocal microscopy. Chem Phys 247:69–83

6. Talaga DS, Lau WL, Roder H et al (2000) Dynamics and folding of single two-stranded coiled-coil peptides studied by fluorescent energy transfer confocal microscopy. Proc Natl Acad Sci U S A 97:13021–13026

7. Deniz AA, Laurence TA, Beligere GS et al (2000) Single-molecule protein folding: diffusion fluorescence resonance energy transfer studies of the denaturation of chymotrypsin inhibitor 2. Proc Natl Acad Sci U S A 97:5179–5184

8. Schuler B, Lipman EA, Eaton WA (2002) Probing the free-energy surface for protein folding with single-molecule fluorescence spectroscopy. Nature 419:743–747

9. Schuler B, Eaton WA (2008) Protein folding studied by single-molecule FRET. Curr Opin Struct Biol 18:16–26

10. Schuler B, Haran G (2008)) In: Rigler R, Vogel H (eds) Single molecules and nanotechnology, vol 12. Springer, Berlin, pp 181–216

11. Ferreon AC, Moran CR, Gambin Y et al (2010) Single-molecule fluorescence studies of intrinsically disordered proteins. Methods Enzymol 472:179–204

12. Van Der Meer BW, Coker G III, Chen SYS (1994) Resonance energy transfer: theory and data. VCH Publishers, Inc., New York, NY

13. Deniz AA, Laurence TA, Dahan M et al (2001) Ratiometric single-molecule studies of freely diffusing biomolecules. Annu Rev Phys Chem 52:233–253

14. Eggeling C, Berger S, Brand L et al (2001) Data registration and selective single-molecule analysis using multi-parameter fluorescence detection. J Biotechnol 86:163–180

15. Schuler B, Lipman EA, Steinbach PJ et al (2005) Polyproline and the "spectroscopic ruler" revisited with single molecule fluorescence. Proc Natl Acad Sci U S A 102:2754–2759

16. Böhmer M, Enderlein J (2003) Fluorescence spectroscopy of single molecules under ambient conditions: methodology and technology. ChemPhysChem 4:793–808

17. Michalet X, Kapanidis AN, Laurence T et al (2003) The power and prospects of fluorescence microscopies and spectroscopies. Annu Rev Biophys Biomol Struct 32:161–182

18. Wahl M, Koberling F, Patting M et al (2004) Time-resolved confocal fluorescence imaging and spectrocopy system with single molecule sensitivity and sub-micrometer resolution. Curr Pharm Biotechnol 5:299–308

19. Hoffmann A, Kane A, Nettels D et al (2007) Mapping protein collapse with single-molecule fluorescence and kinetic synchrotron radiation circular dichroism spectroscopy. Proc Natl Acad Sci U S A 104:105–110

20. Nettels D, Gopich IV, Hoffmann A et al (2007) Ultrafast dynamics of protein collapse from single-molecule photon statistics. Proc Natl Acad Sci U S A 104:2655–2660

21. Felekyan S, Kuhnemuth R, Kudryavtsev V et al (2005) Full correlation from picoseconds to seconds by time-resolved and time-correlated single photon detection. Rev Sci Instrum 76:083104

22. Wahl M, Rahn H-J, Röhlicke T et al (2008) Scalable time-correlated photon counting system with multiple independent input channels. Rev Sci Instrum 79:123113

23. Wahl M, Rahn HJ, Gregor I et al (2007) Dead-time optimized time-correlated photon counting instrument with synchronized, independent timing channels. Rev Sci Instrum 78:033106

24. Sisamakis E, Valeri A, Kalinin S et al (2010) Accurate single-molecule FRET studies using multiparameter fluorescence detection. Methods Enzymol 475:455–514

25. Nettels D, Hoffmann A, Schuler B (2008) Unfolded protein and peptide dynamics investigated with single-molecule FRET and correlation spectroscopy from picoseconds to seconds. J Phys Chem B 112:6137–6146

26. Gopich IV, Nettels D, Schuler B et al (2009) Protein dynamics from single-molecule fluorescence intensity correlation functions. J Chem Phys 131:095102

27. Mukhopadhyay S, Krishnan R, Lemke EA et al (2007) A natively unfolded yeast prion monomer adopts an ensemble of collapsed and rapidly fluctuating structures. Proc Natl Acad Sci U S A 104:2649–2654

28. Hoffmann A, Nettels D, Clark J et al (2011) Quantifying heterogeneity and conformational dynamics from single molecule FRET of diffusing molecules: recurrence analysis of single particles (RASP). Phys Chem Chem Phys 13:1857–1871

29. Dunsby C, Lanigan PMP, McGinty J et al (2004) An electronically tunable ultrafast laser source applied to fluorescence imaging and fluorescence lifetime imaging microscopy. J Phys D Appl Phys 37:3296–3303

30. Nettels D, Müller-Späth S, Küster F et al (2009) Single molecule spectroscopy of the temperature-induced collapse of unfolded proteins. Proc Natl Acad Sci U S A 106:20740–20745

31. Murphy CJ (2002) Optical sensing with quantum dots. Anal Chem 74:520A–526A

32. Dawson PE, Kent SB (2000) Synthesis of native proteins by chemical ligation. Annu Rev Biochem 69:923–960

33. Ratner V, Kahana E, Eichler M et al (2002) A general strategy for site-specific double labeling of globular proteins for kinetic FRET studies. Bioconjug Chem 13:1163–1170

34. Kapanidis AN, Weiss S (2002) Fluorescent probes and bioconjugation chemistries for single-molecule fluorescence analysis of biomolecules. J Chem Phys 117:10953–10964

35. David R, Richter MP, Beck-Sickinger AG (2004) Expressed protein ligation. Method and applications. Eur J Biochem 271:663–677

36. Schuler B, Pannell LK (2002) Specific labeling of polypeptides at amino-terminal cysteine residues using Cy5-benzyl thioester. Bioconjug Chem 13:1039–1043

37. Yamaguchi J, Nemoto N, Sasaki T et al (2001) Rapid functional analysis of protein-protein interactions by fluorescent C-terminal labeling and single-molecule imaging. FEBS Lett 502:79–83

38. Cropp TA, Schultz PG (2004) An expanding genetic code. Trends Genet 20:625–630

39. Zhou HX (2002) Dimensions of denatured protein chains from hydrodynamic data. J Phys Chem B 106:5769–5775

40. Müller-Späth S, Soranno A, Hirschfeld V et al (2010) Charge interactions can dominate the dimensions of intrinsically disordered proteins. Proc Natl Acad Sci U S A 107:14609–14614

41. Doose S, Neuweiler H, Sauer M (2005) A close look at fluorescence quenching of organic dyes by tryptophan. Chemphyschem 6:2277–2285

42. Hillisch A, Lorenz M, Diekmann S (2001) Recent advances in FRET: distance determination in protein–DNA complexes. Curr Opin Struct Biol 11:201–207

43. Zondervan R, Kulzer F, Orlinskii SB et al (2003) Photoblinking of rhodamine 6G in poly(vinyl alcohol): Radical dark state formed through the triplet. J Phys Chem A 107:6770–6776

44. Vogelsang J, Kasper R, Steinhauer C et al (2008) A reducing and oxidizing system minimizes photobleaching and blinking of fluorescent dyes. Angew Chem Int Ed Engl 47:5465–5469

45. Widengren J, Schwille P (2000) Characterization of photoinduced isomerization and back-isomerization of the cyanine dye Cy5 by fluorescence correlation spectroscopy. J Phys Chem A 104:6416–6428

46. Chung HS, Louis JM, Eaton WA (2009) Experimental determination of upper bound for transition path times in protein folding from single-molecule photon-by-photon trajectories. Proc Natl Acad Sci U S A 106:11837–11844

47. Englander SW, Calhoun DB, Englander JJ (1987) Biochemistry without oxygen. Anal Biochem 161:300–306

48. Rasnik I, McKinney SA, Ha T (2006) Nonblinking and long-lasting single-molecule fluorescence imaging. Nat Methods 3:891–893

49. Vogelsang J, Cordes T, Forthmann C et al (2009) Controlling the fluorescence of ordinary oxazine dyes for single-molecule switching and superresolution microscopy. Proc Natl Acad Sci U S A 106:8107–8112

50. Laurence TA, Kong XX, Jager M et al (2005) Probing structural heterogeneities and fluctuations of nucleic acids and denatured proteins. Proc Natl Acad Sci U S A 102: 17348–17353

51. Hillger F, Nettels D, Dorsch S et al (2007) Detection and analysis of protein aggregation with confocal single molecule fluorescence spectroscopy. J Fluoresc 17:759–765

52. Laurence TA, Kwon Y, Yin E et al (2007) Correlation spectroscopy of minor fluorescent species: signal purification and distribution analysis. Biophys J 92:2184–2198

53. Tjernberg LO, Pramanik A, Bjorling S et al (1999) Amyloid beta-peptide polymerization studied using fluorescence correlation spectroscopy. Chem Biol 6:53–62

54. Kapanidis AN, Lee NK, Laurence TA et al (2004) Fluorescence-aided molecule sorting: Analysis of structure and interactions by alternating-laser excitation of single molecules. Proc Natl Acad Sci U S A 101:8936–8941

55. Müller BK, Zaychikov E, Bräuchle C et al (2005) Pulsed interleaved excitation. Biophys J 89:3508–3522

56. Amirgoulova EV, Groll J, Heyes CD et al (2004) Biofunctionalized polymer surfaces exhibiting minimal interaction towards immobilized proteins. ChemPhysChem 5:552–555

57. Groll J, Amirgoulova EV, Ameringer T et al (2004) Biofunctionalized, ultrathin coatings of cross-linked star-shaped poly(ethylene oxide) allow reversible folding of immobilized proteins. J Am Chem Soc 126:4234–4239

58. Boukobza E, Sonnenfeld A, Haran G (2001) Immobilization in surface-tethered lipid vesicles as a new tool for single biomolecule spectroscopy. J Phys Chem B 105:12165–12170

59. Sakmann B, Neher E (1995) Single channel recording. Plenum, New York, NY

60. Talaga DS (2007) Markov processes in single molecule fluorescence. Curr Opin Colloid Interface Sci 12:285–296

61. Watkins LP, Yang H (2005) Detection of intensity change points in time-resolved single-molecule measurements. J Phys Chem B 109:617–628

62. Brown FLH (2006) Generating function methods in single-molecule spectroscopy. Accounts Chem Res 39:363–373

63. Gopich IV, Szabo A (2003) Statistics of transitions in single molecule kinetics. J Chem Phys 118:454–455

64. Baba A, Komatsuzaki T (2007) Construction of effective free energy landscape from single-molecule time series. Proc Natl Acad Sci U S A 104:19297–19302

65. Gopich IV, Szabo A (2003) Single-macromolecule fluorescence resonance energy transfer and free-energy profiles. J Phys Chem B 107:5058–5063

66. Gopich IV, Szabo A (2005) Theory of photon statistics in single-molecule Förster resonance energy transfer. J Chem Phys 122:1–18

67. Gopich IV, Szabo A (2009) In: Barkai E, Brown FLH, Orrit M, Yang H (eds) Theory and evaluation of single-molecule signals. World Scientific Pub. Co., Singapore, pp 1–64

68. Gopich IV, Szabo A (2007) Single-molecule FRET with diffusion and conformational dynamics. J Phys Chem B 111:12925–12932

69. Nir E, Michalet X, Hamadani KM et al (2006) Shot-noise limited single-molecule FRET histograms: comparison between theory and experiments. J Phys Chem B 110:22103–22124

70. Antonik M, Felekyan S, Gaiduk A et al (2006) Separating structural heterogeneities from stochastic variations in fluorescence resonance energy transfer distributions via photon distribution analysis. J Phys Chem B 110:6970–6978

71. Kalinin S, Felekyan S, Antonik M et al (2007) Probability distribution analysis of single-molecule fluorescence anisotropy and resonance energy transfer. J Phys Chem 111:10253–10262

72. Kalinin S, Felekyan S, Valeri A et al (2008) Characterizing multiple molecular states in single-molecule multiparameter fluorescence detection by probability distribution analysis. J Phys Chem 112:8361–8374

73. Best R, Merchant K, Gopich IV et al (2007) Effect of flexibility and cis residues in single molecule FRET studies of polyproline. Proc Natl Acad Sci U S A 104:18964–18969

74. Gopich IV, Szabo A (2009) Decoding the pattern of photon colors in single-molecule FRET. J Phys Chem B 113:10965–10973

75. Hofmann H, Hillger F, Pfeil SH et al (2010) Single-molecule spectroscopy of protein folding in a chaperonin cage. Proc Natl Acad Sci U S A 107:11793–11798

76. Gopich I, Szabo A (2005) Fluorophore-quencher distance correlation functions from single-molecule photon arrival trajectories. J Phys Chem B 109:6845–6848

77. Fleury L, Segura JM, Zumofen G et al (2000) Nonclassical photon statistics in single-molecule fluorescence at room temperature. Phys Rev Lett 84:1148–1151

78. Hillger F, Hänni D, Nettels D et al (2008) Probing protein–chaperone interactions with single molecule fluorescence spectroscopy. Angew Chem Int Ed 47:6184–6188

79. Gopich IV, Szabo A (2007) Single-molecule FRET with diffusion and conformational dynamics. J Phys Chem B 111:12925–12932

80. Chung HS, Gopich IV, McHale K et al (2011) Extracting rate coefficients from single-molecule photon trajectories and FRET efficiency histograms for a fast-folding protein. J Phys Chem A 115(16):3642–3656

81. Merchant KA, Best RB, Louis JM et al (2007) Characterizing the unfolded states of proteins using single-molecule FRET spectroscopy and molecular simulations. Proc Natl Acad Sci U S A 104:1528–1533

82. Rigler R, Elson ES (2001) Flourescence correlation spectroscopy: theory and applications. Springer, Berlin

83. Lipman EA, Schuler B, Bakajin O et al (2003) Single-molecule measurement of protein folding kinetics. Science 301:1233–1235

84. Hamadani KM, Weiss S (2008) Nonequilibrium single molecule protein folding in a coaxial mixer. Biophys J 95:352–365

85. Pfeil SH, Wickersham CE, Hoffmann A et al (2009) A microfluidic mixing system for single-molecule measurements. Rev Sci Instrum 80:055105

86. Gambin Y, Vandelinder V, Ferreon AC et al (2011) Visualizing a one-way protein encounter complex by ultrafast single-molecule mixing. Nat Methods 8:239–241

87. Gambin Y, Deniz AA (2010) Multicolor single-molecule FRET to explore protein folding and binding. Mol Biosyst 6:1540–1547

88. Kapanidis AN, Laurence TA, Lee NK et al (2005) Alternating-laser excitation of single molecules. Acc Chem Res 38:523–533

89. Muller BK, Zaychikov E, Brauchle C et al (2005) Pulsed interleaved excitation. Biophys J 89:3508–3522

Single-Molecule Force Spectroscopy of Chimeric Polyprotein Constructs Containing Intrinsically Disordered Domains

Marco Brucale, Isabella Tessari, Luigi Bubacco, and Bruno Samorì

Abstract

Here, we describe the single molecule force spectroscopy (SMFS)-based experimental protocol we have recently used to single out different classes of conformations in a chimeric multimodular protein containing an intrinsically disordered (human Alpha Synuclein) domain. Details are provided regarding cloning, expression and purification of the chimeric polyprotein constructs, optimal surface preparation, SMFS data collection and filtering. Although the specificity of the issue and the ensemble of nonstandard techniques needed to perform the described procedures render this a rather unorthodox protocol, it is relatively straightforward to adapt it to the study of other protein domains.

Key words: Single molecule, Force spectroscopy, Conformational equilibria, Amyloidogenesis, Proteopathies, Alpha Synuclein, Neurodegenerative diseases

1. Introduction

In physiological conditions, intrinsically disordered proteins (IDPs) are characterized by a relatively shallow conformational free energy landscape with multiple minima, corresponding to different, rapidly interconverting structures (1–3). During the last decade, IDPs attracted a steadily increasing amount of interest from the scientific community, generating a virtuous feedback of technical and theoretical advances that in turn made the study of IDPs increasingly accessible (4). Since IDPs are involved in extremely diverse biochemical pathways and can interact with a plethora of molecular partners in vivo (5–7), the first step towards understanding their behavior in a given environment means to characterize their conformational equilibria. However, obtaining a detailed characterization

Vladimir N. Uversky and A. Keith Dunker (eds.), *Intrinsically Disordered Protein Analysis:*
Volume 2, Methods and Experimental Tools, Methods in Molecular Biology, vol. 896,
DOI 10.1007/978-1-4614-3704-8_3, © Springer Science+Business Media New York 2012

of the fast-interconverting population of structures assumed by an IDP in a given condition is not a trivial task: very few techniques can give insights into these equilibria, most notably single molecule fluorescence resonance energy transfer (SM-FRET) (8), nuclear magnetic resonance (NMR) (9), and few other spectroscopic techniques (10).

Recently, we have found that, somewhat surprisingly, single molecule force spectroscopy (SMFS) measurements can discern several different classes of conformations in an intrinsically disordered domain (having the sequence of human Alpha Synuclein, aSyn) inserted in a chimeric multimodular polyprotein (11). The insertion of aSyn into a chimeric polyprotein in which it is flanked by several globular domains, acting both as "molecular handles" and as internal mechanical gauges, is absolutely necessary to perform sufficiently clean SMFS experiments (11). This requirement limits the general usefulness of our technique, since the needed flanking modules unavoidably influence the energy landscape of the aSyn domain via steric and electrostatic effects, and its resulting conformational equilibria are thus different from those of WT aSyn in the same conditions (12). Nonetheless, this technique can be used to assess the impact of a single factor of choice on the (perturbed) conformational distribution of a disordered domain. For example, we used this approach to show that aSyn point mutations linked to familial Parkinson increase the propensity of the aSyn domain to acquire compact structures in the tested conditions (12).

Herein, we describe the full procedure we used to apply this SMFS experimental approach to the study of aSyn. Although the specificity of the issue and the ensemble of nonstandard techniques needed to perform the described procedures render this quite an unorthodox protocol, it is relatively straightforward to adapt it to the study of other protein domains.

2. Materials

As remarked above, some of the materials needed to follow this protocol are not available commercially (Subheading 2.1) and are described mainly to provide a basis for customization. Moreover, the practicalities of the SMFS experiments are intrinsically specific to the type of microscope employed. We decided to describe the procedures needed to perform the experiments using a widely commercially available instrument (Subheading 2.3).

2.1. Cloning, Expression, and Purification of Chimeric Polyprotein Constructs

1. pAFM1-4 and pAFM5-8 expression vectors described in detail in Steward et al. (13) and kindly provided by Prof. Jane Clarke (Cambridge University). Briefly, in each plasmid four I27 modules are cloned in tandem, separated by unique restriction sites

(respectively, BamH I, Sac I, BssH II, Kpn I, Nhe I/EcoR I and Nhe I Xba I, Spe I, Mlu I, and EcoR I).

2. WT and mutant (A30P, E46K, and A53T) alpha-synuclein cDNA cloned in a pET28 expression plasmid.

3. Luria–Bertani (LB) broth: 10 g/l bacto tryptone, 5 g/l bacto yeast extract, 5 g/l NaCl; add 15 g/l bacto agar for solid medium for plates.

4. Buffer for proteins purification:

   Resuspension buffer: 20 mM phosphate buffer pH 8, 150 mM NaCl.

   Buffer A: 20 mM phosphate buffer pH 8, 500 mM NaCl.

   Buffer B: 20 mM phosphate buffer pH 8, 500 mM NaCl, 20 mM imidazole.

   Buffer C: 20 mM phosphate buffer pH 8, 500 mM NaCl, 150 mM imidazole.

5. Co^{2+}-affinity resin (Histidine-Select Cobalt Affinity Gel, Sigma Aldrich).

2.2. Surface Preparation

1. Round microscope borosilicate glass coverslips, 15 mm diameter (Glaswarenfabrik Karl Hecht KG, Sondheim, Germany).

2. Bunsen burner.

3. mQ water (see Note 1).

4. Medium Quality (V4–V6) Muscovite Red Mica Sheets (Electron Microscopy Sciences, Hatfield, PA, USA).

5. 99.99% Gold wire, 0.1 mm diameter (Alpha Aesar, Ward Hill, MA, USA).

6. Two-component epoxydic adhesive with low viscosity and high glass transition temperature, such as EPO-TEK 377 (Epoxy Technology, Billerica, MA, USA).

7. High-vacuum thermal evaporator.

8. Oven.

2.3. Constant-Velocity SMFS Experiments

1. 10 μl of the chimeric polyprotein construct solution (see Subheadings 2.1 and 3.1) at a concentration of around 100 μg/ml.

2. Template-Stripped Gold (TSG) surfaces prepared as described in Subheading 3.2.

3. Multimode picoforce atomic force microscope with nanoscope controller (Bruker AXS, Mannheim, Germany).

4. Silicon nitride V-shaped cantilevers, DNP model (Bruker AFM probes, Camarillo, CA, USA).

5. Around 1 ml of the buffer of choice (see Note 2).

2.4. SMFS Data Filtering and Handling

1. "Hooke," an open source software platform (14) for SMFS data handling and automatic filtering, available at http://code.google.com/p/hooke/.

---

3. Methods

3.1. Cloning, Expression, and Purification of Chimeric Polyprotein Constructs

3.1.1. Cloning

1. First, as described elsewhere (13), taking advantage of the *EcoR* I site four bases downstream from the *Nhe* I site in pAFM1-4, clone the 5-8 module fragment from pAFM5-8 into pAFM1-4, obtaining an eight modules plasmid named pAFM-8m in which the eight I27 domains are separated by *BamH* I, *Sac* I, *Bss*H II, *Kpn* I, *Nhe* I *Xba* I, *Spe* I, *Mlu* I, and *EcoR* I.

2. Amplify coding sequences of WT and mutant alpha-synuclein by PCR using mutagenic primers containing respectively *Kpn* I and *Xba* I restriction sites (Syn-pAFM8m-KpnI-FOR: 5′-GAGATCT**GGTACC**ATGGATGTATTC-3′; Syn-pAFM8m-XbaI-REV: 5′-TATTAGT**CTAGA**GGCTTCAGGTTC-3′).

3. Digest both plasmid pAFM-8m and PCR products with *Kpn* I and *Xba* I restriction endonucleases.

4. Dephosphorylate digested pAFM-8m by calf intestinal phosphatase (CIP).

5. Ligate vector and inserts using T4 DNA ligase and use the product of the reaction to transform chemical competent DH5α *Escherichia coli* cells.

6. Screen colonies obtained after overnight growth at 37°C for the presence of alpha-syn sequences and sequence positive clones in order to confirm the accuracy of the sequence and of the insertion.

7. The resulting plasmids are named pAFM-3s3(WT), pAFM-3s3 (A30P), pAFM-3s3(E46K), pAFM-3s3(A53T).

3.1.2. Expression

Expression of recombinant chimeric proteins is achieved as described in Steward et al. (13). Briefly:

1. Transform the expression plasmids into *E. coli* C41 cells (15).

2. Subsequently, inoculate a single colony in 250 ml of rich medium (Luria–Bertani, LB) and grow cells in at 37°C to an OD_{600} of 0.4–0.6.

3. Induce expression by addition of IPTG to 0.2 mM.

4. Grow cells over night at 28°C.

3.1.3. Purification

1. Harvest cells by centrifugation and then resuspend them in 20 ml of resuspension buffer and sonicate on ice (typically six cycles of 30 s at 0.8 Hz).

2. Following centrifugation, incubate the supernatant for 10 min at 60°C and then centrifuge again.

3. Perform on the soluble fraction, containing the protein of interest, a two-step (20 and 45%) ammonium sulfate precipitation.

4. Resuspend the precipitate of the second step in 5 ml of Buffer A and dialyze extensively against the same buffer.

5. Incubate the sample with 1 ml of preequilibrated Co^{2+}-affinity resin in agitation for 1 h at 4°C (see Note 3).

6. Harvest the resin by centrifugation (according to manufacturer's indications), collect supernatant, and save it for reincubation.

7. Wash the resin with 5 volumes of Buffer A and collect the supernatant.

8. Wash the resin with 2 volumes of Buffer B and collect the supernatant.

9. Finally, detach the residual binding protein with 2 volumes of Buffer C and collect the supernatant.

10. Reequilibrate the resin and reincubate with the flow-through saved previously.

11. Repeat the whole procedure for a second and, in some cases, for a third time.

12. Check all the fractions collected by SDS-PAGE, put together the ones containing a significant amount of pure protein, and dialyze extensively against resuspension buffer.

13. Finally, quantify the protein sample and store at −80°C in small aliquots, with addition of glycerol to a final concentration of 15% (v/v) and sodium azide to a final concentration of 0.02% (v/v) as preservatives (see Note 4).

3.2. Surface Preparation

This section describes the practical operations needed to prepare TSG surfaces suited to be used in a typical SMFS experiment. The procedure closely follows that originally described by Wagner et al. (16).

1. From a muscovite mica sheet, cut 15 × 15 mm squares and cleave them using a sharp razor to expose two clean crystalline planes (see Note 5).

2. Place the mica squares on the substrate holder of the thermal evaporator and protect them from the source with the shutter. We usually use a substrate–source distance of approximately 15 cm.

3. Cut a segment of gold wire of appropriate length to obtain a gold layer around 250 nm thick on the substrate (the exact amount depends on the configuration of the metal evaporator) and place it in the thermal evaporator source.

4. Heat the substrate holder to 320°C, bring the thermal evaporator chamber to high vacuum ($<10^{-6}$ torr) and wait for complete thermal equilibration, then allow the mica squares to be degassed for 120 min.

5. Evaporate the gold pellet with a deposition rate of 0.5–1.5 nm/s, protecting the substrates with the shutter until a constant rate is reached.

6. When evaporation is complete, shut off substrate holder heating and allow thermal equilibration at room temperature under high vacuum overnight.

7. Briefly expose several glass cover slides to the Bunsen flame and then rinse with mQ water and dry with a clean nitrogen flow.

8. Mix epoxy adhesive components as specified by the manufacturer.

9. Apply 5 μl of the adhesive mixture to each gold-coated mica square and then gently put one glass slide on each trying to make sure that the entire gold surface under the glass slide is covered by the glue.

10. Put the freshly glued TSGs in an oven at 120°C for 2 h.

11. Remove the TSGs from the oven. To use them, mechanically separate the glass coverslide from the mica support with tweezers immediately before use. The thin gold layer will stick to the glass slide, exposing its flat (mica-templated) side.

3.3. Constant-Velocity SMFS Experiments

The SMFS experimental strategy described here mostly relies on the recognition of different unfolding mechanical behaviors in different force curves. Due to this, the practical precautions normally needed in SMFS experiments to ensure accurate force measurements are less crucial than usual, while it is very important for the TSG surface and the analyte solution to be extremely clean.

1. Glue one freshly cleaved TSG surface (see Subheading 3.2) on an AFM metal support disk.

2. Deposit 10 μl of the analyte polyprotein solution (see Subheading 3.1) on the gold surface and incubate at room temperature for 20 min (see Note 6).

3. Rinse the top of the polyprotein-functionalized TSG with the same buffer that will be used during the SMFS experiment.

4. Assemble the AFM fluid cell, the analyte-bearing TSG, and the cantilever chip in the AFM head (see Note 7).

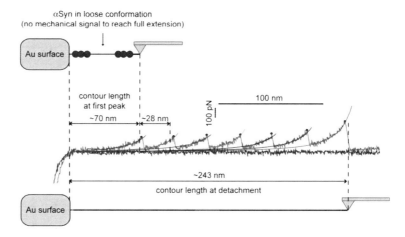

Fig. 1. Mechanical unfolding trace of a 3S3 polyprotein (see Subheading 3.1) containing a WT aSyn domain showing six mechanical unfolding events. *Dark grey trace* = approach, *light grey trace* = retraction. The retraction force curve shows six clean rupture peaks separated by ~28 nm, corresponding to the unfolding of the six I27 domains. The average WLC contour length fitted on the first rupture event is ~70 nm. This distance is compatible with the sum of six folded I27 modules (~4.5 nm each = 4.5 × 6 = 27 nm) and one fully unfolded α-Syn moiety (see cartoon at the *top* of the figure). The WLC contour length fitted on the last peak (detachment from surface) is ~243 nm, compatible with the expected length of a fully unfolded 3S3 construct (680 AA).

5. Flood the cell with the buffer of choice (see Note 8) and focus the laser on the softest cantilever of the chip (in this case, the one having a nominal elastic constant of 0.06 N/m).

6. Determine cantilever spring constant in liquid (see Note 9). We routinely employ the thermal tune procedure as specified by the AFM manufacturer.

7. Start collecting force/distance curves: Apply around 2–3 nN of pressure during contact phase (see Note 10), make sure that the cantilever velocity does not exceed 2 μm/s to avoid hydrodynamic drag artifacts, and occasionally retune the cantilever to ensure maximal consistency of the collected data. see Fig. 1 for an example of a typical force curve obtained following this protocol.

3.4. Data Filtering and Handling

1. Load your raw SMFS data in a Hooke playlist, as described by the developers elsewhere. The software automatically converts raw force/distance curves into force/extension curves.

2. Filter curves not containing clear unfolding events using the software's features.

3. Apply further filtering as needed to arrive at a set of force curves containing clear, single-molecule events in which the mechanical unfolding of the whole polyprotein construct is captured, i.e., with a detachment distance corresponding to the extension of a fully stretched polyprotein, and showing all the expected unfolding events of the globular domains.

4. A completely disordered domain will not give rise to any deviation from the ideal Worm-Like-Chain (WLC) entropic extension behavior. Any repeatedly and consistently observed deviation from this behavior could correspond to a set of mechanically similar configurations assumed by the disordered domain during stretching.

4. Notes

1. The term "mQ water" is used throughout the text to refer to ultrapure water having a resistivity of 18.2 MΩ/cm at 25°C and total organic contaminants content inferior to 5 parts per billion.

2. The overall readability of data gathered in an SMFS experiment is highly dependent on the absence of impurities in the buffer used in the AFM fluid cell. All the buffers should be prepared with ultrapure mQ water and filtered with 0.20-μm syringe filters, e.g., Millex-HPF (Millipore, Billerica, MA, USA) immediately prior to use.

3. All the chimeric recombinant proteins are expressed in frame with a Histidine tag (at the N terminus) suitable for purification but, probably because of its high molecular weight, they have a low affinity for the NTA-matrix. So the protocol has been optimized for high recovery and high purity by incubating two or three times over with the resin the crude sample containing the protein.

4. Protein aliquots prepared in this way can be used for SMFS experiments even after several months of storage at −80°C. The number of curves showing clean mechanical events involving the whole length of a construct molecule will gradually decrease in time.

5. An alternative (and cheaper) procedure is to stick some adhesive tape on top of the mica, then peel it away to expose a fresh layer. In this way, it is easy to reuse mica squares several times.

6. If you are not working in a cleanroom, take care to protect the TSG surface from dust contamination during analyte deposition, for example by placing the TSG in a closed dessicator. In this case, however, do not use TSGs that were freshly glued to the AFM metal support disks, because the cyanoacrilate glue vapors can form contaminating thin films on the exposed drop of analyte.

7. It is mandatory to ensure that the AFM head is protected from mechanical high-frequency vibrations in the surroundings with some sort of vibration-dampening apparatus. A simple and cheap approach that gives excellent results is to place the AFM head on a heavy concrete block suspended via rubber

bands to a chassis. Using this approach, a reasonable signal to noise level can be obtained with no extra precautions.

8. It is critical to avoid the inclusion of air bubbles into the AFM fluid cell when preparing the experiment, and to prevent their formation when data collection started. Take extra care to ensure that that all fluid cell connections are air-tight.

9. We have found that in our experimental setup, a pronounced thermal drift is observed in the first 30-45 min of the SMFS experiment, during which the laser heats the fluid cell to around 28°C when room temperature is around 20°C. Since optical path proportionality factors of the AFM and the spring constant value of the cantilever can be influenced by the varying temperature, it is definitely advantageous to wait for this drift to stop before collecting SMFS data. The thermalization time can of course be reduced by preheating the buffer to around 28°C.

10. During the experiment, it is often useful to empirically fine-tune the amount of pressure exerted by the cantilever tip on the surface and the time the tip spends in contact with the surface. This is because the higher the pressure reached during the contact phase, and the longer the time spent by the cantilever at this pressure, the higher the probability that one or more molecular interactions will be established between the tip and the surface and/or the analyte molecules. By increasing both the exerted pressure and surface delay, it is possible to enhance the occurrence rate of force curves actually containing mechanical signals, at the price of increased aspecific signals and multiple-molecule stretching events.

Acknowledgments

We warmly acknowledge Dr. Massimo Sandal, Dr. Francesco Valle, Dr. Fabrizio Benedetti, Dr. Francesco Musiani, Dr. Sabrina Bonuso, Dr. Laura Civiero, and Dr. Elisa Belluzzi for their contribution to constant refinement of the laboratory procedures during their respective stay in BS's and LB's groups.

References

1. Uversky VN (2003) Protein folding revisited. A polypeptide chain at the folding–misfolding–nonfolding cross-roads: which way to go? Cell Mol Life Sci 60:1852–1871

2. Fink AL (2005) Natively unfolded proteins. Curr Opin Struct Biol 15:35–41

3. Uversky VN, Dunker AK (2010) Understanding protein non-folding. Biochim Biophys Acta 1804:1231–1264

4. Uversky VN, Longhi S (eds) (2009) Instrumental analysis of intrinsically disordered proteins: assessing structure and conformation. Wiley, Hoboken, NJ

5. Dyson HJ, Wright PE (2005) Intrinsically unstructured proteins and their functions. Nat Rev 6:197–208

6. Uversky VN (2010) Multitude of binding modes attainable by intrinsically disordered

proteins: a portrait gallery of disorder-based complexes. Chem Soc Rev 2011 Mar; 40(3): 1623–34. Epub 2010 Nov 3

7. Uversky VN (2010) Targeting intrinsically disordered proteins in neurodegenerative and protein dysfunction diseases: another illustration of the D(2) concept. Expert Rev Proteomics 7:543–564

8. Ferreon AC, Moran CR, Gambin Y, Deniz AA (2010) Single-molecule fluorescence studies of intrinsically disordered proteins. Methods Enzymol 472:179–204. http://www.ncbi.nlm.nih.gov/pubmed/20580965

9. Eliezer D (2007) Characterizing residual structure in disordered protein states using nuclear magnetic resonance. Methods Mol Biol (Clifton, NJ) 350:49–67

10. Eliezer D (2009) Biophysical characterization of intrinsically disordered proteins. Curr Opin Struct Biol 19:23–30

11. Sandal M, Valle F, Tessari I, Mammi S, Bergantino E, Musiani F, Brucale M, Bubacco L, Samori B (2008) Conformational equilibria in monomeric alpha-synuclein at the single-molecule level. PLoS Biol 6:99–108

12. Brucale M, Sandal M, Di Maio S, Rampioni A, Tessari I, Tosatto L, Bisaglia M, Bubacco L, Samori B (2009) Pathogenic mutations shift the equilibria of alpha-synuclein single molecules towards structured conformers. Chembiochem 10:176–183

13. Steward A, Toca-Herrera JL, Clarke J (2002) Versatile cloning system for construction of multimeric proteins for use in atomic force microscopy. Protein Sci 11:2179–2183

14. Sandal M, Benedetti F, Brucale M, Gomez-Casado A, Samori B (2009) Hooke: an open software platform for force spectroscopy. Bioinformatics (Oxford, England) 25:1428–1430

15. Miroux B, Walker JE (1996) Over-production of proteins in *Escherichia coli*: mutant hosts that allow synthesis of some membrane proteins and globular proteins at high levels. J Mol Biol 260:289–298

16. Wagner P, Hegner M, Guntherodt HJ, Semenza G (1995) Formation and in-situ modification of monolayers chemisorbed on ultraflat template-stripped gold surfaces. Langmuir 11:3867–3875

Chapter 4

Visualization of Mobility by Atomic Force Microscopy

Toshio Ando and Noriyuki Kodera

Abstract

Intrinsically disordered regions (IDRs) of proteins are very thin and hence hard to be visualized by electron microscopy. Thus far, only high-speed atomic force microscopy (HS-AFM) can visualize them. The molecular movies identify the alignment of IDRs and ordered regions in an intrinsically disordered protein (IDP) and show undulation motion of the IDRs. The visualized tail-like structures contain the information of mechanical properties of the IDRs. Here, we describe methods of HS-AFM visualization of IDPs and methods of analyzing the obtained images to characterize IDRs.

Key words: High-speed atomic force microscopy, AFM, High-speed AFM, Visualization, Dynamic imaging, Mobility, Mechanical properties

1. Introduction

There are several methods to analyze the structure of IDPs as described in this book. However, visualization of IDRs is very difficult by conventional methods. As IDRs are very thin and highly mobile (and hence take a huge number of conformations), electron microscopy techniques are incapable of visualizing them (1). IDPs are hardly crystallized and hence X-ray crystallography is ineffective for IDPs. AFM can visualize individual nanometer-scale objects under various environments (2). However, its imaging rate is very low to capture highly mobile IDRs in aqueous solutions. Of course, the sample can be dried and immobilized on a substrate surface. However, the structures become thinner upon being dried, resulting in infeasibility of the visualization. Moreover, the drying process very likely alters the IDR structure, which should be avoided. Thus far, visualization of IDRs immobilized on a substrate surface in aqueous solutions has not been successful, because chemical treatments of a substrate surface are apt to increase the surface roughness, which prevents discerning IDRs (1). Thus, only HS-AFM can

Vladimir N. Uversky and A. Keith Dunker (eds.), *Intrinsically Disordered Protein Analysis: Volume 2, Methods and Experimental Tools*, Methods in Molecular Biology, vol. 896, DOI 10.1007/978-1-4614-3704-8_4, © Springer Science+Business Media New York 2012

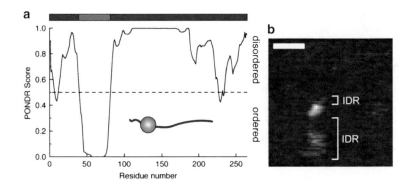

Fig. 1. Prediction by PONDR (Predictor Of Naturally Disordered Regions, http://www.pondr. com/) (3) for the ordered and disordered regions of an IDP (**a**) and an AFM image of the IDP weakly attached to a mica surface in a low ionic strength solution (**b**). Imaging rate, 23.8 fps (42 ms/frame); scan range, 80×80 nm^2; scale bar, 20 nm.

visualize IDRs that are weakly attached to a highly flat surface in buffer solutions, without chemical immobilization (Fig. 1) (1).

HS-AFM (4, 5) has already been applied to the observation of dynamic behavior of several proteins in action (6–9), including the walking behavior of a motor protein myosin V along actin filaments (6) and the structural changes in bacteriorhodopsin in response to light illumination (7). However, the dynamics of these proteins are much slower than that of IDRs on a substrate surface. To visualize highly mobile IDRs with very thin structures using HS-AFM, we need to consider several factors, which are described in this chapter. The molecular movies of IDPs reveal several characteristics of IDPs including the location of the ordered and disordered regions in the molecules, mechanical properties of IDRs, and dynamics of order–disorder transitions of IDRs. Methods to analyze the molecular movies are also described.

2. Materials

2.1. Substrate Surface

Mica (natural muscovite or synthetic fluorophlogopite) has frequently been used as a substrate owing to its surface flatness at the atomic level over a large area (10). It has a net negative charge and is therefore quite hydrophilic. For the following procedures, see Fig. 2.

1. Prepare mica disks (1–2 mm in diameter) from a mica sheet with thickness of <0.1 mm by cutting holes using a sharp puncher (see Note 1).

2. Glue a mica disk to the top surface of a sample stage (a glass rod with 2 mm diameter and 2 mm height) using epoxy and wait until it has dried (~1 to 2 h) (Fig. 2a).

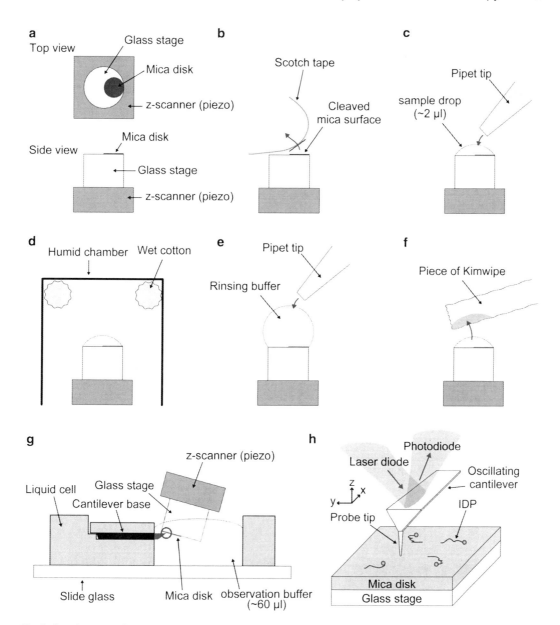

Fig. 2. Sample preparation procedures for HS-AFM imaging. For details, see the text.

3. Glue the sample stage onto the z-scanner of the HS-AFM apparatus using nail enamel and leave it for 5–10 min (Fig. 2a) (see Note 2).

4. Press a Scotch tape to the surface of the mica disk and then smoothly remove the tape from the mica (Fig. 2b). The top layer of the mica will be removed with the tape, which can be checked by inspecting the surface of the removed tape.

5. Place a sample solution on the freshly cleaved mica disk surface for 1–3 min (Fig. 2c, d) and then rinse using an appropriate

buffer solution and a piece of Kimwipe cleaning paper (Fig. 2e, f) (see Note 3).

2.2. Small Cantilevers

Small cantilevers (BL-AC10DS-A2, Olympus: f_c in air 1.5 MHz, f_c in water 600 kHz, $k_c \sim 0.1$ N/m) are commercially available (Atomic Force F&E GmbH, Manheim, Germany). The small cantilevers with a sharp tip made by electron bean deposition (EBD) are also available as an option (this is highly recommended). Although not yet commercialized, small cantilevers will soon be available also from NanoWorld AG (Neuchâtel, Switzerland). When a scanning electron microscope (SEM; a low vacuum type is recommended) is available, sharp EBD tips can be easily made and the expensive small cantilever chips can be used repeatedly (5).

1. Prepare a small container with small holes (~0.1 mm diameter) in the lid.
2. Put a piece of phenol crystal (sublimate) in the container.
3. Put the container in a SEM chamber (if available, under a low vacuum condition).
4. Put cantilevers on the lid of the container immediately above the small holes.
5. Wait for a while until mechanical drift ceases.
6. Irradiate a spot-mode electron beam onto the original tip of the cantilever. Under a low vacuum condition, the tip grows at a rate of ~1 μm/min, while under a high vacuum condition it grows a few times slower (see Note 4).
7. Sharpen the tip (apex radius, ~25 nm) using a plasma etcher (e.g., PE200E; South Bay Technology, California, USA) in argon or oxygen gas. The apex radius can be reduced to ~0.5–1 nm in the best case (see Note 5).

2.3. Buffer Solutions

Prepare all solutions using pure water. Keep all solutions in glass bottles, not in plastic bottles. Pure water used for cleaning several instrument elements such as a cantilever holder, sample stage, etc. should also be kept in a glass bottle (see Notes 6 and 7).

There is no unique choice for the buffer solution used for HS-AFM imaging. However, it is recommended to begin with using a low ionic strength solution such as (10 mM Tris–HCl, pH 7.5, 1 mM $MgCl_2$), because IDRs, which are often rich in both positively and negatively charged amino acids (11), are only weakly adsorbed onto a bare mica surface. When the mobility of IDRs is too high to be imaged, the following changes are recommended for the buffer solution.

1. Change the concentration of divalent cations (Mg^{2+} or Ca^{2+}). Increase in the concentration facilitates binding of negatively charged amino acids of IDPs to a negatively charged mica surface.

2. Reduce the concentration of K^+ (if contained) which binds to a mica surface much more strongly than Na^+ and inhibits the electrostatic protein–surface interaction.

3. Change pH within an allowable range.

Once IDRs are imaged in a low ionic strength solution, the ionic strength should be increased gradually by increasing the concentration of NaCl to examine whether or not the structure of IDRs are influenced by the low ionic strength (see Notes 8 and 9).

2.4. HS-AFM Apparatus   The HS-AFM apparatus will be commercially available worldwide in 2011 from Research Institute of Biomolecule Metrology, Co. Ltd., Tsukuba, Japan.

3. Experimental Methods

3.1. IDP Samples   1. Dilute a stock solution of an IDP (usually ~10 μM order) to 100–200 nM using an appropriate buffer solution containing protease inhibitors (e.g., 1 mM PMSF, 0.5 mM benzamidine, 20 μg/ml TPCK).

2. Divide the diluted sample into small aliquots (~10 μl) and quickly freeze them using liquid nitrogen.

3. Just before AFM experiments, thaw an aliquot of the frozen sample and dilute it to 1–5 nM using an appropriate buffer solution containing protease inhibitors.

4. Store the further diluted sample on ice and use it within 6 h to avoid possible proteolysis.

Ordered domains in an IDP usually attach to a mica surface stably. However, when an IDP is comprised entirely of IDRs, its mobility on a mica surface is too fast to be imaged. In this case, the IDP with a small protein tag, such as GFP, chitin binding protein (CBP), glutathione-S-transferase (GST), or poly-His tag, has to be constructed. Introducing a small protein tag to either the N- or C-terminus of the IDP is recommended because it facilitates the identification of the termini of the imaged elongate molecule. The protein tags attach to a mica surface stably and thereby the mobility of IDRs is significantly reduced.

3.2. HS-AFM Imaging   The procedures for tapping-mode HS-AFM imaging, which are described below, are basically the same as that for conventional tapping-mode AFM imaging. However, note that in the HS-AFM apparatus (Fig. 3), the cantilever and the sample are arranged upside-down so that the cantilever tip points upward to the sample (Fig. 2g) (4, 5).

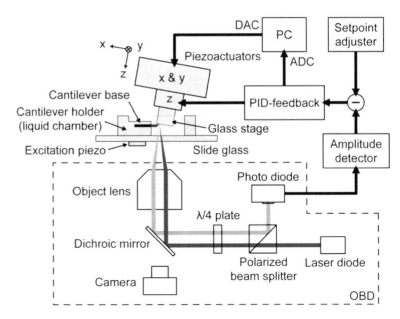

Fig. 3. Schematic of HS-AFM apparatus.

1. Place a buffer solution (~60 μl) around the cantilever installed in its holder of the HS-AFM apparatus (see Note 10).

2. Install the scanner, which has a sample on the z-scanner through the mica-attached sample stage, in the HS-AFM apparatus so that the sample is immersed in the buffer solution placed around the cantilever (Fig. 2g).

3. Adjust the positions of the cantilever and the sample stage by observing them through an objective lens of the optical beam deflection (OBD) detector, optical components, a CCD camera, and a video display, which are implemented in the HS-AFM apparatus.

4. Adjust the position of the incident laser beam outgoing from the OBD detector so that it is focused onto the cantilever.

5. Find the cantilever resonant frequency by measuring the power spectrum of thermal fluctuations in the cantilever deflection.

6. Excite the cantilever by applying AC voltage to a piezoactuator attached to the cantilever holder. Adjust the frequency of the AC voltage at the resonant frequency of the cantilever.

7. Finely adjust the laser beam position so that the maximum amplitude signal from the OBD detector appears (thus, the optimum sensitivity is attained) (see Note 11).

8. Start tip–sample approach by driving a stepper motor attached to the scanner.

9. Once the cantilever tip makes contact with the sample, slightly move the sample stage upward to break the tip–sample contact

by changing the off-set DC voltage for the z-scanner and then measure an amplitude–distance curve to estimate the sensitivity of the amplitude signal to the tip–sample interaction.

10. Slightly move the sample stage upward to break the tip–sample contact again and then readjust the AC voltage for cantilever excitation to attain appropriate free oscillation amplitude of the cantilever (A_0).

11. Adjust the amplitude set point (A_s) at ~$0.95 \times A_0$.

12. Set the values of parameters for the scan size, number of scan lines, and scan speed.

13. Switch on the proportional-integral-derivative (PID) feedback circuit (see Note 12).

14. Start imaging (Fig. 2h).

15. Readjust the amplitude set point so that clear images are obtained (see Note 12).

16. Move the imaging area by changing the off-set DC voltages for the x- and y-scanners. By repeating this operation and video imaging, we can quickly find IDP molecules and record dynamic behavior of many IDP molecules.

3.3. HS-AFM Visualization of IDRs

In HS-AFM, the alteration of imaging parameters is quickly reflected in the images. Therefore, the operation of HS-AFM is easier than conventional AFM. However, we have to take the following considerations into account to successfully visualize IDRs.

1. Cantilever oscillation amplitude: To minimize the tip–sample interaction force, A_0 and A_s should be adjusted to ~1 nm and 0.9–$0.95 \times A_0$, respectively.

2. Scan range: Within a scan range of 100×100 nm^2, the entire of one or two IDP molecules appears when a 1–5 nM sample is applied to a mica surface. This scan size or a slightly smaller size is appropriate for high-speed imaging of IDRs.

3. Imaging rate: Mobile IDRs are usually captured on video at an imaging rate of 10–15 fps. When a higher imaging rate is necessary to record highly undulating IDRs, the imaging rate can be increased only by reducing the scan range and the number scan lines. Although a higher rate of imaging with a wider imaging area is possible, the resulting insufficient feedback operation will damage or fillip the IDP molecules due to too strong tip–sample interaction.

3.4. Image Analysis

3.4.1. Determination of N- and C-Termini

When an IDP is comprised of ordered and disordered regions, we first estimate which end of the imaged elongate molecule is the N-terminus by comparing the image with theoretical predictions for the ordered and disordered regions of the protein (an example is shown in Fig. 1). We can thereby estimate the locations of these regions in terms of the amino acid sequence. However, this estimation sometimes needs to be checked by the below-mentioned additional experiments when the arrangement of the ordered and disordered regions of imaged molecules differs from that provided by the theoretical predictions (see Note 13). When the arrangement is roughly symmetric along the elongated molecule, we also need to conduct the additional experiments.

1. Take images of partially proteolyzed samples of the IDP when the proteolysis products are already characterized (see Note 14).

2. Take images of an IDP construct with a small protein tag at either the N- or C-terminus.

3.4.2. Height and Width of IDR

AFM images can provide the information of sample height with sub-nanometer accuracy. The identification of IDRs can be made by measuring the sample height.

1. Take cross-section profiles of the imaged tail-like structure at various positions along the structure.

2. Subtract the average height of portions at sample-free substrate surface from the peak heights of the cross-section profiles.

3. Make a histogram of the subtracted peak heights.

The histogram usually shows a Gaussian distribution, and therefore, the height value corresponding to the peak of the distribution is considered as the height of the IDR. The height of IDRs is 0.4–0.5 nm, which can be used as a criterion for judging the imaged tail-like structure to be an IDR (see Note 15). The width $(2w)$ of an imaged IDR is largely affected by the tip radius R: $w = (R\rho)^{1/2}$, where ρ is the radius of an IDR. When R is inspected by electron micrographs of the tip, the value of ρ can be estimated using the equation $w = (R\rho)^{1/2}$. However, the mobility of IDRs is often very high, so that they move during capturing one image. In this case, the width is only roughly estimated.

3.4.3. Mechanical Properties of IDR

The tail-like structure of an IDR can be viewed as a macroscopic or microscopic one (Fig. 4) (1). An AFM image of an IDR only provides its macroscopic view because the spatial resolution of AFM is insufficient to resolve the polypeptide chain. The stiffness of a macroscopically viewed IDR string is described by the persistence length p of the string. In two dimensions, the mean square point-to-point distance of the string is given by

$$<r^2(l)>_{2D} = 4pl\left[1 - \frac{2p}{l}(1 - e^{-l/2p})\right],\qquad(1)$$

where l is the contour length between two points on the string (12). The persistence length p represents the stiffness of a macroscopic string structure (13). We can estimate the value of p from the result of best fitting of Eq. (1) to the plot $<r^2(l)>_{2D}$ vs. l (Fig. 4c).

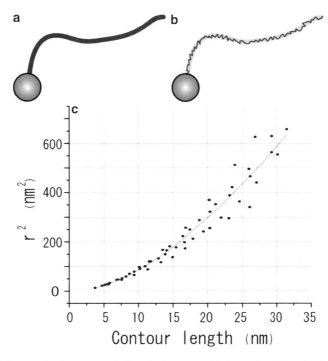

Fig. 4. Schematic of macroscopic (**a**) and microscopic (**b**) views of an IDR and a plot of the mean-square point-to-point distance as a function of the macroscopic contour length of an imaged IDR (1). The *solid line* represents the best-fit curve of Eq. (1).

The value of p is usually 11–12 nm for IDPs (1) (see Note 16) and thus can be used as a criterion for judging the imaged tail-like structure to be an IDR.

When the one-to-one correspondence is cleared between the imaged IDR and its amino acid sequence, we can estimate the microscopic persistence length L_p of the IDR. L_p does not represent the stiffness but represents how loosely the polypeptide chain is folded; when L_p is small, its polypeptide chain is well folded. For an ordered globular protein, L_p is known to be 0.3–0.5 nm (14, 15). The microscopic contour length L_c (i.e., polypeptide chain length) is larger than the macroscopic contour length. Therefore, for the relationship between the end-to-end distance of the IDR (L_{E-E}) and L_c, we can approximately use an equation $<L_{E-E}^2> = 4L_pL_c$ which is obtained from Eq. (1) for $l \to \infty$. L_c can be estimated by $N_{aa} \times 0.34$ nm, where N_{aa} represents the number of amino acids contained in the IDR and "0.34 nm" is the average distance between the nearest neighbor amino acids. When either end of an IDR adjoins an ordered region, we first measure the direct distance between the free end of the IDR and the center of the ordered region and then subtract the half of the height value of the ordered region from the measured distance to estimate the L_{E-E}. When an IDR locates between two ordered regions, we first measure the direct distance between the centers of the ordered regions and then subtract half of the sum of the height values of the ordered regions to estimate the L_{E-E}. Interestingly, the value of L_p is usually in a small range of 1.2–1.3 nm for IDRs (1) (see Note 16). Therefore, this is also used as a criterion for judging the imaged tail-like structure to be an IDR.

3.4.4. Order–Disorder Transition

An IDR is not necessarily always in disordered conformations. Some IDRs show reversible transitions between the ordered and disordered states, which clearly appear in the AFM images; the ordered conformation gives a bright image while the disordered conformation gives a dark image. By measuring the average lifetimes of the ordered state ($<\tau_O>$) and the disordered state ($<\tau_D>$), we can estimate the free energy difference ($\Delta E = E_O - E_D$) between the two states using the equation (16)

$$\frac{\tau_O}{\tau_D} = \exp\left(\frac{-\Delta E}{k_B T}\right) \tag{2}$$

4. Notes

1. Serrated edge formation, which often accompanies partial cleavage of interlayer contacts in the mica disk, should be avoided. Hydrodynamic pressure produced by rapid scanning of the sample stage induces vibrations of the disk through movement of the cleaved sites. For high-speed imaging, the disk should be small (1–2 mm in diameter) to avoid generation of too large a hydrodynamic pressure (17).

2. The sample stage should be removed from the z-scanner soon after performing experiments, using ethanol or acetone. When the stage is tightly glued to the z-scanner after letting it stand for a long time, the removal becomes difficult and its forced removal often breaks the z-scanner.

3. Do not tear a cleaning paper to pieces. Avoid the cleaning paper being touched with the mica surface. The cleaning paper should be touched briefly with a sample solution on the surface. Otherwise, dust particles will be contaminated.

4. The total tip length (original tip + EBD tip) should be longer than 2.5 μm. Otherwise, a so-called "squeeze effect" becomes significant. When an oscillating cantilever is close to the substrate surface, the solution confined between them is squeezed, which damps the cantilever oscillation and lowers the quality of AFM images.

5. The plasma etcher also can be used for cleaning used cantilevers, which lengthens the lifetime of expensive small cantilevers.

6. When water or buffer solutions are kept in plastic bottles for a long time, nano-particles ooze out into the solutions.

7. The cantilevers, cantilever holder and sample stage should be cleaned after performing AFM experiments and stored in a clean container.

8. Since not many IDPs are thus far imaged by HS-AFM, the identification of the ordered and disordered regions by AFM may be claimed to be artifacts due to the surface–sample interaction or the low ionic solutions used. To remove the possible artifact arising from the use of a low ionic solution, it is best to use a higher ionic solution if it does not increase the mobility too much.

9. For a limited numbers of IDPs examined thus far by HS-AFM, we have not found discrepancy between the AFM and NMR results in the identification of the ordered and disordered regions of the IDPs.

10. In HS-AFM, the z-piezoactuator is installed close to the sample stage immersed in a solution. When the volume of buffer solution exceeds a certain level, the solution touches the z-piezoactuator, leading to irreparable damage of the z-piezoactuator.

11. Because the cantilevers for high-speed imaging are very small, precisely aligning the laser beam position relative to a small cantilever cannot be made by the optical view through a $20\times$ objective lens installed in the OBD detector. Therefore, the best alignment is judged by the optimum sensitivity of the OBD detection.

12. When the amplitude set point A_s is larger than $0.95 \times A_0$, the tip often completely detaches at steep downhill regions of the sample (parachuting). During parachuting, bright streaks leaving long tails in the x-direction appear. By gradually getting A_s smaller, these bright streaks disappear and clear images are obtained. However, avoid setting A_s smaller than $0.9 \times A_0$ even when clear images appear. Otherwise, the sample will be damaged. The PID controller implemented in the HS-AFM apparatus is specially designed so that parachuting does not occur as far as A_s is smaller than ~0.9 to $0.95 \times A_0$ (5, 18).

13. When predicted ordered and disordered regions have relatively small numbers of amino acids, the accuracy of the prediction seems low. We have to consider the prediction only as a reference.

14. Analysis of proteolysis products of IDPs has often been carried out to identify the locations of the ordered and disordered regions in the amino acid sequence, particularly in NMR analysis of IDPs. However, we have to keep in mind that even partial proteolysis possibly removes segments critical for forming disordered or ordered regions.

15. A tail-like flexible structure of an imaged molecule does not necessarily means that it is an IDR.

16. This statement is made based on HS-AFM imaging of a few numbers of IDPs. Therefore, we are not yet sure whether this is the case for any IDRs. However, if this is the case, it means that there is no stable state with an intermediate level of disorder (or order) and the intermediate level of disorder may only appear temporarily during transitioning between the highly ordered and completely disordered structure. This issue remains an open question.

Acknowledgment

This work was supported by Grant-in-Aid for Basic Research (S) from JSPS, Knowledge Cluster/MEXT—Japan, and Grant-in Aid for Scientific Research on Innovative Areas (Research in a Proposed Research Area)/MEXT—Japan.

References

1. Miyagi A, Tsunaka Y, Uchihashi T, Mayanagi K et al (2008) Visualization of intrinsically disordered regions of proteins by high-speed atomic force microscopy. Chem Phys Chem 9:1859–1866
2. Binnig G, Quate CF, Gerber Ch (1986) Atomic force microscopy. Phys Rev Lett 56:930–933
3. Romero P, Obradovic Z, Kissinger CR et al (1997) Identifying disordered proteins from amino acid sequences. Proc IEEE Int Conf Neural Networks 1:90–95
4. Ando T, Kodera N, Takai E et al (2001) A high-speed atomic force microscope for studying biological macromolecules. Proc Natl Acad Sci U S A 98:12468–12472
5. Ando T, Uchihashi T, Fukuma T (2008) High-speed atomic force microscopy for nanivisualization of dynamic biomolecular processes. Prog Surf Sci 83:337–437
6. Kodera N, Yamamoto D, Ishikawa R, Ando T (2010) Video imaging of walking myosin V by high-speed atomic force microscopy. Nature 468:72–76
7. Shibata M, Yamashita H, Uchihashi T et al (2010) High-speed atomic force microscopy shows dynamic molecular processes in photo-activated bacteriorhodopsin. Nat Nanotechnol 5:208–212
8. Yamamoto D, Uchihashi T, Kodera N, Ando T (2008) Anisotropic diffusion of point defects in two-dimensional crystal of streptavidin observed by high-speed atomic force microscopy. Nanotechnology 19:384009 (9 pp)
9. Milhiet P-E, Yamamoto D, Berthoumieu O et al (2010) Deciphering the structure, growth and assembly of amyloid-like fibrils using high-speed atomic force microscopy. PLos One 5: e13240 (8 pp)
10. Yamamoto D, Uchihashi T, Kodera N et al (2010) High-speed atomic force microscopy techniques for observing dynamic biomolecular processes. Methods Enzymol 475(B): 541–564
11. Uversky VN, Dunker AK (2010) Review understanding protein non-folding. Biochim Biophys Acta 1804:1231–1264
12. Strobl GR (1996) The physics of polymers. Springer, Berlin
13. Manning GS (2006) The persistence length of DNA is reached from the persistence length of its null isomer through an internal electrostatic stretching force. Biophys J 91: 3607–3616
14. Dietz H, Rief M (2004) Exploring the energy landscape of GFP by single-molecule mechanical experiments. Proc Natl Acad Sci U S A 101:16192–16197
15. Müller DJ, Baumeister W, Engel A (1999) Controlled unzipping of a bacterial surface layer with atomic force microscopy. Proc Natl Acad Sci U S A 96:13170–13174

16. Yamashita H, Voïtchovsky K, Uchihashi T et al (2009) Dynamics of bacteriorhodopsin 2D crystal observed by high-speed atomic force microscopy. J Struct Biol 167:153–158

17. Ando T, Kodera N, Maruyama D et al (2002) A High-speed atomic force microscope for studying biological macromolecules in action. Jpn J Appl Phys 41:4851–4856

18. Kodera N, Sakashita M, Ando T (2006) Dynamic proportional-integral-differential controller for high-speed atomic force microscopy. Rev Sci Instrum 77:083704 (7 pp)

Chapter 5

Unequivocal Single-Molecule Force Spectroscopy of Intrinsically Disordered Proteins

Javier Oroz, Rubén Hervás, Alejandro Valbuena and Mariano Carrión-Vázquez

Abstract

Intrinsically disordered proteins (IDPs) are predicted to represent about one third of the eukaryotic proteome. The dynamic ensemble of conformations of this steadily growing class of proteins has remained hardly accessible for bulk biophysical techniques. However, single-molecule techniques provide a useful means of studying these proteins. Atomic force microscopy (AFM)-based single-molecule force spectroscopy (SMFS) is one of such techniques, which has certain peculiarities that make it an important methodology to analyze the biophysical properties of IDPs. However, several drawbacks inherent to this technique can complicate such analysis. We have developed a protein engineering strategy to overcome these drawbacks such that an unambiguous mechanical analysis of proteins, including IDPs, can be readily performed. Using this approach, we have recently characterized the rich conformational polymorphism of several IDPs. Here, we describe a simple protocol to perform the nanomechanical analysis of IDPs using this new strategy, a procedure that in principle can also be followed for the nanomechanical analysis of any protein.

Key words: Single-molecule force spectroscopy, Atomic force spectroscopy, Intrinsically disordered proteins, Conformational plasticity, Protein nanomechanics

Abbreviations

ΔL_c	Increase in contour length
AFM	Atomic force microscope
DTT	Dithiothreitol
FPLC	Fast protein liquid chromatography
F_u	Average unfolding force
IDP	Intrinsically disordered protein
IPTG	Isopropyl β-D-1-thiogalactopyranoside
LB	Lysogeny broth
MOPS	3-(N-Morpholino) propanesulfonic acid
MPTS	Mercaptopropyl trimethoxysilane
NTA	Nitrilotriacetic acid
OD_{595}	Optical density at 595 nm

Vladimir N. Uversky and A. Keith Dunker (eds.), *Intrinsically Disordered Protein Analysis: Volume 2, Methods and Experimental Tools*, Methods in Molecular Biology, vol. 896, DOI 10.1007/978-1-4614-3704-8_5, © Springer Science+Business Media New York 2012

p	Persistence length
pFS	Plasmid for force spectroscopy
SDS	Sodium dodecyl sulfate
SDS-PAGE	Polyacrylamide gel electrophoresis in the presence of SDS
SMFS	Single-molecule force spectroscopy
WLC	Worm-like chain

1. Introduction

IDPs have not a well defined structure when isolated in solution. Instead, it has been recently shown that these proteins can adopt several conformations, some of which may be scarcely populated and fast fluctuating (1–4). Thus, the characterization of these conformers remains a challenging task for bulk biophysical techniques (5). As a single-molecule technique, SMFS provides a unique means of analyzing this conformational polymorphism, relating protein conformation to mechanical stability.

In SMFS, the protein of interest is stretched in order to measure its mechanical resistance, typically. This resistance is usually unique and characteristic in folded proteins. However, the different conformations of IDPs may also exhibit diverse mechanical properties. Hence, SMFS techniques provide a way to analyze the conformational plasticity of these proteins. In the most common SMFS technique used, AFM, a protein attached to a substrate and the tip of a cantilever (the force sensor) is stretched (usually in N-C direction) by a piezoelectric device and the resistance forces are measured (6). Several approaches have been developed to unambiguously identify and select single-molecule recordings. Most of these approaches are based on polyproteins, tandem repeats of proteins or protein modules, which are easily identified in SMFS force-extension recordings based on the periodicity of equally spaced peaks seen when the length-clamp mode of the AFM is used (6). In this so-called saw-tooth pattern each peak typically originates from the unfolding of an individual protein structure (6–11). The height of each force peak is used to calculate the mechanical stability of the protein (F_u, defined as the average unfolding force) while the distance between peaks reflects the length of the protein region that was previously hidden to the force. By determining the so-called increase in contour length of the molecule (ΔL_c, obtained after fitting the force-extension recordings to the worm-like chain (WLC) model of polymer elasticity, 7) the number of amino acids contained in the force-hidden region of the protein can be calculated and thus, the position of the mechanical barriers can be assigned allowing to infer the type of mechanical structures involved (9, 12–14).

The mechanical unfolding of proteins is a hierarchical process such that a less mechanostable structure will unfold prior a more mechanostable one. However, the proximal region of the force-extension recording is frequently contaminated with nonspecific interactions (15). This spurious contamination can sometimes mask the mechanical unfolding pattern of the protein under study, particularly if it has weak mechanostability and/or a complex ΔL_c pattern (as is the case for some IDPs, (16)).

To circumvent this particular drawback, we have developed a new family of vectors for general use in protein nanomechanics (pFS for *p*lasmid for *F*orce *S*pectroscopy, (15)) that contains a polypeptide with undetectable mechanical resistance by SMFS, as a spacer to bridge the proximal region of the force-extension recordings (pFS-1 version). A second version of the vector (pFS-2) that carries a multi-cloning site in a tolerant loop of a ubiquitin repeat (or the I27 module) has also been developed, which allows the sequence of the protein of interest to be lodged inside its fold, an approach we have termed "carrier-guest" strategy (15, 16). Using this approach, the protein of interest will always be stretched after the unfolding of the carrier protein, and thus far from the problematic proximal region of the recordings. This strategy has recently enabled the unambiguous mechanical analysis of a variety of IDPs (both amyloidogenic and non-amyloidogenic) at the single-molecule level (16). However, this strategy only guarantees that our SMFS data originate from stretching a single IDP molecule, which could still be involved in a series of interactions (with the AFM elements, substrate or cantilever tip, or other IDP molecules forming dimers or even oligomers). We have already described strategies to control each of these interactions (16). For instance for the case of amyloidogenic IDPs, since they have a tendency to oligomerize, we used an inhibitor of the oligomerization process to confirm (by comparison with the untreated sample) that our SMFS data originate from stretching single monomeric molecules (i.e., intramolecular interactions). Ideally, one could perform refolding experiments in a buffer devoid of IDPs in solution as a more general control that would rule out possible interactions between the IDP molecules.

As amyloidogenic IDPs (and in particular, neurotoxic proteins, which are causally related to neurodegenerative diseases) exhibit a well-known amyloidogenic behavior (i.e., formation of toxic oligomers and amyloid fibers), it is of great interest to determine the aggregation state of the sample in order to select the species of potential mechanical interest (i.e., monomers, soluble oligomers, insoluble aggregates, or fibrils). In its original configuration, our custom-made AFM setup was unable to allow the determination of the topography of a given sample (17). However, we have recently added imaging capabilities to this system by integrating commercial imaging hardware (Dulcinea control unit) and software (WSxM) both from Nanotec Electrónica S.L., (http://www.nanotec.es; (18, 19)). This system is now capable of obtaining an AFM image of a region, in contact or dynamic mode, and then selecting the area of interest in order to record a force curve *quasi*-simultaneously with the image. However, since there are conflicting technical requirements between both modes of AFM, applying this new approach to the combined analysis of single molecules would require the development of a new functionalization protocol preserveing the integrity of the single molecules attached to the substrate (18).

Here, we describe a protocol for the use of the pFS-2 vector to analyze the mechanical properties of IDPs using the carrier-guest strategy (see Note 1). This protocol can also be applied in principle to the nanomechanical analysis of any protein (or protein region) using either the pFS-1 or pFS-2 vectors. We detail the steps involved, including the cloning, expression and purification procedures, as well as SMFS data acquisition and analysis.

2. Materials

The materials used for the cloning, expression, and purification of the recombinant proteins are all commercially available, and where relevant, the information of the provider is specified. The materials related to SMFS data acquisition and analysis are specific to our AFM setup, which was first described in ref. 17, and then added with imaging capacities in ref. 18. As such, our setup can perform *quasi*-simultaneous imaging-pulling analysis, although the protocol described here focuses on pulling. While the protocol we describe is based on our specific setup, the use of pFS-1/pFS-2 vectors, their mechanical properties and the criteria for nanomechanical analysis are independent of the AFM apparatus used (see Note 2).

2.1. Cloning, Expression, and Purification of IDPs in pFS-2

1. pFS-1 and pFS-2 vectors have been described elsewhere in detail (Fig. 1a, b; (15, 16)). In brief, these vectors contain a fragment of around 200 amino acids from the N2B polypeptide of human cardiac titin (UniProtKB/Swiss-Prot code Q8WZ42), which unfolds without detectable mechanical resistance and that therefore acts as a spacer, bridging the problematic proximal region of the force-extension recordings (Fig. 1c, d; (20)). In addition, they contain several human ubiquitin repeats (UniProtKB/Swiss-Prot code P0CG47), a protein with putative chaperone activity and well-characterized mechanical properties (21, 22). Both vectors contain a series of interdomain restriction sites (BamHI, XbaI, SalI, NotI, SpeI, BssHII, XhoI, and KpnI, from the N- to C-terminus, Fig. 1a, b) some of which can be used for the directional cloning of any protein of interest (NotI, SpeI, BssHII, and

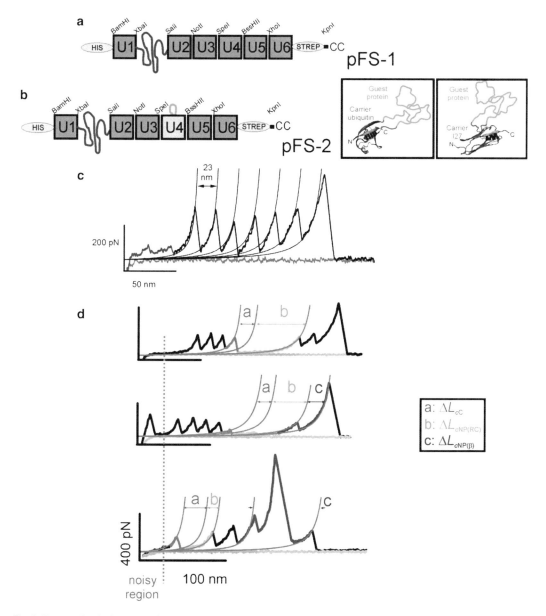

Fig. 1. Nanomechanical analysis of IDPs using pFS polyproteins. (**a**) Schematic representation of the pFS-1 polyprotein. The ubiquitin repeats are represented by *grey boxes* and the N2B fragment is represented as a non-folded polypeptide. (**b**) Schematic representation of pFS-2. The ubiquitin (or I27) repeat containing the multi-cloning site is located at position 4 of pFS. On the *right* is a representation of the carrier-guest construction using both carrier modules: ubiquitin (*left*, with the multi-cloning site located between residues T9 and G10) and I27 (*right*, with the multi-cloning site located between A42 and A43). The mechanical clamps of both carrier modules are indicated (6, 22), demonstrating that the guest IDP is "force-hidden" and that the carrier must unfold prior to stretching the grafted IDP. (**c**) Typical force-extension recording of the pFS-2 polyprotein. The extension gained by the stretching of the N2B fragment, at the beginning of the force-extension recording, serves as a spacer to avoid the usually contaminated proximal region of the force-extension recordings. The ubiquitin force peaks are shown in *black*. (**d**) Representative force-extension recordings of the pFS-2 + Sup35NM. By using this vector we can unambiguously resolve a variety of conformations adopted by Sup35NM (NP in the figure), ranging from mechanically undetectable conformations (putatively random coil, RC, b in the figure, *top trace*) to different mechanostable conformations that exhibit different degrees of mechanical stability (putatively β-structured, shown as C in the recordings). Modified from refs. 15 and 16.

XhoI). In pFS-2, a multi-cloning site is positioned inside the fold of either a ubiquitin repeat (containing AgeI, BsiWI, SmaI, and MluI sites, from the N- to C-terminus, located in loop AB between residues T9-G10, Fig. 1b; (15, 16)) or an I27 module (AgeI and SmaI restriction sites located between residues A42 and A43 in the CD loop, Fig. 1b; (16)) to clone proteins following the carrier-guest strategy. Both multicloning sites are located behind the resistance region of the modules (a.k.a. mechanical clamp). The pRSETA vector was used as the basic platform to construct these vectors (Invitrogen).

2. The cloning steps are performed in the *E. coli* XL1-Blue strain (Stratagene). The culture medium used is Lysogeny Broth (LB, 10 g/l Bacto Tryptone, 5 g/l Bacto Yeast extract, 10 g/l NaCl) with antibiotic added according to the plasmid vector's antibiotic resistance.

3. The recombinant protein is expressed in the *E. coli* C41(DE3) strain (23) by 1 mM IPTG using LB medium (see above).

4. The purification of recombinant proteins can be performed using many different approaches, although we use an FPLC apparatus (ÄKTA Purifier, GE Healthcare). Depending on the purity achieved at each step, several chromatography purification steps can be performed alternatively or sequentially:

   - Ni^{2+}-affinity chromatography: use Histrap HP FPLC columns (GE Healthcare).
     - Binding buffer: 50 mM sodium phosphate/500 mM NaCl/50 mM imidazole [pH 7.4].
     - Elution buffer: 50 mM sodium phosphate/500 mM NaCl/500 mM imidazole [pH 7.4].
   - Strep-tag affinity chromatography: use Streptrap HP FPLC columns (GE Healthcare).
     - Binding buffer: PBS (137 mM NaCl, 10 mM Na_2HPO_4, 2 mM KH_2PO_4, 2.7 mM KCl, [pH 7.4]).
     - Elution buffer: PBS/2.5 mM desthiobiotin.
   - Size exclusion chromatography: use HiLoad 16/60 Superdex TM 200 column (GE Healthcare).
     - Buffer: 100 mM Tris–HCl [pH 7.5]/1.25–1.5 M guanidinium chloride (this concentration does not denature ubiquitin or I27 domains allowing the removal of contaminants that may co-elute with the recombinant protein; (24, 25)).

5. All concentration and buffer-exchange steps are performed by ultrafiltration using Amicon 10K filters (Millipore).

2.2. Preparation of AFM Substrates

Two SMFS substrates have been used with pFS-based recombinant polyproteins: gold-coated substrates, which permit covalent attachment of pFS polyproteins through their C-terminal cysteine residues (26), and NTA-Ni^{2+} functionalized glass coverslips, which attach pFS polyproteins via their N-terminal His-tag (27). Although gold-coated coverslips can be custom-made by thermal deposition (13), commercially available gold substrates (Arrandee) yield acceptable results in SMFS. The following materials are used in the preparation of NTA-Ni^{2+} substrates:

1. Round microscope borosilicate glass coverslips, 14 mm diameter (Thermo Scientific).

2. An oven.

3. Solutions:
   - 20N KOH.
   - MilliQ water.
   - 2% 3MPTS (Sigma-Aldrich)/0.02% acetic acid.
   - 100 mM DTT.
   - 3 mg/ml maleimide-C3-NTA (Dojindo Laboratories) dissolved in 10 mM MOPS [pH 7.0].
   - 10 mM NiCl$_2$.

2.3. Data Acquisition in SMFS

All experiments with IDPs using the pFS-2 vector were performed in the length-clamp mode of the AFM (see below; (6)). Force-clamp mode and "refolding" protocols to analyze structure formation can also be used (28). A basic protocol for length-clamp nanomechanical analysis of IDPs is presented here focusing on the "unfolding" process to analyze the breakage of structures.

1. 10–20 μl of the pFS-2 polyprotein carrying the IDP (see Subheading 2.1) at a concentration of 2–3 μM.

2. AFM substrates, as described in Subheading 2.2.

3. The AFM apparatus:
   - Multimode AFM head: TVOH-MMAFMLN (Veeco).
   - Laser System (Schäfter + Kirchhoff): power supply SK9732C, laser diode collimator 50BM, laser-beam coupler 60SMS-1-4-A8-07, single mode fiber cable SMC630-5-NA010-3APC-0-50, collimator lens 60 FC-4-M12.
   - Fluid cell (Veeco).
   - Multiaxis closed-loop PicoCube P-363.3CD piezoelectric positioner (Physik Instrumente). This piezoelectric positioner is equipped with a capacitive sensor (PZT-Servocontroller E-509.C3A, Physik Instrumente) that allows subnanometer resolution in its displacement.

- AFM controller (JRC Instruments).
- Dulcinea high voltage unit (Nanotec Electrónica S.L.).
- Data acquisition boards (PCI-6052E and PCI-6703, National Instruments) mounted into a personal computer.
- Commanding software: Igor Pro (Wavemetrics) and WSxM (Nanotec Electrónica S.L.).

4. Si_3N_4 AFM cantilevers: Biolever (Olympus) or MLCT-AUNM (Veeco).

5. UV/Ozone ProCleaner™ Plus lamp (Bioforce Nanosciences Inc.).

6. Experimental buffer (0.22 μm filtered).

7. Vibration isolation table (Nano-K 25BM-4; Minus K Technology).

2.4. SMFS Data Analysis   All SMFS data were collected and analyzed in Igor Pro (Wavemetrics) using home-made protocols. The analysis is based on the fitting of the recordings to the WLC, which models the elasticity of polymers (7). The following parameters are typically analyzed: p (persistence length), F_u (unfolding force), ΔL_c (increase in contour length), total length of the molecule, and extension at which the force peak appears.

---

3. Methods

3.1. Cloning, Expression, and Purification of IDPs in the pFS-2   As an example of the use of the pFS-2 vector to analyze IDPs, we focus on the nanomechanical analysis of the Sup35NM prion from *Saccharomyces cerevisiae* (residues 1–253, UniProtKB/Swiss-Prot code P05453; (29)), which was cloned into the pFS-2 using the carrier-guest strategy (16).

Cloning

1. The codifying sequence for Sup35NM was PCR-cloned using the pJCSUP35 plasmid (Addgene) as the template. As the sites chosen for the cloning of this sequence into the multi-cloning site inside the I27 module were AgeI and SmaI, the synthesized oligonucleotides primers should contain these sites immediately flanking the Sup35NM sequence of the oligonucleotide. The sequences of the primers were (in 5′→3′):

Forward oligonucleotide: *ACCGGT*ATGTCGGATTCAAAC-CAAGGC

Reverse oligonucleotide: *CCCGGG*ATCGTTAACAACTTCG TCATCC

The chosen restriction sites are shown in italics (*ACCGGT* for AgeI and *CCCGGG* for SmaI).

2. The AgeI-Sup35NM-SmaI sequence is amplified by conventional PCR using Taq DNA polymerase (New England Biolabs) and the PCR product is then purified by electrophoresis in a 1.5–2% agarose gel to clone it into a convenient vector to verify its sequence before subcloning it into the expression vector.

3. The gel-purified PCR product (AgeI-Sup35NM-SmaI) is ligated into the pCR2.1 vector (Invitrogen) using T4 DNA ligase (Fermentas).

4. Transform competent ("Z-competent", Zymo Research) XL1-blue cells with the ligation mix, plate them on LB + agar plates containing antibiotic (carbenicillin), and incubate overnight at 37°C.

5. Using a toothpick, isolate single colonies and grow them overnight in 5 ml of LB + carbenicillin at 37°C with agitation (280 rpm) to screen the colonies for the correct insert sequence. The plasmid DNA is then obtained from the colonies using the conventional mini-prep method (30), and it is digested with appropriate enzymes to confirm the presence of the insert and size by gel electrophoresis. Both strands of the insert are sequenced to verify it has the correct sequence.

6. Once the sequence of the insert is confirmed, repeat steps 2–5 in order to subclone AgeI-Sup35NM-SmaI (already cloned into pCR2.1) into dephosphorylated AgeI-pFS-2-SmaI. It is now not essential to confirm the correct sequence of the insert as the probability of introducing errors in the sequence with this additional cloning step is extremely low. The resulting expression plasmid will be pFS-2 + Sup35NM.

Expression

1. Transform Z-competent C41(DE3) cells (23) with pFS-2 + Sup35NM, plate the bacteria on LB + agar Petri dishes containing antibiotic (carbenicillin) and incubate the inverted dishes overnight at 37°C.

2. The following day, inoculate three colonies in separate tubes containing 5 ml of LB medium + carbenicillin and incubate them overnight at 37°C with agitation.

3. The following morning, inoculate fresh media with a small volume of the overnight cultures and incubate them until an OD_{595} of 0.6–0.8 is reached. Induce protein expression by adding 1 mM IPTG and incubating for 3–4 additional hours at 37°C with agitation.

4. Harvest cells by pelleting at $4,000 \times g$ for 10 min and lyse them for 4 min at 98°C in Laemmli Sample Buffer (5× LSB: 156.25 mM Tris–HCl [pH 6.8], 5% (w/v) SDS, 25% (v/v) glycerol, and 0.1% (w/v) bromophenol blue). Run an aliquot of the lysed cultures in 8% SDS-PAGE gels and determine which colony yields the stronger over-expression of the full-length

pFS-2 + Sup35NM recombinant polyprotein with the least degradation, by Coomassie Blue staining or Western blotting. Then, take 500 μl of this clone (as the expression clone stock), pellet the cells, add fresh medium with 10% glycerol, and freeze at −80°C for storage.

5. Inoculate 5 ml of LB + carbenicillin with this clone and incubate overnight at 37°C with agitation. The following day, inoculate 500 ml of LB + carbenicillin with the overnight culture and repeat step 3.

Purification

1. Harvest the cells by centrifugation ($6,000 \times g$ for 10 min) and then resuspend them in 20 ml of Ni^{2+}-affinity binding buffer (see Subheading 2.1). Add a 1:1,000 dilution of protease inhibitor cocktail (Calbiochem) and then snap freeze (liquid N_2) and snap thaw (42°C) the suspension to facilitate cell lysis.

2. Add lysozyme (Calbiochem) to a final concentration of 1 mg/ml and incubate the cells for 30 min at 4°C. Add Triton X-100 to a final concentration of 1%, and DNase I and RNase A (Sigma-Aldrich) to a final concentration of 5 μg/ml each.

3. Incubate for 30 min in a rocking platform at 4°C.

4. Remove the insoluble debris by centrifugation at $18,100 \times g$ for 20 min and filter the supernatant through a 0.45 μm filter to prevent clogging of the resin. Store the pellet at −80°C as this may contain the recombinant protein insoluble in the form of inclusion bodies.

5. Perform FPLC Ni^{2+}-affinity chromatography purification using Histrap HP columns (see the manufacturer's instructions for the column specifications; GE Healthcare).

6. Separate the eluted fractions by 8% SDS-PAGE to verify the purification procedure and monitor the proteins in the fractions using Coomassie Blue staining or Western blotting (using antibodies against the N-terminal His-tag or the C-terminal Strep-tag).

7. Pool the correct elution fractions and perform ultrafiltration using Amicon 10K filters (Millipore). Repeat the ultrafiltration step several times to obtain the sample in a convenient buffer and volume for the next step.

8. Repurify the sample by size exclusion chromatography using the buffer and column specified in Subheading 2.1. Select the appropriate flow rate based on the molecular weight of the recombinant protein of interest.

9. Extensively dialyze the fractions containing the protein of interest against the final SMFS buffer: Tris–HCl 10 mM [pH 7.5]. Repeat steps 6–7, and leave the sample in the experimental buffer.

10. Determine the protein concentration by absorbance at 280 nm using its theorethical molar extinction coefficient.

11. Add 5 mM DTT to avoid disulphide bonding through the C-terminal cysteine residues present in the pFS-2, and divide the sample into small aliquots (~100 μl). Snap freeze (if the recombinant protein of interest tolerates freezing/thawing) in liquid N_2 and store at −80°C. Aliquots should be snap thawed (at 42°C) before use.

3.2. Preparation of NTA-Ni^{2+} Functionalized Glass Coverslips

As mentioned above, two different substrates can be used with the pFS polyproteins. The following is the protocol used to prepare NTA-Ni^{2+} functionalized glass coverslips (27), as gold substrates are commercially available (Arrandee).

1. Immerse the coverslips overnight in a 20N KOH solution.

2. Place the coverslips under a MilliQ water flow for 1 h and then transfer them to a solution of 2% 3MPTS/0.02% acetic acid for 1 h at 90°C.

3. Wash the coverslips in a MilliQ water flow for 1 h and then cure them for 15 min in an oven at 120°C. Then, cool them at room temperature for about 10 min.

4. Next, transfer the coverslips to a 100 mM DTT solution for 15 min and wash under a MilliQ water flow for 1 h.

5. To each coverslip, add a ~50 μl drop of a solution containing 3 mg/ml maleimide-C3-NTA dissolved in 10 mM MOPS [pH 7.0]. Incubate the coverslips for 30 min. Keep the orientation of the coverslips from now on.

6. Wash quickly in MilliQ water each coverslip while holding it. To each, add a drop (60 μl) of 10 mM NiCl$_2$ and incubate them for 10 min.

7. Wash coverslips briefly in MilliQ water (as in step 6) prior to storage.

3.3. SMFS of pFS-2 + Sup35NM (see Notes 3 and 4)

1. Glue one NTA-Ni^{2+} functionalized coverslip (see Subheading 3.2) on a metal support disc using double-sided tape.

2. Wash the coverslip briefly with 20 μl of experimental buffer and repeat a few times (previously filtered through a 0.22 μm filter).

3. Mount the disc onto the AFM head and place a drop of the buffer (20 microlitersl) and ensure that the laser spot is focused directly on the center of the drop. Be careful when mounting the disc as the magnets on top of the piezoelectric device can displace the disc and break the coverslip.

4. Switch off the laser and then add 15–20 μl of the sample (protein concentration 2–3 μM) to the drop of buffer. Incubate for 15–30 min at room temperature.

5. In the meantime, switch on the JRC controller and the oscilloscope.

6. Introduce a cantilever (we typically use Biolever cantilevers, Olympus) into the UV cleaner and switch on the UV for 15–30 s. Leave the cleaner closed as the ozone formed will further clean the cantilever.

7. Meanwhile, fill two 1 ml syringes with filtered buffer (0.22 μm filter) and insert them into the ports of the fluid cell. Once ready, quickly place the clean correctly positioned cantilever into the fluid cell. Keep the cantilever hydrated with a drop of buffer.

8. Recover the unbound sample from the substrate. Wash gently three times with buffer, trying to leave the least amount of liquid possible on the coverslip.

9. Mount the fluid cell. Remove the buffer from the chamber to avoid overflow of fluid from the o-ring when pressing the fluid cell onto the coverslip. Once firmly adjusted, gently apply pressure to the syringes to fill the chamber up.

10. Switch on the laser and reposition the mirrors in order to focus the laser directly onto the tip of the selected cantilever and to obtain the most intense signal in the photodiode.

11. Open Igor Pro and compile the appropriate data acquisition and analysis procedures. In our case, these are custom-made.

12. A spectrum of thermal fluctuations must be acquired. We select here the first resonance peak of the cantilever (for Biolever cantilevers the nominal value is 37 kHz in air and a lower value for our experiments in liquids) in order to calibrate its elastic constant by the equipartition theorem (31).

13. Switch on the Dulcinea and Physik Instrumente units and open the WSxM program (19). Activate the external controllers, which will send a voltage signal to the Dulcinea controller. The latter will amplify both voltage signals (the signal from the external controllers and that generated by the Dulcinea itself) before sending them to the piezoelectric device.

14. Switch on the motor and approach the sample until the piezo-electric positioner enters in range. Then, the position detected by WSxM will be taken as 0 and the SMFS experiments can start.

15. Calibrate the cantilever. After measuring the thermal spectrum (step 12) of the cantilever, its sensitivity must be measured (Sens [nm/V]) by recording an F–z curve. This should be performed in a clean zone of the substrate choosing a region of the recording where the tip is pressing the surface and the forward and backward traces of the piezoelectric movement overlap. The elastic constant of the cantilever detected by Igor must be close to the nominal value specified by the manufacturer (close to 30 pN/nm

for Biolever cantilevers). Otherwise, the cantilever may be defective and should be replaced.

16. Once calibrated, choose the SMFS mode (*length-clamp, force-clamp* or its variant *force-ramp*). For length-clamp set the following parameters: pulling speed, extension of pulling, and the fraction of contact between tip and substrate (see Note 3).

17. Thermal drift must be controlled, particularly at the beginning of the experiment, as the temperature inside the fluid cell will steadily increase by the action of the laser heating until it stabilizes. Withdraw and approach with the motor (or retract the piezoelectric positioner) each time the position in z is corrected, to avoid breaking the cantilever.

18. On completing the experiment, retract the piezoelectric positioner and withdraw it with the motor. Restore the force and position offsets (button "Zero") before switching off the controlling programs.

19. Switch off all the electronics, the laser, and the oscilloscope.

3.4. Data Analysis

1. The data must be analyzed peak by peak, setting the *zero* position as close as possible to the *zero* measured by Igor.

2. The adjustable parameters that can be modified to best fit the curve to the WLC are the total length, p, and the ΔL_c.

3. Create a table and record in the analysis procedure all the following parameters for every force peak: p, F_u, ΔL_c, total length of the molecule, and extension at which the force peak appears.

4. Once the table is complete, a histogram for any variable can be constructed, and it can be normalized and fitted to any function of interest (Gaussian, Log-Normal, etc.).

   The protocol above is suitable for the SMFS analysis of proteins with defined folds, the mechanical properties of which are usually fixed (6). However, analyzing IDPs is not that simple particularly when they display conformational polymorphism (e.g., amyloidogenic IDPs like neurotoxic proteins), as these proteins do not exhibit a single and reproducible mechanical feature ("signature" or "fingerprint") in force-extension recordings but rather, multiple mechanical events. In order to characterize and unambiguously quantify such polymorphisms, we designed the carrier-guest strategy (15, 16).

5. Select good single-molecule IDP recordings (see Note 4). The following criteria must be applied in order to select good single-molecule IDP recordings when using this strategy for the nanomechanical analysis of IDPs with conformational polymorphism:

   (a) The spacer present in the pFS-2 (N2B fragment) unfolds without detectable mechanical resistance and will appear in the proximal region of the force-extension recordings

(~70 nm if the polyprotein is being pulled from its termini). This allows us to avoid the noisy proximal region of the force-extension spectra. As such, any recording where the putative IDP force peak appears in this region is discarded (Fig. 1d).

(b) The recording should show several (less than 6) equally spaced force peaks as markers, attributable to the unfolding of the ubiquitin repeats present in the pFS-2 vector, the F_u and ΔL_c of which are characteristic and well described in the literature: $F \approx 200$ pN and $\Delta L_c \approx 23$ nm (22).

(c) The polyprotein should not exhibit more force peaks than the number expected based on the construction of the pFS-2 protein (excluding those derived from the unfolding of the IDP due to its mechanical plasticity).

(d) The total length of the unfolded molecule should not be greater than that of the extended polypeptide (considering a gain in length of 0.4 nm per stretched amino acid; (12)).

(e) In the carrier-guest strategy, the guest IDP is "force hidden" inside the carrier module (Fig. 1b) and therefore, the force peak originated from the unfolding of the carrier module should always precede (although not necessarily immediately) the force peaks corresponding to the unfolding of the grafted IDP (Fig. 1d).

(f) Any force peak that appears at an extension shorter than that corresponding to the complete unfolding of the carrier module (29.5 nm for the carrier I27 and 25.5 nm for the carrier ubiquitin) is excluded from our analyses as, in principle, it may originate from spurious interactions between the IDP and the carrier module.

(g) Only the force data with an ΔL_c value that, when summed, coincides exactly with the ΔL_c of the carrier-guest construction are included in our analyses. In the specific case of I27 + Sup35NM, the latter value is 29.5 nm from the unfolding of the carrier I27 ("a" in Fig. 1d) and 101 nm from the stretching of the grafted Sup35NM ("b" and "c" in Fig. 1d). This ensures an exact measurement of the expected ΔL_c to detect the force events, and allows these events to be observed far from the problematic proximal region of the force-extension recordings. Alternatively, if a classical hetero-polyprotein strategy (10) were used, in which the protein of interest is placed in series with the repeats of the marker, many of the force events of the protein of interest could be hidden by nonspecific interactions between the tip and the sample (particularly those lower than the mechanical stability of the markers), introducing false positive data (contamination) into the analyses.

4. Notes

1. As the carrier-guest strategy described here for unequivocal SMFS of neurotoxic proteins implies the insertion of these proteins into a carrier, several structural controls for the carrier-guest protein should be performed in order to rule out the possibility that artifactual effects may occur in and between both proteins (e.g., ^{1}H monodimensional nuclear magnetic resonance or circular dichroism), which could induce structural changes in the grafted neurotoxic protein. Furthermore, controls must be included to ensure that the neurotoxic protein maintains its amyloidogenic properties when hosted in the carrier (16), such as amyloid aggregation (turbidometry, congo Red or thioflavin binding assays) and fibrillogenesis (imaging AFM or transmission electron microscopy).

2. The specific configuration of our custom-made AFM is protected by an international patent (PCT/ES2008/070130) licensed to Nanotec Electrónica S.L. while the use of pFS vectors is protected by another international patent (PCT/ES2011/070867). The procedures for data acquisition and data analysis were modified (for connecting with Dulcinea control unit) from those originally developed by Prof. Julio M. Fernández (http://fernandezlab.biology.columbia.edu).

3. The protocol described in this section is specific for the experiments performed using our AFM setup. As mentioned above, our AFM combines three electronics: the AFM controller, the piezoelectric sensors and the high-voltage unit, which itself can be used also to control the AFM. To that end, several command lines were added to the original custom-made software written in Igor to allow crosstalk with WSxM (18, 19).

4. The procedure for refolding (used as a control for unwanted interactions) includes the following steps: unfolding without detachment, a series of cycles of limited approach/extension (to relax and unfold the same molecule away from the substrate several times), and the final complete stretching. We perform it using a home-made procedure.

Acknowledgements

We thank the members of the laboratory for their critical reading of the manuscript. This work was funded by grants from the Ministerio de Ciencia e Innovación (BIO2007-67116), the Consejería de Educación de la Comunidad de Madrid (S-0505/MAT/0283), and the Consejo Superior de Investigaciones Científicas

(200620F00). J.O. and R.H. are recipients of fellowships from the Consejería de Educación de la Comunidad de Madrid and the Fundación Ferrer (Severo Ochoa's fellowship), respectively.

References

1. James LC, Tawfik DS (2003) Conformational diversity and protein evolution—a 60-year-old hypothesis revisited. Trends Biochem Sci 28:361–368

2. Chiti F, Dobson CM (2006) Protein misfolding, functional amyloid and human disease. Annu Rev Biochem 75:333–366

3. Ferreon AC et al (2010) Single-molecule fluorescence studies of intrinsically disordered proteins. Methods Enzymol 472:179–204

4. Uversky VN, Dunker AK (2010) Understanding protein non-folding. Biochim Biophys Acta 1804:1231–1264

5. Receveur-Bréchot V et al (2006) Assessing protein disorder and induced folding. Proteins 62:24–45

6. Carrión-Vázquez M et al (2000) Mechanical design of proteins studied by single-molecule force spectroscopy and protein engineering. Prog Biophys Mol Biol 74:63–91

7. Bustamante C et al (1994) Entropic elasticity of lambda-phage DNA. Science 265:1599–1600

8. Yang G et al (2000) Solid-state synthesis and mechanical unfolding of polymers of T4 lysozyme. Proc Natl Acad Sci U S A 97:139–144

9. Li H et al (2001) Multiple conformations of PEVK proteins detected by single-molecule techniques. Proc Natl Acad Sci USA 98:10682–10686

10. Steward A, Toca-Herrera JL, Clarke J (2002) Versatile cloning system for construction of multimeric proteins for use in atomic force microscopy. Protein Sci 11:2179–2183

11. García-Manyes S et al (2007) Force-clamp spectroscopy of single-protein monomers reveals the individual unfolding and folding pathways of I27 and ubiquitin. Biophys J 93:2436–2446

12. Ainavarapu SR et al (2007) Contour length and refolding rate of a small protein controlled by engineered disulfide bonds. Biophys J 92:225–233

13. Valbuena A et al (2009) On the remarkable mechanostability of scaffoldins and the mechanical clamp motif. Proc Natl Acad Sci USA 106:13791–13796

14. Oroz J et al (2011) Nanomechanics of the cadherin ectodomain: "canalization" by Ca^{2+} binding results in a new mechanical element. J Biol Chem 286:9405–9418

15. Oroz J, Hervás R, Carrión-Vázquez M (2012) Unequivocal single-molecule force spectroscopy of proteins by AFM using pFS vectors. Biophys J 102:682–690

16. Hervás R, et al (In Press) Common features at the start of the neurodegeneration cascade. PLoS Biol

17. Schlierf M, Li H, Fernández JM (2004) The unfolding kinetics of ubiquitin captured with single-molecule force-clamp techniques. Proc Natl Acad Sci USA 101:7299–7304

18. Valbuena A et al (2007) *Quasi*-simultaneous imaging/pulling analysis of single polyprotein molecules by atomic force microscopy. Rev Sci Instrum 78:113707

19. Horcas I et al (2007) WSXM: a software for scanning probe microscopy and a tool for nanotechnology. Rev Sci Instrum 78:013705

20. Li H et al (2002) Reverse engineering of the giant muscle protein titin. Nature 418:998–1002

21. Finley D, Bartel B, Varshavsky A (1989) The tails of ubiquitin precursors are ribosomal proteins whose fusion to ubiquitin facilitates ribosome biogenesis. Nature 338:394–401

22. Carrión-Vázquez M et al (2003) The mechanical stability of ubiquitin is linkage dependent. Nat Struct Biol 10:738–743

23. Miroux B, Walker JE (1996) Over-production of proteins in *Escherichia coli*: mutant hosts that allow synthesis of some membrane proteins and globular proteins at high levels. J Mol Biol 260:289–298

24. Carrión-Vázquez M et al (1999) Mechanical and chemical unfolding of a single protein: a comparison. Proc Natl Acad Sci USA 96:3694–3699

25. Went HM, Benítez-Cardoza CG, Jackson SE (2004) Is an intermediate state populated on the folding pathway of ubiquitin? FEBS Lett 567:333–338

26. Rief M et al (1997) Reversible unfolding of individual titin immunoglobulin domains by AFM. Science 276:1109–1112

27. Hossain MD et al (2006) The rotor tip inside a bearing of a thermophilic F1-ATPase is dispensable for torque generation. Biophys J 90:4195–4203

28. Oberhauser AF et al (2001) Stepwise unfolding of titin under force-clamp atomic force microscopy. Proc Natl Acad Sci USA 98:468–472

29. Glover JR et al (1997) Self-seeded fibers formed by Sup35, the protein determinant of [*PSI+*], a heritable prion-like factor of *S. cerevisiae*. Cell 89:811–819

30. Sambrook J, Russel DW (2001) Molecular cloning: a laboratory manual. Cold Spring Harbor Laboratory Press, New York

31. Florin EL et al (1995) Sensing specific molecular interactions with the atomic force microscope. Biosens Bioelectron 10:895–901

Part II

Methods to Assess Protein Size and Shape

Chapter 6

Sedimentation Velocity Analytical Ultracentrifugation for Intrinsically Disordered Proteins

Andrés G. Salvay, Guillaume Communie, and Christine Ebel

Abstract

The size of intrinsically disordered proteins (IDPs) is large compared to their molecular mass and the resulting mass-to-size ratio is unusual. The sedimentation coefficient, which can be obtained from sedimentation velocity (SV) analytical ultracentrifugation (AUC), is directly related to this ratio and can be easily interpreted in terms of frictional ratio. This chapter is a step-by-step protocol for setting up, executing and analyzing SV experiments in the context of the characterization of IDPs, based on a real case study of the partially folded C-terminal domain of Sendai virus nucleoprotein.

Key words: Intrinsically disordered proteins, Hydrodynamic radius, Frictional coefficient, Sedimentation velocity, Analytical ultracentrifugation, SEDFIT, Sendai virus nucleoprotein

1. Introduction

The size of intrinsically disordered proteins (IDPs) is large compared to their molecular mass and the resulting mass to size ratio is unusual (1, 2). Sedimentation velocity (SV) experiments in analytical ultracentrifugation (AUC) is a powerful technique to determine the molar mass, M, and hydrodynamic (Stokes) radius, R_H, of macromolecules in solution (3–6). SV studies macromolecules in solution that are subjected to a large centrifugal field. SV combines the separation of the macromolecules and the analysis of their transportation in view of a rigorous thermodynamics. The transport is determined by the sedimentation coefficient, s, and by the diffusion coefficient, D, which are directly related, for noninteracting species, to M/R_H and to $1/R_H$, respectively. AUC is complementary to the methods based on size determination (e.g., size exclusion chromatography, dynamic light scattering) that probe R_H. As an illustration, an IDP (monomer) that appears large in SEC can be hardly

Vladimir N. Uversky and A. Keith Dunker (eds.), *Intrinsically Disordered Protein Analysis: Volume 2, Methods and Experimental Tools*, Methods in Molecular Biology, vol. 896, DOI 10.1007/978-1-4614-3704-8_6, © Springer Science+Business Media New York 2012

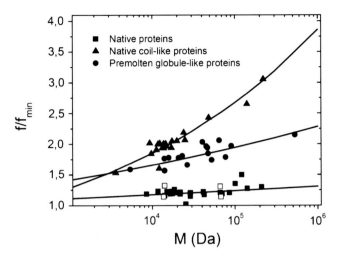

Fig. 1. Frictional ratios of native globular proteins, PMG-like IDPs and coil-like IDPs as a function of their molar masses. *Lines* are derived from M/R_H relationships given in Uversky (2). Data for PMG-like IDPs (*filled circles*) and coil-like IDPs (*filled triangles*) are from the same work. Data for native globular proteins (*filled* and *empty squares*) are from Tcherkasskaya and Uversky (24) and from Cantor and Schimmel (25), respectively. The Figure is adapted from Manon and Ebel (7), with permission.

distinguished from a dimer. The same IDP will sediment more slowly than a folded monomer while a dimer would sediment faster, leading to an easy experimental diagnostic of the extended shape and association state of the protein (7).

In the last decade, SV data analysis based on numerical solutions of the transport were successfully developed (8–10). In particular, the $c(s)$ analysis allows deciphering sample homogeneity, evidencing equilibrium of association, and characterizing species in a very easy way in terms of their s-values (11). It gives indication of the average shape of the macromolecules present in solution. It allows determining if further data analysis may provide independently determined reliable values for D, thus M and R_H.

The frictional ratio, f/f_{min}, is a very useful parameter for describing the shape of the macromolecules in solution. It is the ratio of the hydrodynamic radius to the radius of the anhydrous volume. The former tells about the size (dimension) of the molecule in solution, the later depends on the cubic root of the molar mass. The frictional ratio f/f_{min} depends on the hydration, surface roughness, shape, and flexibility of the particle. For globular compact macromolecules, its value is nearly constant, while very slightly increasing, from 1.15 to 1.3, for M from 5 to 1,000 kDa. Uversky has classified IPDs in two classes: coil-like and pre-molten globule (PMG)-like IPDs, according to their M-R_H characteristics (2, 12, 13). We have used the derived M-R_H relationships to derive f/f_{min}-values (7). Figure 1 shows that f/f_{min}-values of IDPs are significantly larger being compared to compact folded globular proteins and increase significantly when the protein has a larger molar mass. f/f_{min} increases from 1.5 to 2 (PMG-like IDPs) and from 1.6 to 3 (coil-like IPDs) for M increasing from 5 to 200 kDa.

The present chapter represents a detailed protocol for retrieving and analyzing SV data for IDPs. This protocol does not contain AUC sedimentation equilibrium experiments, which lead to the determination of M-values, but requires more homogeneous and stable solutions. The protocol is based on a real case study of the partially folded C-terminal domain of Sendai virus nucleoprotein: Ntail.

Sendai virus belongs to the Paramixoviridae family. Its RNA genome is tightly packed by many copies of the nucleoprotein, forming a helical nucleocapsid (14). This nucleocapsid is the matrix for the polymerase complex during the viral replication. Ntail domain, 125 residues in length, is known to mediate the interaction with the polymerase complex by a folding upon binding mechanism. The protein is predominantly unfolded but contains a transient α-helical motif in its molecular recognition element (15).

2. Materials

1. Analytical buffer: 50 mM Na/K phosphate, 500 mM NaCl, pH 6.

2. The protein solution, in volume and concentration required for SV measurements at three dilutions (see Notes 1 and 2). The solvent (10 mL) should be well defined and preferentially contain salt above 100 mM (see Note 3). Our study used 233 µL of Ntail protein of Sendai virus (15) at 1.16 mg/mL in 50 mM Na/K phosphate pH 6, 500 mM NaCl.

3. An analytical ultracentrifuge (Optima XLI Beckman) with associated program Beckman XL-I. A rotor (4-hole AnTi 60 or 8-hole AnTi-50, Beckman) (see Note 4).

4. AUC cell assemblies equipped with sapphire windows and 2-channel Titane, aluminum or Epon centerpieces of 1.5, 3, and/or 12 mm optical path length (Nanolytics or Beckman) (see Notes 2 and 4) with manual for cell montage (Beckman, provided with the rotor).

5. The program SEDNTERP created by D. Hayes, T. Laue, J. Philo and available free (http://www.jphilo.mailway.com/), for calculating the parameters relevant to SV analysis (see Note 5).

6. The program SEDFIT created by P. Schuck and available free (http://www.analyticalultracentrifugation.com), for the analysis of AUC experiments (see Note 6).

7. An UV spectrophotometer for absorbance measurements and determination of protein concentration in the solution samples (DU 7400 Beckman).

8. The amino acid sequence as a text file.

3. Methods

3.1. Theoretical Background

Sedimentation is a transport method akin to diffusion and sedimentation. Sedimentation velocity measures in a rotor spinning at high angular velocity, ω, in the centrifuge, the evolution of the weight concentration, c, with time, t, and radial position, r. For each homogeneous ideal solute, and given the sector shaped cells used in AUC, the transport is described by the Lamm equation:

$$(\partial c / \partial t) = -1/r \; \partial/\partial r[r(cs\omega^2 r - D \times \partial c/\partial r)], \qquad (1)$$

where s and D are the sedimentation and diffusion coefficients of the macromolecule. s is defined as the ratio of the macromolecule velocity (cm/s) to the centrifugal field ($\omega^2 r$ in cm/s^2). s is expressed in Svedberg unit S ($1\,S = 10^{-13}$ s). s and D are functions of the molar mass M, the hydrodynamic radius R_H (also referred to as the Stokes radius R_S) and the partial-specific volume $\bar{v}$ of the macromolecule. s and D also depend on the solvent density ρ and viscosity η.

The Svedberg equation relates s to R_H (or D), M and $\bar{v}$:

$$s = M(1 - \rho\bar{v})/(N_A 6\pi\eta R_H) = M(1 - \rho\bar{v})D/RT \qquad (2)$$

N_A is Avogadro's number and T the absolute temperature. The sedimentation coefficient is generally expressed as $s_{20,w}$ after correction for solvent density and viscosity in relation to the density and viscosity of water at 20°C ($\rho_{20,w} = 0.99832$ g/mL; $\eta_{20,w} = 1.022$ mPa/s):

$$s_{20,w} = s[(1 - \rho_{20,w}\bar{v})/(1 - \rho\bar{v})](\eta/\eta_{20,w}) \qquad (3)$$

The Stokes–Einstein equation relates D to R_H:

$$D = RT/(N_A 6\pi\eta R_H) \qquad (4)$$

D is often expressed as $D_{20,w}$ after correction for temperature and solvent viscosity compared to the conditions of water at 20°C ($T_{20} = 293.45$ K, $\eta_{20,w} = 1.002$ mPa/s):

$$D_{20,w} = D(T_{20}/T)(\eta/\eta_{20,w}) \qquad (5)$$

Non-ideality effects in concentrated samples influences s and D. The sedimentation and diffusion at infinite dilution, s_0 and D_0, be derived from the linear approximations:

$$s^{-1} = s_0^{-1}(1 + k_s c) \qquad (6)$$

$$D = D_0(1 + k_D c) \qquad (7)$$

In the favorable case of an ideal solution comprising one non-interacting solute (eventually two), D (thus R_H) can be measured experimentally from SV experiments from the Lamm equation. The ratio of R_H to the minimum theoretical hydrodynamic radius R_{min} of non-hydrated volume, V, of the particle defines the frictional ratio f/f_{min}:

$$V = (4/3)\pi R_{min}^{3} = M\bar{v}/N_A \qquad (8)$$

$$R_H = (f/f_{min})R_{min} \qquad (9)$$

3.2. Prior to the Experiment

1. In order to choose the experimental conditions and prepare the analysis, we use the program SEDNTERP for calculating from the amino acid composition of the protein, $M = 13,388$ Da, $\bar{v} = 0.6989$ mL/g and the extinction coefficient at 280 nm, $E_{0.1\%,280} = 0.933$ mg/mL cm. From the solvent composition: $\rho = 1.023$ g/mL, $\eta = 1.067$ mPa/s (1 mPa/s = 1 centipoise (cP)).

2. Equations 2, 4, 8, and 9 are used to calculate estimates for s and D considering a monomeric protein with different shapes. We use homemade excel sheets. We can also use the calculation facilities in the program SEDFIT for calculating $s_{spherical}$, R_{min}, and $D_{spherical}$ for the anhydrous spherical protein: *option/calculator/calculate s(M) for spherical particle/*. Input values are M, $\bar{v}$, temperature, ρ and η. Output values are $s_{spherical} = 2.03$ S, $D_{spherical} = 1.3 \times 10^{-6}$ cm²/s, and $R_{min} = 1.55$ nm. Then, estimates for a globular compact hydrated shape ($f/f_{min} = 1.25$) are: $R_H = 1.25 \times R_{min} = 1.9$ nm, $D = D_{spherical}/1.25 = 1.0 \times 10^{-6}$ cm²/s and $s = s_{spherical}/1.25 = 1.6$ S. For a rather extended shape ($f/f_{min} = 2$): $R_H = 2 \times R_{min} = 3.1$ nm, $D = D_{spherical}/2 = 0.6 \times 10^{-6}$ cm²/s, and $s = s_{spherical}/2 = 1$ S.

3. Select the temperature of measurement: here, 20°C, because our sample is stable at that temperature (see Note 7).

4. Use SEDFIT to simulate the sedimentation in our standard rotor speed for SV experiments of 42,000 rpm (revolution per minute), i.e. 130,000×*g*. Open SEDFIT; *generate/single/fit M and s instead for s and D*: no/*dr* = *1e-3*: OK/*rotor speed*: change to 42,000 rpm/*simulate...*: yes/*acceleration...*: OK/*Time interval of scan (sec)*: write 600 (i.e., 10 min)/*number of simulated scans*: write 30 (i.e., 5 h)/*std of noise = 1e-2*: OK. In the table, check component 1 box is marked, write as input value $c = 0.5$, $D = 0.6$, and $s = 1$ corresponding to our hypothetical extended Ntail, meniscus = 6.0, then: OK; create a folder to temporary save the generated data; a window opens

showing in the top panel the superimposition of simulated sedimentation profiles: the vertical and horizontal axis are the concentration (stated as absorbance) and radial position, respectively. Drag the vertical green line at right to $r = 7$ cm. Change the scale: *display/data range*. Copy the screen in a word file if desired. The SV profiles show at $130,000 \times g$. important boundary spreading, because the protein is relatively small; particularly the concentration hardly reaches zero upon centrifugation. This may decrease the robustness of analysis. Thus, make the same steps for a simulation at a rotor speed of 55,000 rpm. Boundary spreading is decreased. Thus, $(220,000 \times g)$ is selected (see Note 8). It imposes the choice of the rotor (AnTi 60) and centerpieces (see Note 4).

5. Estimate or measure the absorbance A of stock protein solution at 280 nm, to decide the dilutions to be made, the AUC centerpieces to be used, given their path length (l) and the sample volume (V) (see Note 2). We measured $A_{280} = 1.24$. We decided to measure SV for the stock solution with $l = 0.3$ cm, $V = 100$ μL providing $A_{init} = 0.3 \times 1.24 = 0.372$, dilution 4 with $l = 1.2$ cm, $V = 400$ μL providing $A_{init} = 1.2 \times 1.24/4 = 0.372$ and dilution 12 with $l = 1.2$ cm, $V = 400$ μL providing $A_{init} = 1.2 \times 1.24/12 = 0.124$, respectively.

3.3. Experiment

1. Prepare the diluted samples (see previous paragraph) and store the samples at the appropriate temperature (here 4°C).

2. Equilibrate the centrifuge at 20°C.

3. Prepare the AUC cells assemblies. Details concerning cell assemblies and practical steps for cell montage and rotor preparation are described as videos that can be downloaded ((http://www.beckman.com/resourcecenter/labresources/sia/cellassy_video.asp?pf = 1) (16)). We prepared three cells equipped with Ti-centerpieces with $l = 0.3$ cm (for the stock sample) and 1.2 cm (for the diluted ones), and comprising sapphire windows (see Note 9). We filled the sample compartments with 400 or 100 μL of sample, and the solvent compartment with the same volume of solvent.

4. Once the cells are closed, check that cells that will be placed in opposite positions in the rotor have the same weights: the two 12 mm cells, and the 3 mm cell and the counterbalance (required for instrumentation calibration). Place the cells in the rotor.

5. Put the rotor into the analytical ultracentrifuge, and initiate vacuum and temperature equilibration at 20°C. Wait for temperature equilibration before the SV experiment, but at least 30 min, irrespective of the temperature reading (see Note 10).

6. During temperature equilibration, start the program Beckman XL-I. Give a title for each cell and a common localization for the generated data. Start the AUC at 3,000 rpm ($665 \times g$) and acquire for each cell a wavelength scan between 240 and 400 nm at a radial position of 6.5 cm (the mean radial position of the compartments) (see Note 11). Check the absorbance corresponds to expectation. As a second control, start for all cells a radial scan at 280 nm (see Note 12).

7. Start the SV experiments overnight: In our study, the sedimentation profiles, i.e., A_{280} as a function of the radial position, r, were measured for each cell every 5 min for 14 h at 55,000 rpm (see Note 13).

8. Stop the data acquisition and centrifugation after night. After breaking the vacuum, remove the rotor, set off AUC and remove the cells from the rotor. Rotate the cells to provoke the displacement of an air bubble within the sample compartment if the samples are to be recovered. Dismount and clean the cell assemblies.

9. Zip the raw data and duplicate them for saving and analysis.

3.4. Data Analysis

1. Open the program SEDFIT (see Note 6) and select, starting with the more concentrated sample, a set of 20–30 SV profiles corresponding to the whole sedimentation process (see Fig. 2, top). We select 1 over 7 profiles for a total of 140 profiles (time interval: ≈ 40 min, total time 14 h). Fix the meniscus (air-sample interface) and bottom position and the radial limits for the fit, avoiding the pellet region at the bottom of the cell.

2. Analyze first our data in the model of the $c(s)$ analysis. The analysis considers that the solution contains a continuous distribution of a large number of types of particles, which are characterized by s and are quantified by their absorbencies at 280 nm. The $c(s)$ method deconvolutes the effects of diffusion broadening for obtaining resolution distributions. This is approximately but efficiently done by assuming that all proteins have the same shape, which gives a relationship between s and D. The relation is established through the inputs given for $\bar{v}, f/f_{\min}$ (which can be fitted), ρ and η (see Note 14). We considered first 50 particles, in the range 0.1–20 S without regulation procedure (see Note 15), with $f/f_{\min} = 1.25$ corresponding to compact proteins. Since no aggregates were detected above 4 S, a second analysis was done in the range 0.1–4 S (see Fig. 2a). A third analysis was done fitting $f/f_{\min}$ leading to $f/f_{\min} = 2.0$ (see Fig. 2b). The quality of the fit increased significantly as attested by the better superposition of the fitted and experimental SV profiles (top panels) and the observation of the residuals (middle panels), the significant decrease of the root mean square

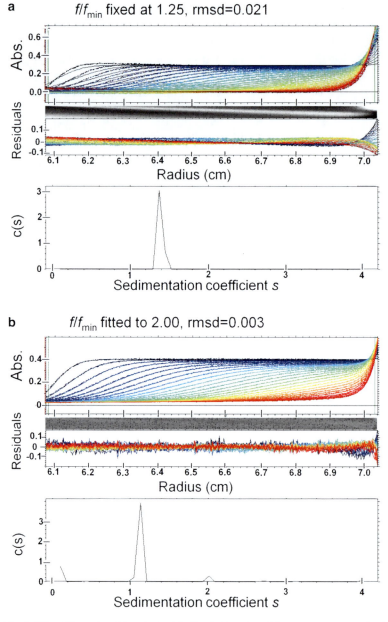

Fig. 2. Fitting f/f_{min} in the $c(s)$ analysis of Ntail at 1.16 mg/mL. Print screens were obtained using the program SEDFIT. *Panels* **a** and **b** show the results of the fit when f/f_{min} is fixed at 1.25 and fitted, respectively.

deviation (rmsd). The details of the distributions (bottom panels) are different, but the general features of $c(s)$ are similar attesting the robustness of this type of analysis.

3. The last $c(s)$ analysis is made with 300 particles and a regularization (F-ratio of 0.95) leading to a more regular $c(s)$ distribution and avoiding possible irrelevant details (see Fig. 3a–c).

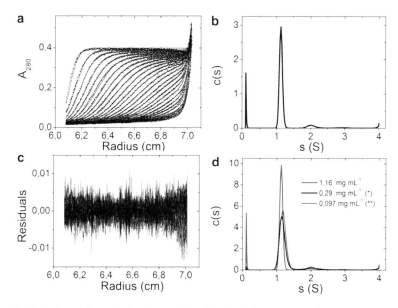

Fig. 3. Sedimentation velocity $c(s)$ analysis of Ntail. (**a**) Superposition of selected experimental (*dots*) and fitted (*continuous line*) SV profiles corrected for all systematic noises, for Ntail at 1.16 mg/mL measured at 280 nm in a 3 mm optical path length centerpiece. The last profile corresponds to 14 h of sedimentation at 55,000 rpm (220,000 × g) at 20°C. (**b**) Superposition of the difference between the experimental and fitted curves. (**c**) $c(s)$ distribution in the range of 0.1–4 S. (**d**) Superposition of $c(s)$ for Ntail at 1.16 mg/mL and diluted 4 and 12 times. For clarify, $c(s)$ are normalized for optical path length and for sample dilution.

Integrate the main peaks to obtain mean values for s, $s_{20,w}$, and signals in absorbance units. The peak at $s_{20,w} = 0.15$ S results most probably from a poor deconvolution of the baseline and will not be considered further. The peak at $s_{20,w} = 4.4$ S (less than 2%) may be related to a very small amount of larger species or to a poor deconvolution of the data. Copy the $c(s)$ data in a spreadsheet software (see Note 16).

4. Apply steps 2 and 3 for the other cells. Compare the results by superposing the $c(s)$ (see Fig. 3d). Summarize the data (see Table 1).

5. Analyze the rmsd: it is always comparable to the systematic noise in absorbance data, indicating the fit is nice.

6. Analyze the fitted f/f_{min}: they are much larger than 1.25 corresponding to globular compact particles. This is what we expect for an IDP (see Note 17).

7. Analyze the changes in s with concentration: all Ntail samples have one main contribution (>86%) at $s_{20,w} = 1.3$ S and a minor species at $s_{20,w} = 2.5$ S. The invariance of the s-values means the solution may be considered as composed of noninteracting species: a peak in the $c(s)$ represents a species (see Note 18).

Roni Wilbur

Table 1
c(s) analysis of Ntail

				Main species				Minor species		
Concentration (mg/mL)	rmsd	f/f_{min}	s (S)	$s_{20,w}$ (S)	Abs_{z1cm}	%	s (S)	s_{20w} (S)	%
1.16	0	2	1.13	1.28	1.12	92	2.1	2.4	8
0.29	0.01	1.99	1.17	1.32	0.26	90	2.2	2.5	10
0.097	0	1.81	1.16	1.34	0.10	99	2.5	2.8	1

rmsd, f/f_{min}, s and $s_{20, w}$ are obtained from the $c(s)$ analysis; Abs_{1cm} is normalized by the optical path length; %s are given considering 100% for the two types of species

This is fortunate because the analysis of the value of s in terms of R_H and f/f_{min} (presented below) is only valid in that case.

8. Analyze the changes of the percentage of the different species with concentration. If the percentages are constant with dilution, the minor species is a contaminant. An increase of the percentage of the larger species when increasing the concentration would indicate a slow equilibrium of association (see Note 18). Our data are not enough precise to discriminate between these two cases. Ntail is from SDS-PAGE >95% pure, which does not solve the ambiguity. In case of interest, complementary experiments would be needed to characterize this species.

9. Analyze the slight changes in the s-values for the main peak with concentration. Non-ideality related to inevitable excluded volume effect leads to a decrease of s when increasing concentration. This is indeed observed for Ntail (because the determination of s is accurate). The linear extrapolation of $1/s$ to infinite dilution (see Eq. 6) gives s_0, which can be used in the Svedberg equation (see Eq. 2). We extrapolate $s_{20w,0} = 1.34$ S (see Note 19).

10. Attempt to analyze the data in SEDFIT with the noninteracting species model for determining independent values of s and D, thus M and R_H. Following $c(s)$ analysis, we have to consider two noninteracting species for Ntail. Figure 4a, b shows the starting and fitted, respectively, values for the fit. Clearly, in our case, the deconvolution of the signal for the two species is not straightforward, and the approach is here inappropriate.

11. Use the Svedberg equation (Eq. 2) combined with Eqs. 8 and 9 to calculate R_H and f/f_{min} from $s_{0,20,w}$ with different hypothesis on the association state for the main species. We use homemade excel sheets. We can also use the calculation facilities proposed in the program SEDFIT: *option/calculator/calculate axial and frictional coefficient ratios/*. Input values are $M = 13,388$ Da (for a monomer), $s_{20w,0} = 1.34$ S, $\bar{v} = 0.6989$ mL/g,

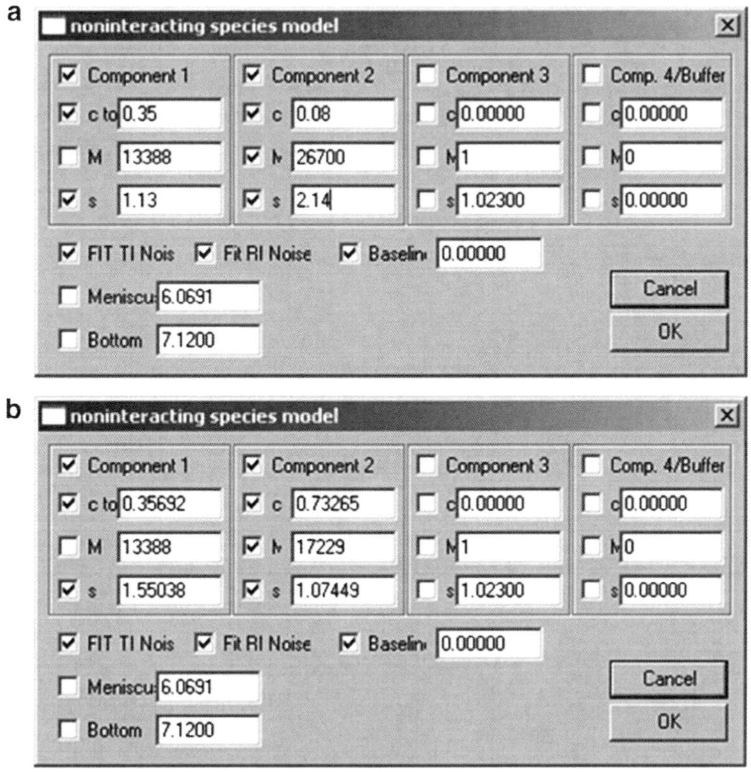

a

noninteracting species model				☒
☑ Component 1	☑ Component 2	☐ Component 3	☐ Comp. 4/Buffer	
☑ c to 0.35	☑ c 0.08	☐ c 0.00000	☐ c 0.00000	
☐ M 13388	☑ M 26700	☐ M 1	☐ M 0	
☑ s 1.13	☑ s 2.14	☐ s 1.02300	☐ s 0.00000	

☑ FIT TI Nois ☑ Fit RI Noise ☑ Baselin 0.00000

☐ Meniscu 6.0691   Cancel

☐ Bottom 7.1200   OK

b

noninteracting species model				☒
☑ Component 1	☑ Component 2	☐ Component 3	☐ Comp. 4/Buffer	
☑ c to 0.35692	☑ c 0.73265	☐ c 0.00000	☐ c 0.00000	
☐ M 13388	☑ M 17229	☐ M 1	☐ M 0	
☑ s 1.55038	☑ s 1.07449	☐ s 1.02300	☐ s 0.00000	

☑ FIT TI Nois ☑ Fit RI Noise ☑ Baselin 0.00000

☐ Meniscu 6.0691   Cancel

☐ Bottom 7.1200   OK

Fig. 4. Noninteracting species analysis of the sedimentation velocity of Ntail at 1.16 mg/mL. *Panel a* shows the selected input values. They were chosen according to the result of the $c(s)$ analysis. We considered Component 1 as Ntail monomer, the "concentration" for component 1 represents the total signal at 280 nm for this cell. Component 2 is the minor species. The "concentrations" for components 2, 3, and 4 represent the fraction of the total signal at 280 nm. *Panel b* gives the result of the fit. The result for relative proportion of the two species is obviously wrong, meaning the two species are not properly distinguished.

hydration = 0 (see Note 20), $T = 20$ C, $\rho = 0.99828$ g/mL, and $\eta = 0.01002$ P (see Note 21). The outputs are $R_H = 2.66$ nm and $f/f_{min} = 1.72$. The hypothesis of a dimer leads to $R_H = 5.31$ nm and $f/f_{min} = 2.72$.

12. Analyze the f/f_{min} results: Fig. 1 shows that f/f_{min}-values of 1.2 ± 0.05, 1.7 ± 0.1, and 1.95 ± 0.1 are expected for an $M = 13,388$ Da protein (i.e., Ntail monomer) that would be globular compact, PMG-like, and coil-like, respectively. The experimental $s_{20w,0} = 1.34$ S is thus compatible with Ntail as a PMG-like monomer. On the other hand, Fig. 1 shows that f/f_{min}-values of 1.2 ± 0.05, 1.8 ± 0.1, and 2.2 ± 0.1 are expected for an $M = 27,776$ Da protein (i.e., Ntail dimer) that would be globular compact, PMG-like, and coil-like, respectively. The experimental $s_{20w,0} = 1.34$ S is not compatible with Ntail as a dimer, whatever its shape.

13. Conclusion: we conclude that the studied solution comprises essentially of a monomeric protein characterized by f/f_{min} = 1.7. Ntail thus belongs to the PMG-like IDP family. This is in agreement with the fact that the protein is predominantly unfolded but contains a transient α-helical element.

4. Notes

1. The measurement at three concentrations is made for evaluating nonideal effects and for attesting that the protein is not undergoing association equilibrium (see Note 18). We typically use a dilution series between 1, 2, 4 and 1, 4, 16.

2. An absorbance between 0.1 and 1.2 in the ultracentrifuge is optimal. Absorbance can be measured between 230 and 650 nm (typically 280 nm). The absorbance and volume of the stock sample determines the choice of the optical path length and the extent of the dilution series. The centerpieces with optical path lengths of 1.2, 0.3, and 0.15 cm require typically sample volumes of 420, 110, and 55 μL, respectively.

3. Lowering solvent salt below 100 mM may lead to significant hydrodynamic non-ideality for charged proteins—which is the usual case. The non-ideality is related to the differential sedimentation of the counter-ions and the macromolecule and to electrostatic repulsion between proteins and/or to excluded volume effects (17, 23).

4. Rotor 8-hole AnTi 50 is used for AUC experiments performed at angular velocity up to 50,000 rpm ($185,000\times g$), while 4-hole AnTi 60 rotor tolerates up to 60,000 rpm ($265,000\times g$). Epon centerpieces are not to be used above 42,000 rpm ($130,000\times g$), while Al and Ti centerpieces can be used up to 60,000 rpm ($265,000\times g$).

5. The program SEDNTERP allows evaluating, from amino acid composition, M, $\bar{v}$ and extinction coefficients of proteins, and, from the solvent composition, ρ and η. For unusual solvents, ρ and η can be measured experimentally: we use the density-meter DMA50 and a viscosity-meter AMVn (Anton Paar).

6. The program SEDFIT uses numerical solution of the Lamm equation for analyzing SV experiments in different ways. Detailed help, including step-by-step tutorials is available online (http://www.analyticalultracentrifugation.com). SEDFIT incorporates the possibility of accounting for the systematic noise of the experimental data, a procedure we routinely apply. Note the program SEDPHAT, which will not be described here and is available free on SEDFIT Web page

(http://www.analyticalultracentrifugation.com), may be used for analyzing globally different sets of data, e.g., SV data obtained at different concentrations.

7. SV measurements can be done between 4 and 20°C on our AUC. The temperature of 20°C is the less time consuming for temperature equilibration of the rotor prior SV experiments.

8. The maximum rotor speed of 60,000 rpm ($265,000 \times g$) should be even more appropriate but was not selected because our AnTi rotor is aged.

9. Sapphire windows allow the detection of concentration changes using interference optics in our ultracentrifuge (XLI). Interference is particularly appropriate when concentrations are too large (absorbance optics would be saturated), or for the study or multicomponents systems (see, e.g., refs. 18, 19). In the present case, the same information is obtained at 280 nm or using interference, which requires more care in terms of sample design (solvent has to be rigorously the same in the sample and reference solvent compartments) and cell montage (the sample and solvent compartments have to be filled with the same volume) for the analysis to be easy. Since interference measurements are done without loss of time, we however generally acquire the two sets of data in case. The modifications of the present protocols for using interference are not detailed here for clarity but details can be found in, e.g., ((http://www.analyticalultracentrifugation.com) (19).

10. Temperature equilibration is very important because sedimentation depends on solvent viscosity which changes significantly with temperature.

11. In addition to angular velocity 3,000 rpm ($665 \times g$) and temperature (20°C), parameters to be defined for each cell are: the number of measurements to be averaged (2); the wavelength increment (1 nm).

12. Radial scan at $665 \times g$ allows checking the proper filling of the cells, since the meniscus, i.e., the interface between air and the solutions (sample and solvent), are easily detected. It will measure A_{280} as a function of r at time "zero." A flat profile is expected, since the concentration is the same at all radial positions (at $665 \times g$, sedimentation is not significant unless for very large assemblies). The quality of the signal is also checked at that step (noise should be below 0.01 absorbance unit). The parameters to be defined for each of the cell are: the wavelength (280 nm); the range of radial position (5.8–7.2 cm); the number of A_{280} measurements to be averaged for each radial position (2); the radial step (0.003 cm); the mode of acquisition (continuous).

13. Global parameters for SV experiments: 55,000 rpm ($220,000 \times g$), 20°C. The duration for scanning the absorbance of the three

cells is typically 5 min (for seven cells in the 8-hole rotor, it would be 12 min). In the method, we define: the interval between scans (1 min, much below 5 min in such a way the successive measurements are done as soon as possible); the total number of scans (typically 200 scans for exceeding overnight and stopping manually the AUC in the morning -other protocols are possible); an overlay of four scans (for each of the cells, the superposition of the last four scans is displayed upon centrifugation). The parameters for each of the cells are those given above.

14. The *continuous* $c(s, ffo)$ model of SEDFIT allowing fitting a distribution of s and of f/f_{min} will not be described here (20).

15. Use F-ratio = 0.5 for the analysis without regularization procedure. Use F-ratio of 0.68 or 0.95 for analysis with regularization.

16. For making Fig. 3, we also copied the raw and fitted data as well as the residuals.

17. f/f_{min} determination is only valid in theory if the sample is composed of noninteracting (and ideal) species. In that case, f/f_{min} is determined as a mean for all species in solution. It may be related to the extended shape of the particle but also may artificially arise from the non-ideality of the solution, which perhaps would explain in our data the larger apparent values of f/f_{min} at the largest concentrations.

18. For a rapid equilibrium between monomer and dimer, one peak at s-value intermediate between monomer and dimer is expected (see, e.g., ref. 21, 22).

19. The same equation in principle allows determining k_s, whose value is related to the particle dimension (17, 23). Our determination is not precise enough for such analysis.

20. This is because f_{min} refers to the non-hydrated protein.

21. $\rho = 0.99828$ g/mL and $\eta = 0.01002$ P correspond to water at 20°C, since $s_{20w,0}$ value is reported as the input value. Alternatively, experimental s-value, temperature, solvent density, and viscosity can be chosen as input values.

Acknowledgments

We thank the IBS platform of the Partnership for Structural Biology and the Institut de Biologie Structurale in Grenoble (PSB/IBS), for the assistance and access to the instrument of AUC.

References

1. Receveur-Brechot V, Bourhis JM, Uversky VN et al (2006) Assessing protein disorder and induced folding. Proteins 62:24–45

2. Uversky VN (2002) What does it mean to be natively unfolded? Eur J Biochem 269:2–12

3. Ebel C (2007) Analytical ultracentrifugation. State of the art and perspectives. In: Uversky VN, Permyakov EA (eds) Protein structures: methods in protein structure and stability analysis. Nova, New York

4. Lebowitz J, Lewis MS, Schuck P (2002) Modern analytical ultracentrifugation in protein science: a tutorial review. Protein Sci 11:2067–2079

5. Ebel C (2004) Analytical ultracentrifugation for the study of biological macromolecules. Prog Colloid Polym Sci 127:73–82

6. Howlett GJ, Minton AP, Rivas G (2006) Analytical ultracentrifugation for the study of protein association and assembly. Curr Opin Chem Biol 10:430–436

7. Manon F, Ebel C (2010) Analytical ultracentrifugation, a useful tool to probe intrinsically disordered proteins. In: Uversky VN, Longhi S (eds) Instrumental analysis of intrinsically disordered proteins: assessing structure and conformation. Wiley, Hoboken

8. Demeler B, Behlke J, Ristau O (2000) Molecular parameters from sedimentation velocity experiments: whole boundary fitting using approximate and numerical solutions of Lamm equation. Methods Enzymol 321:38–66

9. Schuck P (1998) Sedimentation analysis of noninteracting and self-associating solutes using numerical solutions to the Lamm equation. Biophys J 75:1503–1512

10. Stafford WF, Sherwood PJ (2004) Analysis of heterologous interacting systems by sedimentation velocity: curve fitting algorithms for estimation of sedimentation coefficients, equilibrium and kinetic constants. Biophys Chem 108:231–243

11. Schuck P (2000) Size-distribution analysis of macromolecules by sedimentation velocity ultracentrifugation and Lamm equation modeling. Biophys J 78:1606–1619

12. Uversky VN (2007) Size-exclusion chromatography in protein structure analysis. In: Uversky VN, Permyakov EA (eds) Protein structures: methods in protein structure and stability analysis. Nova, New York

13. Uversky VN (2010) Analysing intrinsically disordered proteins by size-exclusion chromatography. In: Uversky VN, Longhi S (eds) Instrumental analysis of intrinsically disordered proteins: assessing structure and conformation. Wiley, Hoboken

14. Arnheiter H, Davis NL, Wertz G et al (1985) Role of the nucleocapsid protein in regulating vesicular stomatitis virus RNA synthesis. Cell 41:259–267

15. Jensen MR, Houben K, Lescop E et al (2008) Quantitative conformational analysis of partially folded proteins from residual dipolar couplings: application to the molecular recognition element of Sendai virus nucleoprotein. J Am Chem Soc 130:8055–8061

16. Balbo A, Zhao H, Brown PH et al (2009) Assembly, loading, and alignment of an analytical ultracentrifuge sample cell. J Vis Exp 33:1530

17. Saluja A, Fesinmeyer RM, Hogan S et al (2010) Diffusion and sedimentation interaction parameters for measuring the second virial coefficient and their utility as predictors of protein aggregation. Biophys J 99:2657–2665

18. Balbo A, Minor KH, Velikovsky CA et al (2005) Studying multiprotein complexes by multisignal sedimentation velocity analytical ultracentrifugation. Proc Natl Acad Sci U S A 102:81–86

19. le Maire M, Arnou B, Olesen C et al (2008) Gel chromatography and analytical ultracentrifugation to determine the extent of detergent binding and aggregation, and Stokes radius of membrane proteins using sarcoplasmic reticulum Ca2+-ATPase as an example. Nat Protoc 3:1782–1795

20. Brown PH, Schuck P (2006) Macromolecular size-and-shape distributions by sedimentation velocity analytical ultracentrifugation. Biophys J 90:4651–4661

21. Buisson M, Valette E, Hernandez JF et al (2001) Functional determinants of the Epstein-Barr virus protease. J Mol Biol 311:217–228

22. Schuck P (2003) On the analysis of protein self-association by sedimentation velocity analytical ultracentrifugation. Anal Biochem 320:104–124

23. Solovyova A, Schuck P, Costenaro L et al (2001) Non-ideality by sedimentation velocity of halophilic malate dehydrogenase in complex solvents. Biophys J 81:1868–1880

24. Tcherkasskaya O, Uversky VN (2001) Denatured collapsed states in protein folding: example of apomyoglobin. Proteins 44:244–254

25. Cantor CR, Schimmel PR (1980) Techniques for the study of biological structure and function. Biophysical chemistry. W. H. Freeman, San Franscisco

Chapter 7

Analysis of Intrinsically Disordered Proteins by Small-Angle X-ray Scattering

Pau Bernadó and Dmitri I. Svergun

Abstract

Small-angle scattering of X-rays (SAXS) is a method for the low-resolution structural characterization of biological macromolecules in solution. The technique is highly complementary to the high-resolution methods of X-ray crystallography and NMR. SAXS not only provides shapes, oligomeric state, and quaternary structures of folded proteins and protein complexes but also allows for the quantitative analysis of flexible systems. Here, major procedures are presented to characterize intrinsically disordered proteins (IDPs) using SAXS. The sample requirements for SAXS experiments on protein solutions are given and the sequence of steps in data collection and processing is described. The use of the recently developed advanced computational tools to quantitatively characterize solutions of IDPs is presented in detail. Typical experimental and potential problems encountered during the use of SAXS are discussed.

Key words: Small-angle scattering, Solution scattering, Macromolecular structure, Flexible macromolecules, Functional complexes, Ab initio methods, Rigid body modeling

1. Introduction

Small-angle scattering of X-rays (SAXS) is a powerful method for the analysis of biological macromolecules in solution (1). Over the last decade, major advances in instrumentation and computational methods have led to new and exciting developments in the application of SAXS to structural biology including globular proteins, macromolecular complexes and also flexible systems like IDPs (2–6).

In a SAXS experiment, samples containing dissolved macromolecules are exposed to an X-ray beam and the scattered intensity is recorded by a detector as a function of the scattering angle. Dilute aqueous solutions of proteins give rise to an isotropic scattering intensity I, which depends on the modulus of the momentum transfer s ($s = 4\pi \sin(\theta)/\lambda$, where λ is the wavelength

Vladimir N. Uversky and A. Keith Dunker (eds.), *Intrinsically Disordered Protein Analysis: Volume 2, Methods and Experimental Tools*, Methods in Molecular Biology, vol. 896, DOI 10.1007/978-1-4614-3704-8_7, © Springer Science+Business Media New York 2012

of the beam, and *2θ* is the angle between the incident and scattered beam). The solvent scattering is subtracted and the background corrected intensity is presented as a radially averaged one-dimensional curve *I(s)*. For monodisperse solutions of identical randomly oriented proteins, the SAXS curve is proportional to the scattering of a single particle averaged over all orientations.

Several overall parameters can be directly obtained from a SAXS curve providing information about the size, oligomeric state, and overall shape of the molecule. Moreover, for folded proteins or macromolecular complexes, low-resolution three-dimensional structures can be determined from the scattering data either ab initio or through the refinement using available high-resolution structures and/or homology models (5). SAXS is routinely employed for the validation of structural models, analysis of oligomeric states, and the estimation of volume fractions of components in mixtures. SAXS has been actively used also for flexible systems including solutions of IDPs, but in the past the analysis was often restricted to the determination of simple geometric and weight parameters (7). Novel possibilities in the study of such systems have recently been opened to structural biologists by the advanced approaches that take the conformational flexibility into account (8). These studies provide a novel piece of structural information that has been proven as fundamental to understand biological processes (9–13). In the present chapter the experimental and methodological peculiarities of the use of SAXS for IDPs are presented. We describe the experimental procedures and the analysis techniques including the traditional methods providing overall parameters and also the advanced approaches to characterize the systems in terms of ensemble distributions.

2. Materials

SAXS can probe structure on an extremely broad range of macromolecular sizes, ranging from small proteins and polypeptides (a few kDa) to macromolecular complexes and large viral particles (to several hundred MDa). The proteins can all be measured in solution under near native conditions, and the effect of changes in sample environment (like pH, temperature, ionic strength, ligand addition, etc.) can be easily followed. Still, the protein solutions must be well prepared, thoroughly purified, and characterized before doing SAXS experiments.

2.1. Sample Preparation: General Considerations   In general, SAXS requires a few (typically 1–3) mg of highly pure, monodisperse protein that remains soluble at high concentration. If the sample is aggregated, the scattering data will be difficult or even impossible to interpret (see Note 1). The sample concentrations must be determined as precise as possible, and accuracy better than 10 % is required to appropriately normalize the scattering data and thus to estimate the effective molecular mass of the solute (see Note 2).

A typical sample volume required for a single measurement is about 20–60 µl depending on the SAXS station used (see Note 3). A series of scattering curves is always to be recorded at varying concentration to ensure that the condition of a "dilute" solution is fulfilled. The typical solute concentrations range from about 0.5–1 mg/ml to about 5–10 mg/ml (see Note 4), such that about 1–2 mg of purified material is usually required for a complete SAXS experiment on the given protein or construct at the given conditions (e.g., buffer composition and temperature). Sufficient amount (>10 ml) of matching buffer(s) must be brought to the SAXS station to make the buffer collections and to dilute the samples if necessary. The buffer composition must precisely match the composition of the sample: even small mismatches in the chemical composition of the solvent between the buffer and the sample may lead to difficulties during background subtraction. At best, the last dialysis buffer should be used for the background measurements (see Note 5). The protein solutions are usually centrifuged prior to the measurements for ca 5 min at ca 13,000 rpm and the supernatant is taken for the measurements to remove large aggregates (this procedure is especially useful for the samples stored at −80°C, which were thawed for the SAXS experiments).

3. Methods

3.1. Measurement of the SAXS Profiles: General Considerations

Each measurement of the macromolecular solution typically requires two measurements of the corresponding buffer, before and after the sample. Note that typically all the SAXS measurements for the given experimental session are performed in one and the same measuring cell to ensure that the background scattering remains unchanged allowing for reliable buffer subtraction. The sample compartment (which is either a cell with flat windows or a capillary) is cleaned and refilled after each measurement. In the past, the cleaning and filling procedure was done manually; nowadays, liquid handling robots are becoming more and more popular to facilitate the automatic data collection (14, 15). On some stations (e.g., X33 station of the EMBL in Hamburg) remote operation is possible where the users send the samples and perform the data collection from their laboratory.

On synchrotrons, SAXS experiments are very fast, and the data is typically collected within seconds (see Note 6); on laboratory

X-ray sources, minutes and even hours are required. Using a robotic operation, the user fills in the tray with the samples and corresponding buffers, then specifies the sequence of measurements and monitors the progress. The main problem for the user remains to monitor the bubble-free cell filling (see Note 7). On the stations with the manual operation, each individual sample should be filled (without bubbles), then the data collection started, finished, and the cell compartment cleaned. This cycle is then repeated as many times as needed to measure all the samples and corresponding buffers.

The measurements of the SAXS data are usually made below room temperature (typically, at 5–10°C) to reduce the possible radiation damage effects. However, SAXS stations normally allow for controlled measurement of the temperature series, in the range from ice melting to water boiling points. The temperature measurements are often useful for the studies of the thermodynamic characteristics of IDPs.

3.2. Data Analysis

There are different strategies and different pieces of software available for the analysis of the SAXS data from IDPs. We shall present here the protocols based on the use of the program package ATSAS (16), which allows one to perform major processing and analysis steps. This package and all the programs mentioned below are publicly available for download from http://www.embl-hamburg.de/biosaxs/software.html.

3.2.1. Primary Analysis of SAXS Data

The scattering data are usually collected by two-dimensional detectors, and the images are appropriately processed, corrected, and reduced to one-dimensional scattering profiles by the local software of the SAXS stations. The output is typically an ASCII file containing the subtracted data in columnar format: s, $I(s)$, $\sigma(s)$, where the latter column represents the standard deviation of the processed intensity. The normalization and buffer subtraction are performed following Eq. 1

$$I(s) = (I_{\text{sample}}(s)/T_{\text{sample}} - I_{\text{buffer}}(s)/T_{\text{buffer}})/c, \qquad (1)$$

where $I_{\text{sample}}(s)$ and $I_{\text{buffer}}(s)$ are, respectively, the measured scattering data from the sample and the buffer, T_{sample} and T_{buffer} are the intensities of the transmitted X-ray beam, and c is the solute concentration. This basic operation is also often performed by the station-specific software, scripts, or completely automatically (the latter is the case at the already mentioned synchrotron X33 station of the EMBL in Hamburg). In the following we assume that the radially averaged and subtracted patterns are available and describe the sequence of actions using the interactive data analysis program PRIMUS from the ATSAS package (16).

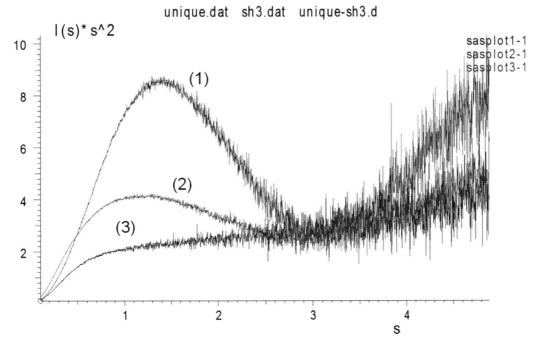

Fig. 1. Kratky representation of SAXS curves measured for different N-terminal constructions of Src-Kinase. (1) The SH3 domain, globular; (2) the Unique-SH3 fragment of the protein, partly disordered; (3) the Unique domain, fully disordered. The features observed in the Kratky representation identify the degree of compactness of the polypeptide. Figure made with the program SASPLOT of the ATSAS package.

1. Load the SAXS curve in PRIMUS.

2. A Kratky representation of the SAXS curve ($I(s)s^2$ vs s) is an excellent tool to qualitatively identify conformational disorder in proteins. Unstructured proteins present a continuous rise of $I(s)s^2$ whereas globular proteins display a peak. Partly disordered proteins present a mixed behavior depending on the relative proportion of each of the parts. The Kratky representation of a SAXS curve can be obtained with the SASPLOT application of PRIMUS that allows several data representations (see Fig. 1).

3. Using Guinier approximation, R_g and the forward scattering, $I(0)$, are derived from the initial part of the scattering profile (17). At very small angles the intensity is represented as:

$$I(s) = I(0)\exp\left(-\left(sR_g\right)^2/3\right). \qquad (2)$$

The R_g and $I(0)$ are obtained by a simple linear fit in logarithmic scale using the *Guinier* option of PRIMUS (see Fig. 2) where the optimal range of points for the analysis can be selected (see Note 8). Note that there is an option *AutoRg* available, which selects the appropriate range and computes the R_g automatically. In some cases, however, it might be useful to cross-validate the results by interactive analysis.

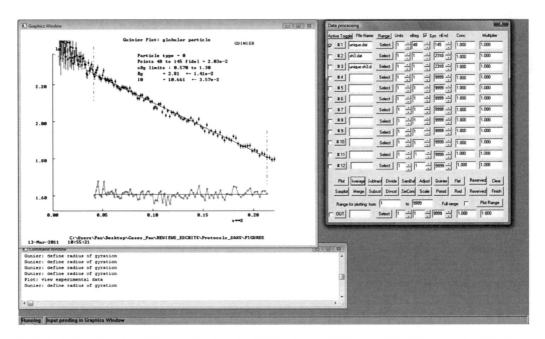

Fig. 2. Derivation of the R_g and the $I(0)$ values from the initial part of the SAXS curve of the Unique domain of Src-Kinase using Guinier approximation with the software PRIMUS of the ATSAS package. Experimental data (*filled dots*) are fitted to Eq. 2 (*straight line*), individual discrepancies are displayed (*empty dots*). The number of points and the R_gs range used, and the resulting R_g and $I(0)$ are displayed on the top of the figure.

4. An initial quantitative estimate of the compactness of an IDP can be obtained from the R_g using the following Flory's relationship as a threshold value (18)

$$R_g^{RC} = (2.54 \pm 0.01) N^{(0.522 \pm 0.01)}. \tag{3}$$

This equation relates the number of amino acids of the protein, N, with the expected radius of gyration of a disordered chain when it behaves as a random coil, R_g^{RC}. Obtaining $R_g < R_g^{RC}$ is an indication of compactness with respect to the random coil (e.g., presence of long-range contacts), if $R_g > R_g^{RC}$ then the IDP is more extended than the random coil (e.g., contains long β-sheet-like regions). Note that this interpretation is not valid for multidomain proteins containing globular domains connected by flexible linkers.

5. The forward intensity, $I(0)$, provides the estimate of the molecular weight (MW) of the protein and hence suggests its oligomeric state. This is done by comparison of the $I(0)$ value of the protein studied with that obtained for a standard protein of known MW and concentration (i.e., Bovine Serum Albumin, with a MW of 66 KDa, is often used as such a standard)

$$MW_{prot} = I(0)_{prot} MW_{stand} / I(0)_{stand}. \tag{4}$$

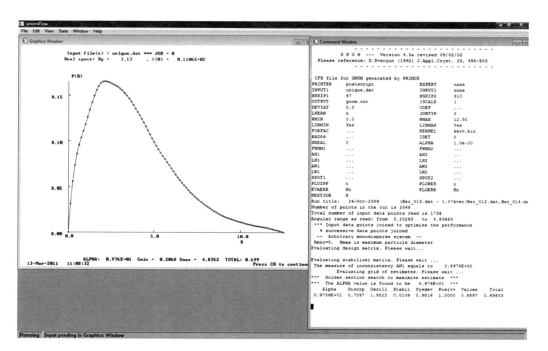

Fig. 3. Distance distribution function, $P(r)$, of the Unique domain of the Src-Kinase obtained with the program GNOM implemented in PRIMUS of the ATSAS package. A D_{max} of 12.5 nm was imposed prior to the indirect transformation. The R_g and the $I(0)$ obtained are displayed on the top of the figure.

To apply this equation experimental $I(0)$ values must be divided by the concentration of the sample in mg/ml, that is why knowledge of the precise solute concentration is important (see Subheading 2.1 and Note 2).

6. The distance distribution function, $P(r)$, is defined as the distribution of distances between volume elements inside the particle. $P(r)$ is obtained from the SAXS curve with the program GNOM (19), which is called directly from PRIMUS or its automated version AUTOGNOM (16). $P(r)$ defines the maximum particle distance, D_{max} beyond which $P(r)$ is equal to zero (see Fig. 3 and Note 9).

7. The scattering data sets recorded on the same protein in the same buffer and temperature but at different concentrations, should be analyzed together to check whether there are concentration-dependent differences between the data sets. Typically, the dependences $R_g(c)$ or $MW(c)$ allow one to judge, whether the concentration dependence is present. If this is the case, the data can be extrapolated to infinite dilution using the *Zerconc* command of PRIMUS or a separately provided *AutoMerge* program (see Note 10). The pattern extrapolated to zero concentration should be employed for a more in-depth analysis in the next section.

Highly flexible proteins such as IDPs probe an astronomical amount of conformations that are in fast equilibrium exchange. Therefore, their accurate structural description requires its definition in terms of ensembles. The following points explain in detail the use of the Ensemble Optimization Method (EOM) which is the original and most used strategy developed for the structural characterization of flexible proteins using SAXS (20) and is included in the ATSAS package. Different variations of the EOM principle have been later reported by other groups (13, 21, 22).

The use of the EOM approach consists in three consecutive steps:

1. Generation of a pool of conformations.

2. Selection of a sub-ensemble of conformations describing the SAXS data.

3. Quantitative description of the structural properties of the selected ensemble.

These most relevant aspects of each one of these steps will be explained in the following points and associated notes. Reader is referred to the EOM manual (available in the ATSAS package) for more specific details about its usage.

1. The three steps of EOM are performed with two different programs: **RanCh** (*Random Chain*) and **GAJOE** (*Genetic Algorithm Judging Optimization of Ensembles*). These programs are ready to use after the installation of the ATSAS suite.

2. The first step is the *generation of a pool of M ≫ conformations* of the protein of interest that must be a good representation of the conformational space sampled. The program **RanCh**, which is the tool to perform this part, distinguishes between the two different scenarios:

   1. Fully disordered protein (*scenario FDP*).

   2. Multidomain protein (*scenario MD*) where several globular domains are connected with (potentially) flexible linkers.

   Conformations can be generated also using alternative methods different than **RanCh** (see Note 11).

3. The files needed to generate the pool of conformations are:

   • A file with the exact amino acid sequence of the protein in a one-letter code FASTA format (*.seq*).

   • The experimental SAXS curve (*.dat*) in ASCII format with three columns corresponding to momentum transfer s, Intensity $I(s)$, and experimental error, $\sigma(s)$, associated to $I(s)$.

   These two files are sufficient for the *FDP scenario*. For the *MD scenario*, the pdb files (*.pdb*) corresponding to each one of the globular domains must be provided.

4. Upon execution of **RanCh** different parameters and file names will be asked by the program (see the EOM manual). It is important to give more detail for two of them:

   - *Type of model*: When modeling disordered proteins [R] option is preferable (see Note 12).

   - *Total number of structures to generate*. A large number ($\approx$10,000) is required in order to have enough survey of sizes and shapes to select in the following step (see Note 13).

   For each of the structures generated the theoretical scattering profile is computed using CRYSOL (23). This step is already incorporated into **RanCh**. Standard parameters for a CRYSOL calculation can be used although a larger order of harmonics (15 is default and 50 maximum) can be used to have more accurate representation of higher angles.

5. Output files of **RanCh** are: (These files must be maintained in the same folder for running following steps).

   - *junX00.int:* This file contains the intensities of the random models created by **RanCh**. *X* is the extension of the new pdb files

   - *RanchX.log:* This is a log file containing the parameters used for the calculation.

   - *Size_listZ.txt:* Contains the individual R_g and D_{max} of all conformers computed with **RanCh**. *Z* is the experimental SAXS data file name.

6. The *selection of a sub-ensemble of conformations* describing the SAXS data is performed with **GAJOE**. The aim of this part is finding in the pool of *M* conformations a subset of *N* (between 20 and 50) theoretical scattering profiles (*chromosome*)

$$I(s) = \frac{1}{N} \sum_{n=1}^{N} I_n(s), \qquad (5)$$

that optimally describes the SAXS data when averaged (with the lowest χ^2 value).

$$\chi^2 = \frac{1}{K-1} \sum_{j=1}^{K} \left[\frac{\mu I(s_j) - I_{exp}(s_j)}{\sigma(s_j)} \right]^2. \qquad (6)$$

This procedure is performed using a genetic algorithm (GA) defined with a number of parameters such as the size of the chromosome, number of chromosomes per generation, maximum number of mutations allowed, the number of generations. The GA is run several times (*cycles*) starting from different randomly selected chromosomes. With this strategy a set of optimal solutions is obtained, one per cycle (see Note 14).

Default values implemented in **GAJOE** ensure the optimal performance of the GA.

7. The output files of **GAJOE** contain information regarding the GA selection procedure and also about the *quantitative description of the structural properties of the selected ensemble* (see step 8). These output files are placed in a subfolder named in the form **GAnum** where **num** is the sequential number for every time the program is run. In each subfolder the following files can be found:

   • *GAnum.log:* The log file that reports on the files, the parameter values used, etc.

   • *best_curvenum.txt:* File showing the number of times each curve/structure was selected in the final generation of all cycles.

   • *selected_ensemnum.txt:* File showing the best 10 ensembles at the final generation of each cycle.

   • *profilesZnum.fit:* The file with the fit of the best selected ensemble, i.e., the cycle with the lowest discrepancy χ to the experimental data **Z.dat**. It also contains the χ values for all cycles. The **.dat* and **.fit* files can be visualized using SASPLOT program.

   The quality of the fit, χ, obtained with EOM should be around unity and the **.fit* file should not display systematic deviation between the experimental and calculated data. If this is not the case, it is an indication that the structures composing the pool are not good representation of the behavior of the protein in solution and other alternatives such as oligomerization must be taken into account (9, 24).

8. The *quantitative description of the structural properties of the selected ensemble* is also performed by the program **GAJOE**. The structural properties that the molecule adopts in solution are described in terms of distributions of low-resolution structural parameters such as R_g, D_{max}, and anisometry. Therefore, EOM is a tool to describe the size and shape distributions sampled by the unfolded molecule in solution (see Note 15). To calculate these distributions, the R_g and D_{max} of the individual conformers belonging to the optimal chromosome (the one that fits best to the SAXS curve) of each cycle are computed and placed in a histogram by **GAJOE**. The output files from **GAJOE** regarding the structural interpretation, also placed in the **GAnum** subfolder, are:

   • *Size_distrZnum.dat/Rg_distrZnum.dat:* Files containing the D_{max} and R_g distribution obtained from the subensembles collected and placed in common histograms. Distributions from the pool are also included in these files for quantitative comparison with the model with complete conformational freedom (see step 9).

- *A subfolder named **pdbs*** which contains the structures that were selected in the cycle with the lowest χ. These structures are intended to offer a visual inspection of the structures, but not a high-resolution picture of the preferred conformations adopted by the protein in solution (see Note 7). In the MD scenario selected structures can provide direct insight into the interdomain distances attained by the molecule in solution (see step 10).

9. Direct comparison of size and shape descriptor distributions obtained for the pool and the EOM selected ensemble obtained (*Size_distr**Znum**.dat/Rg_distr**Znum**.dat*) puts into perspective the conformational sampling of the protein. When pool distributions are a good representation of a random coil model (see Note 16) then structural phenomena such as compaction or enhanced extendedness can be quantified. Sharp distributions of R_g and D_{max} suggest the globularity of the protein (in the MD scenario), while broad distributions indicate large-amplitude motions, see examples in Fig. 4.

10. In the MD scenario, interdomain distance distributions can be derived from the ensemble of selected conformations. The centers of masses of the globular domains of the MD protein can be derived for the individual conformations, and the distribution of distances can be plotted in a histogram. Comparison of these distributions with the random coil model implemented to generate the pool can identify partial ordering in connecting linkers or transient specific interaction between these domains (see Note 17). Examples of this analysis have been reported (20, 25). The specific routines to perform these analyses during the **GAJOE** process are under development.

4. Notes

1. The sample monodispersity (desirably better than 90 %) has to be verified in advance by biophysical methods, e.g., by native gel filtration, ultracentrifugation, and dynamic light scattering. One must keep in mind that having a single band on an SDS gel is not sufficient and only a single band on the native gel suggests that one has a monodisperse solution.

2. To determine concentrations for SAXS measurements, Bradford assays are usually not sufficiently precise and absorption OD measurements at the laser wavelength 280 nm are more reliable. For the proteins lacking aromatic residues the

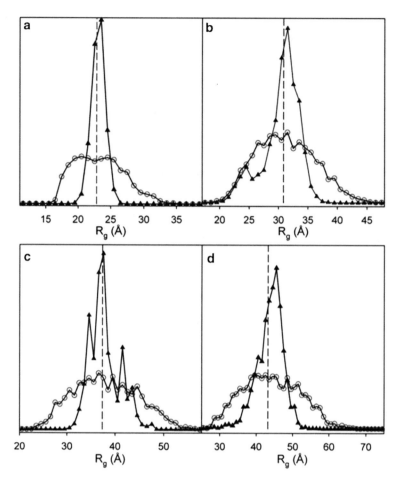

Fig. 4. R_g distributions derived from the EOM analysis of the synthetic SAXS profiles computed for rigid structures (*filled triangles*) and dynamically averaged ensembles of conformations (*empty dots*) for (**a**) di-, (**b**) tri-, (**c**) tetra-, and (**d**) penta-ubiquitin chains, composed by different number of globular ubiquitin domains connected with a 20-residue-long flexible linker. The *vertical dashed lines* represent the R_g of the conformation used to calculate the synthetic curves for the rigid scenario. See ref. 28 for details. The different behavior of the R_g distributions discriminates flexible proteins (broad distributions) from rigid arrangements (sharp distributions).

concentration may also be determined using the refraction index of the solute (i.e., a refractometer can be used).

3. On second generation synchrotrons and on older X-ray laboratory sources, one needs about 50 μl and more; on modern high-brilliance synchrotrons and on new laboratory cameras, about 10 μl is required. The new trends using microfluidic devices (26) are expected to lead to sub-μl sample volumes for SAXS in future.

4. Typically, for each sample a concentration series (e.g., 1, 2, 5, 10 mg/ml) has to be measured. If the sample tends to (or may) aggregate at higher concentrations, it is highly recommended to bring the low-concentration stocks to the SAXS instrument and concentrate the samples prior to the X-ray measurements. If the sample is well behaved, one may bring high-concentration stocks to dilute them before the measurements.

5. To achieve the best SAXS results, the proteins should normally be dissolved in the buffers yielding close-to-ideal solutions. The ionic strength and pH should be selected in such a way that the proteins minimally interact with each other unless it is required by the nature of the experiment. Any additives (salts, small molecules, cosolutes, etc.) diminish the contrast of the protein and increase the X-ray absorption, therefore buffers with excessive amounts of additives (typically, not more than 0.5 M NaCl, not more than 5 % glycerol, not more than 5 mM ATP, etc.) should be avoided unless this is not dictated by the biological or chemical considerations.

6. On high-brilliance synchrotrons, the samples can be damaged by the X-ray radiation even during the seconds and sub-seconds collection time. A reducing agent (e.g., 2 mM DTT or TCEP) or an additive slowing down the free radicals (e.g., up to 5 % glycerol) may be added to the sample before the experiment to diminish the damage/aggregation during the data collection. In any case, synchrotron data should be collected in short individual time frames, which are compared to each other to detect the radiation damage effects.

7. Having a bubble-free cell filling is extremely important: if an edge of a bubble (either of the sample or of the buffer specimen) is illuminated by the X-ray beam, parasitic scattering emerges and data will become unusable.

8. The momentum transfer range used for Guinier's approach must not exceed $s < 1.3R_g$. First points of the curve, if they are influenced by the primary beam and show nonlinear behavior, must be suppressed from Guinier's analysis.

9. For highly flexible proteins such as IDPs or some multidomain proteins, D_{max} derived from the primary analysis of the SAXS curve is smaller than the maximum distance that the molecule can adopt in solution (27, 28). The low population of the highly extended conformations often induces such an underestimation of D_{max}.

10. The simplest procedure often used to obtain the scattering pattern representing the "infinite dilution" case is to take the lowest concentration data at small angles and the highest concentration data at the higher angles and merge the two patterns using the PRIMUS command *Merge*. One should, however,

keep in mind that this procedure is an approximation and is only applicable if the data display negligible concentration dependence at low solute concentrations.

11. The pool of conformations can be generated with alternative strategies that also ensure a reasonable sampling of the conformational space. Molecular Dynamic or Monte-Carlo simulations are potential alternatives. Appropriate scripts and protocols are provided in order to adapt these ensembles to **GAJOE** requirements (see EOM manual for details).

12. **RanCh** distinguishes between Random model [R] and Native-like model [N]. In both cases the disordered regions are represented by self-avoiding C_α chains, where the conformation is defined by the bond and dihedral angles of four consecutive C_α atoms, the so-called quasi-Ramachandran space (see ref. (29) for details). The difference between both models originates in the definition of the conformational sampling forced when building the chain. A Native model, [N], is based on a quasi-Ramachandran space derived from all residues of a large library of high-resolution protein structures. Random model, [R], only take protein residues that are placed neither in α-helices nor in β-strands (30).

13. A large number of conformations is required in the pool in order to provide a vast survey of sizes and shapes for the molecule. In our experience 10,000 is a reasonable number. It is worth noting that, at atomic level, the sampling of the pool cannot be exhaustive as a disordered protein can adopt an astronomical number of conformations. However, SAXS is a low-resolution technique and only descriptors of the overall structural features are relevant.

14. Uniqueness of the solutions obtained in each of the cycles of **GAJOE** in an EOM study is an important aspect. Selected ensembles derived from repeated cycles starting from different random selections normally contain different conformations but they all provide similar low-resolution structural descriptor distributions such as R_g and D_{max}. Therefore, the algorithm is able to find equivalent minima in terms of distributions but of course not in terms of individual molecular configurations: the latter is not identifiable given the low-resolution nature of SAXS data.

15. Conformations created in the pool are described at atomic resolution. However, the structural interpretation of an EOM analysis of a SAXS curve is restricted to low-resolution descriptors of these conformations. This situation is a direct consequence of low-resolution information coded in a SAXS profile. Still, the distributions provided by EOM yield a major improvement over traditional approaches that condense all structural characteristics in the averaged R_g values. Attempts to interpret selected ensembles at atomic or residue level are discouraged.

16. In order to quantify the degree of order of the flexible regions of the protein studied, structures of the pool must represent the conformational sampling in a completely disordered state (random coil). In that sense, **RanCh** is a good program as the coil library used to generate conformations has been proven as a good model of natively disordered regions. Molecular dynamics simulations are often not a good description of the complete conformational space available due to the limited time of simulation and the inaccuracy of force-fields to describe residue-specific Ramachandran's space of natively disordered polypeptides.

Acknowledgments

P.B. acknowledges funds from la Generalitat de Catalunya (grant SGR2009 1352), D.S. acknowledges an EU FP7 e-Infrastructures grant WeNMR (contract number 261572).

References

1. Feigin LA, Svergun DI (1987) Structure analysis by small-angle X-ray and neutron scattering. Plenum, New York/London

2. Svergun DI, Koch MHJ (2002) Advances in structure analysis using small-angle scattering in solution. Curr Opin Struct Biol 12:654–660

3. Petoukhov MV, Svergun DI (2007) Analysis of X-ray and neutron scattering from biomolecular solutions. Curr Opin Struct Biol 17:562–571

4. Putnam CD, Hammel M, Hura GL et al (2007) X-ray solution scattering (SAXS) combined with crystallography and computation: defining accurate macromolecular structure, conformations and assemblies in solution. Q Rev Biophys 40:191–285

5. Mertens HD, Svergun DI (2010) Structural characterization of proteins and complexes using small-angle X-ray solution scattering. J Struct Biol 172:128–141

6. Jacques DA, Trewhella J (2010) Small-angle scattering for structural biology-expanding the frontier while avoiding the pitfalls. Protein Sci 19:642–657

7. Doniach S (2001) Changes in biomolecular conformation seen by small angle X-ray scattering. Chem Rev 101:1763–1778

8. Bernadó P, Svergun DI (2012) Structural Insights into intrinsically disordered proteins from Small-Angle X-ray Scattering. Mol BioSystems 8:151–167.

9. Bernadó P, Pérez Y, Svergun DI et al (2008) Structural characterization of the active and inactive states of Src kinase in solution by small-angle X-ray scattering. J Mol Biol 376:492–505

10. Mylonas E, Hascher A, Bernadó P et al (2008) Domain conformation of Tau protein studied by solution small-angle X-ray scattering. Biochemistry 47:10345–10353

11. Stott K, Watson M, Howe FS et al (2010) Tail mediated collapse of HMGB1 is dynamic and occurs via differential binding of the acidic tail to the A and B domains. J Mol Biol 403:706–722

12. García-Pino A, Balasubramanian S, Wyns L et al (2010) Allostery and intrinsic disorder mediate transcription regulation by condicional cooperativity. Cell 142:101–111

13. Yang S, Blachowicz L, Makowski L et al (2010) Multidomain assembled states of Hck tyrosine kinase in solution. Proc Natl Acad Sci U S A 107:15757–15762

14. Round AR, Franke D, Moritz S et al (2008) Automated sample-changing robot for solution scattering experiments at the EMBL Hamburg SAXS station X33. J Appl Cryst 41:913–917

15. Hura GL, Menon AL, Hammel M et al (2009) Robust, high-throughput solution structural analyses by small angle X-ray scattering (SAXS). Nat Methods 6:606–612

16. Konarev PV, Petoukhov MV, Volkov VV et al (2006) ATSAS 2.1, a program package for small-angle scattering data analysis. J Appl Cryst 39:277–286

17. Guinier A (1939) La diffraction des rayons X aux trés petits angles: application a l'étude de phenomenes ultramicroscopiques. Ann Phys (Paris) 12:161–237

18. Bernadó P, Blackledge M (2009) A self-consistent description of the conformational behavior of chemically denatured proteins from NMR and small angle scattering. Biophys J 97:2839–2845

19. Svergun DI (1992) Determination of the regularization parameter in indirect-transform methods using perceptual criteria. J Appl Crystal 25:495–503

20. Bernadó P, Mylonas E, Petoukhov MV et al (2007) Structural characterization of flexible proteins using small-angle X-ray scattering. J Am Chem Soc 129:5656–5664

21. Pelikan M, Hura GL, Hammel M (2009) Structure and flexibility within proteins as identified through small angle X-ray scattering. Gen Physiol Biophys 28:174–189

22. Rozycki B, Kim YC, Hummer G (2011) SAXS ensemble refinement of ESCRT-III CHMP3 conformational transitions. Structure 19:109–116

23. Svergun DI, Barberato C, Koch MHJ (1995) CRYSOL—a program to evaluate X-ray solution scattering of biological macromolecules from atomic coordinates. J Appl Crystal 28:768–773

24. Mosbaek CR, Nolan D, Persson E et al (2010) Extensive small-angle X-ray scattering studies of blood coagulation factor VIIa reveal interdomain flexibility. Biochemistry 49:9739–9745

25. Bernadó P, Modig K, Grela P et al (2010) Structure and dynamics of ribosomal protein L12: an ensemble model based on SAXS and NMR relaxation. Biophys J 98:2374–2382

26. Toft KN, Vestergaard B, Nielsen SS et al (2008) High-throughput small angle X-ray scattering from proteins in solution using a microfluidic front-end. Anal Chem 80:3648–3654

27. Heller WT (2005) Influence of multiple well defined conformations on small-angle scattering of proteins in solution. Acta Crystallogr D Biol Crystallogr 61:33–44

28. Bernadó P (2010) Effect of interdomain dynamics on the structure determination of modular proteins by small angle scattering. Eur Biophys J 39:769–780

29. Kleywegt GJ (1997) Validation of protein models from Calpha coordinates alone. J Mol Biol 273:371–376

30. Bernadó P, Blanchard L, Timmins P et al (2005) A structural model for unfolded proteins from residual dipolar couplings and small-angle X-ray scattering. Proc Natl Acad Sci U S A 102:17002–17007

Chapter 8

Small Angle Neutron Scattering for the Structural Study of Intrinsically Disordered Proteins in Solution: A Practical Guide

Frank Gabel

Abstract

Small angle neutron scattering (SANS) allows studying bio-macromolecular structures and interactions in solution. It is particularly well-suited to study structural properties of intrinsically disordered proteins (IDPs) over a wide range of length-scales ranging from global aspects (radii of gyration and molecular weight) down to short-distance properties (e.g., cross-sectional analysis). In this book chapter, we provide a practical guide on how to carry out SANS experiments on IDPs and discuss the complementary aspects and strengths of SANS with respect to small angle X-ray scattering (SAXS).

Key words: Small Angle Neutron Scattering, Contrast Variation, Intrinsically Disordered Proteins, Hydration Water, Deuterium labeling

1. Introduction

Small angle neutron (SANS) and X-ray (SAXS) scattering have been used extensively for several decades to study the structural properties of polymers (1, 2) and bio-macromolecules (3–9) in solution. They are particularly well-suited to provide information on several length-scales for unfolded systems (going from radii of gyration over persistence lengths to cross-sectional analysis). An extensive review and discussion of the systems studied is beyond the scope of this book chapter. More recently, SAXS and SANS have been applied to study the structural properties of unfolded proteins (3, 10–19), in part by using the concepts developed in the polymer sciences. Two very interesting and promising recent topics are the presence of residual native structures (20, 21) and the description of the unfolded state in terms of explicit conformational ensembles (10, 22). Both findings were stimulated by complementary information

Vladimir N. Uversky and A. Keith Dunker (eds.), *Intrinsically Disordered Protein Analysis: Volume 2, Methods and Experimental Tools*, Methods in Molecular Biology, vol. 896, DOI 10.1007/978-1-4614-3704-8_8, © Springer Science+Business Media New York 2012

Table 1
Neutron and X-ray scattering lengths of the most common nuclei found in biological macromolecules (5)

Nucleus	b_{coh} (10^{-12} cm)	$f_{X\text{-rays}}$ (10^{-12} cm)
^{1}H	−0.3742	0.28
^{2}H (D)	0.6671	0.28
^{12}C	0.6651	1.69
^{14}N	0.94	1.97
^{16}O	0.5804	2.25
^{31}P	0.517	4.23
^{32}S	0.2847	4.5

from NMR, in particular residual dipolar couplings (RDCs), a fruitful combination with high potential for future studies of disordered proteins (8, 23).

It has been recognized that the study of polymer structures can benefit from contrast variation and specific deuterium-labeling using SANS, in particular regarding the cross-sectional analysis (2, 24). It would be desirable to supplement SAXS with SANS data in a similar way in order to study structural properties of IDPs making use of the concept of contrast variation (H_2O/D_2O ratio in solution and/or deuteration of specific moieties in IDPs). While SAXS and SANS yield similar parameters at "large" distances (such as radii of gyration) for IDPs, particular questions that can be specifically addressed with SANS include the following: (1) how are the interactions of protein and solute organized in solution? (2) solvent properties in the vicinity of unfolded proteins in contrast to folded proteins, (3) properties of side-chain flexibility and its effect on the scattering curve at higher angles, and (4) deviations from the random-coil model, e.g., excluded volume effects.

While X-rays are sensitive to variations of the electronic density between the solutes and the solvent, neutrons are sensitive to variations of the nuclear scattering length density ρ (5). The neutron scattering lengths b_{coh} do not vary in a systematic way with atomic number as do X-rays (Table 1). The most important feature for biological applications is the difference in scattering length between the hydrogen (H) and deuterium (D) nuclei that is the basis for contrast variation of proteins in aqueous solutions (Fig. 1). In contrast to polymers, combined SAXS/SANS data analysis on biological systems is difficult in many cases due to their fragility: both experiments ideally have to be done on the same sample within a minimum time delay which is not always possible.

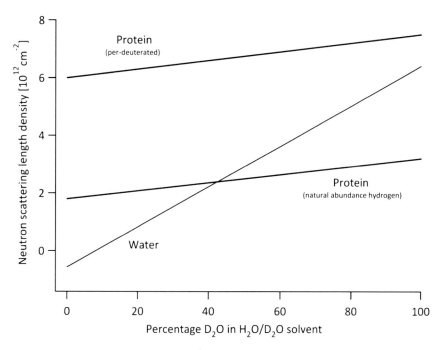

Fig. 1. Neutron scattering length densities ρ (in cm^{-2}) of natural abundance and perdeuterated proteins in aqueous solution. They can be calculated from the chemical composition of the protein, its solvent-excluded volume and the number of exchangeable hydrogen atoms. At a given H_2O/D_2O ratio in solution, "contrast" (=scattering length density difference) $\Delta\rho$ is defined as the difference between the water line and the respective protein line.

2. Material

2.1. Neutron Sources and Instruments

Small angle neutron scattering (SANS) requires cold neutrons (wavelengths $\approx$ 5–10 Å). They are produced either in nuclear fission processes in reactors or by spallation processes in spallation sources and moderated to yield a maximum flux near the desired wavelength. An overview of existing facilities can be found at the following Web site (http://www.neutron.anl.gov/facilities.html). An overview of available SANS instruments is provided under the following link: http://www.ill.eu/instruments-support/instruments-groups/ groups/lss/more/world-directory-of-sans-instruments/. Instrumental access usually requires writing and submitting a proposal that is reviewed by scientific committees of the respective neutron centers which attribute beam-time based on scientific merit.

2.2. Sample State and Experimental Practice

SANS is carried out on solution state samples at typical protein concentrations ranging from about 1 to 10 mg/ml and sample volumes of the order of 100–200 μl. These conditions are very similar to those used in SAXS or NMR experiments. Sample holders are usually glass cells made out of quartz (e.g., Hellma® 100-QS) with optical path-lengths of 1–2 mm (see Note 1). In general, these (spectrophotometer) cells allow an easier sample manipulation and access than the capillary system usually applied in SAXS experiments. It should be noted that cells containing specific chemical elements with high neutron capture rates, e.g., Boron (25) should be avoided. Due to different contrast conditions (as compared to SAXS), SANS can tolerate high concentrations of solutes like salt, sugar, or glycerol without diminishing the signal/noise of the dissolved proteins in the scattering experiment significantly. An additional advantage of SANS is that the protein samples do not suffer from radiation damage and can be recovered for supplemental studies after the experiment. SANS experiments usually take a few minutes to several hours to obtain a good scattering curve, depending on concentration, sample volume, and contrast conditions. The use of D_2O as a solvent can increase the signal/noise of a natural abundance IDP but should be used with care since many proteins tend to oligomerize or aggregate in D_2O (see Note 2). Several SANS instruments offer the possibility to control the sample temperature in a large range during the exposure time. E.g., the instrument D22 at the Institut Laue-Langevin (ILL), Grenoble, France, covers the whole temperature range of interest for biological samples (4–90°C) limited only by practical aspects (freezing and vaporization of the solution). Good general practice of SAXS/SANS experiments on proteins in solution has been reviewed in detail recently (8, 26).

2.3. Sample List for a SANS Experiment on IDPs

1. *IDP in solution.* The protein to be studied should be prepared in the buffer of choice in a concentration series covering the range from about 1 to 10 mg/ml (e.g., 1, 2, 5, and 10 mg/ml) (see Note 3). There are no general restrictions on solvent apart from the fact that it should not contain exotic elements that absorb neutrons strongly like [5]B, [64]Gd, and others (25). The minimum sample volume that should be prepared at each concentration is 200 µl. Sample concentrations should be measured using optical absorbance at 280 nm with the appropriate calibrations.

2. *IDP buffers.* The reference buffers of the samples should match the buffer containing the protein(s) as faithfully as possible, in particular regarding the H_2O/D_2O ratio if contrast variation is used. If the sample preparation is finished by a dialysis step, the buffer used during the last dialysis step should be kept and measured by SANS. If sample preparations imply solubilization of a lyophilisate, the dissolving buffer should be prepared in excess in order to serve as reference buffer for the neutron experiments. The volume of the reference buffer(s) should be the same as the sample volume and measured in an identical cell as the one used for the protein sample(s).

3. *Sample holder reference.* The sample holder reference (e.g., empty quartz cuvette) needs to be measured in the same instrumental conditions as the sample(s) and buffer(s) in order to account for its contribution to the scattered signal.

4. *Detector background measurement.* A strongly neutron-absorbing material (boron composite such as BF_4) is usually measured in order to account for the electronic background signal on the detector once the incoming direct neutron beam is blocked. It is usually provided by the neutron facility.

5. *Water (H_2O) reference.* A pure water sample (same volume as samples and buffers) needs to be measured in order to calibrate the detector efficiency and put data on an absolute scale (27). It has to be measured in the same setup as the other samples/buffers.

---

3. Methods

3.1. Schematic Representation of a SANS Experiment

A SANS experiments consists in following the scattered intensity of a well-defined incoming beam as a function of scattering angle or scattering vector $Q = (4\pi/\lambda)\sin\theta$ (where 2θ is the scattering angle) (Fig. 2). If the solution of bio-macromolecules is isotropic (i.e., the orientation of one molecule does not influence the orientation of the others), the scattered signal is symmetric around the center of the incoming direct beam. The sensitivity for molecular weight ranges from a few kilodalton to several Megadalton, the structural length-scales accessible range from a few Angstrom to several thousand Angstrom.

3.2. Instrument Setup for IDPs (see Note 4)

In contrast to globular, compact proteins, IDPs require a specific instrumental setup in a SANS experiment, in particular a wider angular range (Q-range) has to be covered. Often this cannot be obtained by a single sample-detector distance but requires several ones. The lowest Q-value, Q_{min}, should be chosen in a way to allow a

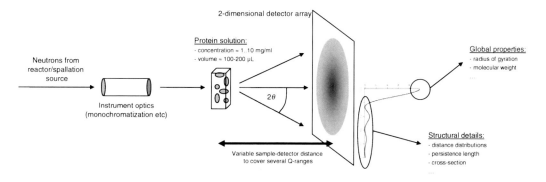

Fig. 2. Schematic representation of a SANS experimental setup.

proper determination of the radius of gyration (see Subheading 3.4) fulfilling the condition $Q_{min} \approx 0.3/R_g$. If not known, the R_g (in Ångström) of the IDP can be estimated in a first approximation from an unfolded protein consisting of the same number N of amino acid residues from a power law or from tables (20). For an IDP, the size of ubiquitin ($N = 76$) the estimation would yield $R_g \approx 25$ Å and require a Q_{min} of smaller or equal to about 0.012 Å^{-1} in order to apply the Guinier approximation properly. The upper limit of the Q-range used, Q_{max}, should ensure a proper analysis of short-distance information (e.g., cross-sectional analysis, Kratky-plots, etc.) as well as an accurate solvent-subtraction. A good compromise between these requirements and a decreasing signal/noise at higher angles restrains the upper limit of the Q-range in practice to about $0.5...1.0$ Å^{-1} for most instruments. If possible, an additional sample with high protein concentration (10–50 mg/ml) should be measured in the high angular range in order to assure a proper solvent subtraction.

3.3. Raw Data Reduction   The raw data consist in two-dimensional sets of intensities, recorded by a detector system (Fig. 2). It is measured both for sample(s) and the corresponding buffer(s). These data need to be corrected for detector background (electronic, terrestrial, and cosmic as well as parasitic radiation from the neutron optics). Background measurements in SANS usually are done using a strongly neutron-absorbing sample (e.g., boron). The empty sample holder (quartz cuvette) and a water (H_2O) sample are also measured to correct for sample holder contribution and detector calibration, respectively.

In addition, all samples and references should be measured in transmission mode in order to determine the respective weight with which they contribute to the signal. The beam center can be determined by a strongly coherently scattering sample such as Teflon in scattering mode or from the transmission experiment of the empty beam. Raw data can be reduced using software developed at neutron facilities (e.g., software developed at ILL Grenoble; http://www.ill. eu/instruments-support/computing-for-science/cs-software/all-software/sans). These programs can also be used to define appropriate masks to cut away the parasitic scattering close to the beam center and beam stop. One-dimensional scattering curves, i.e., intensity vs. scattering vector, $I(Q)$, are determined for both samples and buffers by integrating the 2D pattern in concentric rings around the beam center.

In a final data step, the buffer intensities need to be subtracted from the sample intensities in the one-dimensional curves (see Note 5). This can be done using available data visualization and manipulation programs such as PRIMUS (28). It is in general important to scale the level of both scattering curves at the highest available Q-values prior to the subtraction. Discrepancies beyond about 2% in the two levels are usually indicative of differences in sample and buffer preparations, often due to a slightly different H_2O/D_2O ratio in the solvent.

3.4. Basic Data Analysis   If a solution of IDPs is ideal (sufficiently dilute; no interparticle effects; isotropically oriented particles; see comments in the Subheading 4) and chemically monodisperse (no mixture of several oligomeric states and/or aggregation), basic structural parameters can be extracted from the data and be interpreted as (conformationally averaged) structural properties of the IDP molecules. The two most important ones are the radius of gyration, R_g, and the intensity scattered in the forward direction, $I(0)$. Both can be obtained by a linear fit in an $\ln I(Q)$ vs. Q^2 plot, the so-called Guinier analysis (29):

$$\ln[I(Q)] = \ln[I(0)] - 1/3 R_g^2 Q^2 \tag{1}$$

It should be noted that the range of validity of this approximation for unfolded proteins is only valid up to about $Q_{max}R_g < 0.7...0.8$ rather than $1.0...1.3$ for globular particles (30, 31). As discussed in the instrumental setup, the lower limit for the fit range, $Q_{min}R_g$, should be about 0.3 or lower. An approximation that is more accurate in a wider Q-range ($Q_{max}R_g < 1.2...1.4$) for IDPs is the Debye equation (32):

$$I(Q) = \frac{2I(0)}{R_g^4 Q^4}\left[R_g^2 Q^2 - 1 + \exp(-R_g^2 Q^2)\right] \tag{2}$$

The radius of gyration describes the weighted distribution of scattering length density differences (=contrast with bulk solvent) around the center of scattering length density:

$$R_g^2 = \int \Delta\rho(\vec{r})r^2 dV \Big/ \int \Delta\rho(\vec{r})dV \tag{3}$$

Here, r is the distance between a scattering element (atom or residue if several atoms are grouped together) and the center of scattering length density of an IDP molecule. Please note that R_g can also contain contribution from solvent/solute elements in the proximity of the IDP if they display a different scattering length density than the bulk solvent.

In both SANS and SAXS, the $I(0)$ intensity can be related to the molecular weight of the IDP, M_{IDP}, by a relative comparison to a known standard (e.g., BSA) (33) which has to be measured at the same contrast conditions (=protein with same labeling scheme and in the same buffer as the IDP, in particular at the same H_2O/D_2O ratio):

$$M_{IDP} = M_{BSA} \frac{I_{IDP}(0)}{I_{BSA}(0)} \frac{C_{BSA}}{C_{IDP}} \tag{4}$$

M_{BSA} is the molecular weight of BSA ($\approx$66 kDa) and C the respective protein concentrations in mg/ml. As an alternative, it is possible to determine the molecular weight by SANS in an absolute manner by calibration against water (H_2O) (27):

$$M_{IDP} = \frac{I_{inc}(0)}{I_{IDP}(0)} \frac{f 4\pi \, T_{IDP}}{1 - T_{H_2O}}$$

$$N_A C_{IDP} t 10^{-3} \left[\left(\frac{1}{V_{IDP}} \sum_i b_{coh} - \rho_N^0 \right) V_{IDP} \right]^2 \quad (5)$$

$I_{IDP}(0)$ is the coherent protein scattering in the forward direction, $I_{inc}(0)$ is the incoherent scattering from H_2O in the forward direction, T_{IDP} and T_{H2O} are the transmissions of the sample and of H_2O, respectively, C_{IDP} is the protein concentration in mg/ml, t is the thickness of the quartz cuvette in cm, f is a correction factor for the anisotropicity of the solvent scattering as a function of neutron wavelength (27), Σb_{coh} is the sum of scattering lengths of the protein atoms in cm (5), ρ_N^0 is the solvent scattering length density in cm^{-2}, and V_{IDP} is the solvent-displaced protein volume in cm^3 ((34), and references therein).

The accuracy of both calibrations is about 10–20% and mainly limited by the accuracy of the concentration measurement (usually done by optical absorbance at 280 nm), but also by the accuracy of the solvent-excluded volume of the protein in solution. The determination of molecular weight is very important in order to corroborate the oligomeric state of the protein in solution.

3.5. Sophisticated Data Analysis I: Properties of the Solvent in the Vicinity of IDPs and Solute-IDP Interactions

One of the strengths of SANS is that the contrast conditions of proteins, solvents, and solutes are very sensitive to chemical composition (in particular H/D-exchange) and physical density (e.g., solvent water in the vicinity of protein surfaces (35)). One possibility to monitor these properties is via the $I(0)$ intensity. $I(0)$ is proportional to the number N of particles in solution and the integrated scattering length density differences $\Delta\rho_i$ of their components (protein, hydration shell water, preferentially interacting solute molecules) with respect to the bulk solvent:

$$I(0) \propto N \left(\sum_i \Delta\rho_i V_i \right)^2 \quad (6)$$

Equation 6 has been recently used to study the interaction of urea molecules with ubiquitin by comparing $I(0)$ of native and denatured ubiquitin in a combined SAXS and SANS study (12).

3.6. Sophisticated Data Analysis II: Fit of Complex Mathematical Expressions that Describe the Structural Properties of IDPs Over Several Length-Scales

The entire scattering curve (beyond the Guinier- or Debye-range) can be fitted using mathematical expressions that include structural parameters over several orders of length-scale. These expressions may include some or all of the following parameters: $I(0)$, R_g, L (contour length), b (statistical segment length), R_c, and others (36).

3.7. Sophisticated Data Analysis III: Cross-Sectional Analysis

The scattering curve at higher Q-values reflects structural properties of IDPs at shorter distances, i.e., it is strongly influenced by the distribution of scattering length densities around the central backbone. This distribution can be described in terms of the cross-sectional radius of gyration, R_c:

$$R_c^2 = \int_0^R \Delta\rho(\vec{r})r^2 \mathrm{d}\vec{r} \Big/ \int_0^R \Delta\rho(\vec{r})\mathrm{d}\vec{r} \qquad (7)$$

$\Delta\rho$ is the contrast of the different scattering elements (atoms or grouped moieties such as residues or side-chains) with respect to the bulk solvent, r their distance to the cross-sectional center of scattering length density, and R the maximum distance from this center. R_c can be extracted in a Guinier-like representation of the data by a linear fit in an intermediate Q-range (for not too small or too large angles, $Q_{max} \approx 1/R_c$):

$$\ln[I(Q)Q] = A\ln[I_c(0)] - \frac{1}{2}R_c^2 Q^2 \qquad (8)$$

A and $I_c(0)$ are fit parameters. The experimentally determined cross-sectional term can be interpreted in terms of the scattering length density distribution around the center of scattering length density in the cross-sectional plane (Eq. 7) at different levels of resolution; e.g., if an ensemble of PDB structures is available it can be explicitly calculated from the atomic coordinates (see Subheading 3.9 below). This analysis is also perfectly suited for studying the side-chain conformations and the structural properties of hydration water that may have a different scattering density than the bulk solvent (35).

3.8. Sophisticated Data Analysis IV: Kratky-Plots

Kratky-plots ($I(Q)Q^2$ vs. Q) are a special representation of the scattering data and well-suited to analyze structural properties of unfolded proteins and polymers in solution (37). They contain information on the overall degree of folding of the protein. Comments on their use and interpretation can be found in the following publications (2, 3, 37). However, great care has to be taken in their interpretation since Kratky-plots are particularly sensitive to errors in buffer background subtraction.

3.9. Sophisticated Data Analysis V: Explicit Ensembles of Structural Conformers

The comparison of back-calculated small angle scattering curves from ensembles of structural conformers of IDPs has become very popular over the last 5 years (8, 10, 22). More details are provided in the chapter on Small Angle X-ray Scattering in this book. It can also be applied to SANS data if the specific neutron scattering length densities of protein, solutes, hydration water, and bulk solvent are taken into account.

3.10. Concluding Remarks and Perspectives

While the principles of data analysis and interpretation are relatively similar to SAXS, SANS offers the advantage of contrast variation and specific hydrogen/deuterium-labeling. Contrast variation should be particularly fruitful for IDPs in the analysis of structural properties on shorter length-scales (protein-solvent interface, side chain properties, etc....) by using specific deuterium-labeling schemes as were used in the polymer sciences (2, 24). In addition, segmental labeling schemes as introduced in NMR recently (38)—if applied as H/D-labeling to IDPs—may also provide valuable structural restraints on distance distributions at larger distances ($\approx R_g$). We believe that in both regimes SANS has the potential to provide very useful structural information on IDPs in the near future that can complement SAXS studies.

4. Notes

This section contains a list of typical problems that may be encountered during a SANS experiment and indications on how to avoid them or minimize their impact.

1. *Water condensation on quartz cuvettes.* One of the most common potential sources of error in a SANS experiment can be the condensation of water vapor on quartz cuvettes or in the sample/buffer aqueous solution. This can happen when samples (or cuvettes) are transferred from a cold temperature (cold room) to room temperature environment, and falsify SANS results, in particular for D_2O samples. Therefore, the quartz cuvettes containing samples should be sealed in order to minimize condensation in the solution and condensed water on the outside surface of the sample holder should be carefully wiped off prior to the neutron experiments.

2. *Decreased solubility (attractive interparticle interactions) in solvents containing large amounts of D_2O.* The physicochemical properties of D_2O are different with respect to those of H_2O, in particular the hydrophobic interaction (39). While popular in SANS experiments in order to increase contrast (for hydrogenated proteins) and to reduce the incoherent background from H_2O, D_2O has been observed in several cases to increase the tendency of proteins to oligomerize, aggregate, or precipitate in solution. While a precipitate of very large molecular weight (all linear dimensions large with respect to the protein of interest) may be tolerated, different oligomeric states of proteins in solution will influence the data analysis of the protein of interest. If the aim of the study is a modeling/characterization of a specific protein state (and not the properties of an oligomeric equilibrium), this situation must be avoided. Therefore, the

oligomeric state of the protein should be characterized prior to the SANS experiments with complementary techniques (analytical ultracentrifugation, dynamic/static light scattering, gel filtration, etc.) *in* D_2O. During a SANS experiment, it is in general important to monitor the oligomeric state as a function of protein concentration (both in H_2O and in D_2O) using the approaches described in the Subheading 3.4 (Eqs. 8.4 and 8.5) and the ones presented in the next paragraph.

3. *Repulsive interparticle interactions.* In some cases (mostly charged bio-macromolecules such as DNA or RNA), repulsive interactions can be observed in a SANS experiment, influencing mainly the data interpretation at low angles (radius of gyration, molecular weight, etc.). Usually, these effects tend to increase with increasing protein concentration. They should be avoided in a SANS experiment if they are not themselves the subject of interest. In order to identify their presence in a given IDP sample, it is important to run a concentration series, typically in the range 1–10 mg/ml, as described in the Subheading 2.3 and plot the $I(Q)/C_{IDP}$ series graphically. If all curves superpose, it is generally a strong argument that there are no interaction effects in the sample. If there is a concentration dependence on the shapes of the scattering curves $I(Q)/C_{IDP}$, interaction effects play a role and two remedies are possible: (a) Lowering the sample concentration as far as possible in order to minimize the effects while conserving a good signal/noise. (b) Change the buffer properties (e.g., ionic strength, pH) in order to screen the long-range electrostatic effects that may be at the origin of the interparticle effects.

4. At all stages of a SANS experiment, it is advisable to be in touch with a local contact at the neutron facility (instrument responsible or experienced user) that can help out to (1) optimize the writing of a proposal, (2) discuss sample preparation, (3) carry out the experiment, and (4) help with the data analysis.

5. *Solvent subtraction.* Even in a carefully prepared SANS experiment (see Subheading 2.3 above) the scattering intensity of the sample might not align with the corresponding buffer at the highest measured angles (Q_{max}). Since the absolute level of scattering at high angles in SANS is mainly determined by the incoherent scattering of hydrogen atoms, this discrepancy is in most cases due to slightly different levels of H_2O/D_2O solvent in the sample and buffer or stems from differences in the H/D-density between the proteins and the solvent they displace. The most consistent strategy to adopt when sample and buffer levels differ at high angles is to scale the buffer level to the sample level. If they align a subtraction can be carried out. If the sample curve crosses the buffer line at high angles (=different slope), the highest angles are not sufficient to determine an

accurate subtraction. The most consistent strategy is then to slightly "under-subtract" the buffer assuming that the sample curve will continue to decrease at higher angles. A high-concentration sample ($\approx$10–50 mg/ml) may also be helpful in some cases to accurately subtract the buffer at high angles. In this case, it is important to check the consistency of high-concentration and low-concentration samples ($<$10 mg/ml) in an intermediate Q-range by checking their alignment.

Acknowledgements

The author would like to thank Dr. Giuseppe Zaccaï (Institut Laue-Langevin, Grenoble, France) for critical comments on the manuscript.

References

1. Chu B, Hsiao BS (2001) Small-angle X-ray scattering of polymers. Chem Rev 101:1727–1761

2. Schurtenberger P (2002) Static properties of polymers. In: Lindner P, Zemb T (eds) Neutrons, X-rays and light. Delta Series, North Holland

3. Doniach S (2001) Changes in biomolecular conformation seen by small angle X-ray scattering. Chem Rev 101:1763–1778

4. Heller WT (2010) Small-angle neutron scattering and contrast variation: a powerful combination for studying biological structures. Acta Crystallogr D Biol Crystallogr 66:1213–1217

5. Jacrot B (1976) The study of biological structures by neutron scattering from solution. Rep Prog Phys 39:911–953

6. Koch MH, Vachette P, Svergun DI (2003) Small-angle scattering: a view on the properties, structures and structural changes of biological macromolecules in solution. Q Rev Biophys 36:147–227

7. Lipfert J, Doniach S (2007) Small-angle X-Ray scattering from RNA, proteins, and protein complexes. Annu Rev Biophys Biomol Struct 36:307–327

8. Putnam CD, Hammel M, Hura GL, Tainer JA (2007) X-ray solution scattering (SAXS) combined with crystallography and computation: defining accurate macromolecular structures, conformations and assemblies in solution. Q Rev Biophys 40:191–285

9. Svergun DI, Koch MHJ (2002) Small-angle scattering studies of biological macromolecules in solution. Rep Prog Phys 66:1735–1782

10. Bernado P, Blanchard L, Timmins P, Marion D, Ruigrok RW, Blackledge M (2005) A structural model for unfolded proteins from residual dipolar couplings and small-angle X-ray scattering. Proc Natl Acad Sci USA 102:17002–17007

11. Calmettes P, Durand D, Desmadril M, Minard P, Receveur V, Smith JC (1994) How random is a highly denatured protein? Biophys Chem 53:105–113

12. Gabel F, Jensen MR, Zaccai G, Blackledge M (2009) Quantitative modelfree analysis of urea binding to unfolded ubiquitin using a combination of small angle X-ray and neutron scattering. J Am Chem Soc 131:8769–8771

13. Kataoka M, Nishii I, Fujisawa T, Ueki T, Tokunaga F, Goto Y (1995) Structural characterization of the molten globule and native states of apomyoglobin by solution X-ray scattering. J Mol Biol 249:215–228

14. Kirste RG, Schulz GV, Stuhrmann HB (1969) Die konformationsaenderung des pottwalmesmyoglobins bei der reversiblen denaturierung im pH-bereich 7 bis 1. Z Naturforsch 24b:1385–1392

15. Kohn JE, Millet IS, Jacob J, Zagrovic B, Dillon TM, Cingel N, Dothager RS, Seifert S, Thiyagarajan P, Sosnick TR, Hasan MZ, Pande VS, Ruczinski I, Doniach S, Plaxco KW (2004) Random-coil behavior and the dimensions of chemically unfolded proteins. Proc Natl Acad Sci U S A 101:12491–12496

16. Millet IS, Doniach S, Plaxco KW (2002) Toward a taxonomy of the denatured state:

small angle studies of unfolded proteins. Adv Protein Chem 62:241–262

17. Perez J, Vachette P, Russo D, Desmadril M, Durand D (2001) Heat-induced unfolding of neocarzinostatin, a small all-beta protein investigated by small-angle X-ray scattering. J Mol Biol 308:721–743

18. Petrescu AJ, Receveur V, Calmettes P, Durand D, Desmadril M, Roux B, Smith JC (1997) Small-angle neutron scattering by a strongly denatured protein: analysis using random polymer theory. Biophys J 72:335–342

19. Petrescu AJ, Receveur V, Calmettes P, Durand D, Smith JC (1998) Excluded volume in the configurational distribution of a strongly-denatured protein. Protein Sci 7:1396–1403

20. Plaxco KW, Gross M (2001) Unfolded, yes, but random? never! Nat Struct Biol 8:659–660

21. Shortle D, Ackerman MS (2001) Persistence of native-like topology in a denatured protein in 8 M urea. Science 293:487–489

22. Bernado P, Mylonas E, Petoukhov MV, Blackledge M, Svergun DI (2007) Structural characterization of flexible proteins using small-angle X-ray scattering. J Am Chem Soc 129: 5656–5664

23. Jensen MR, Markwick PRL, Meier S, Griesinger C, Zweckstetter M, Grzesiek S, Bernado P, Blackledge M (2009) Quantitative determination of the conformational properties of partially folded and intrinsically disordered proteins using NMR dipolar couplings. Structure 17:1169–1185

24. Rawiso M, Duplessix R, Picot C (1987) Scattering function of polysterene. Macromolecules 20:630–648

25. Rauch H, Waschkowski W (2001) Neutron scattering lengths. In: Dianoux A-J, Lander G (eds) Neutron data booklet. Institut Laue-Langevin, Grenoble

26. Jacques DA, Trewhella J (2010) Small-angle scattering for structural biology—expanding the frontier while avoiding the pitfalls. Protein Sci 19:642–657

27. Jacrot B, Zaccai G (1981) Determination of molecular weight by neutron scattering. Biopolymers 20:2413–2426

28. Konarev PV, Volkov VV, Sokolova AV, Koch MHJ, Svergun DI (2003) PRIMUS: a windows PC-based system for small-angle scattering data analysis. J Appl Crystallogr 36:1277–1282

29. Guinier A (1939) La diffraction des rayons X aux tres faibles angles: applications a l'Etude des phenomenes ultra-microscopiques. Ann Phys (Paris) 12:161–236

30. Feigin LA, Svergun DI (1987) Structure analysis by small-angle X-Ray and neutron scattering. Plenum Press, New York and London

31. Guinier A, Fournet G (1955) Small angle scattering of X-rays. Wiley, New York

32. Debye P (1947) Molecular-weight determination by light scattering. J Phys Colloid Chem 51:18–32

33. Mylonas E, Svergun DI (2006) Accuracy of molecular mass determination of proteins in solution by small-angle X-ray scattering. J Appl Crystallogr 40(suppl):S245–S249

34. Creighton TE (1993) Proteins: structures and molecular properties. W.H Freeman and Company, New york

35. Svergun DI, Richard S, Koch MHJ, Sayers Z, Kuprin S, Zaccai G (1998) Protein hydration in solution: experimental observation by X-ray and neutron scattering. Proc Natl Acad Sci U S A 95:2267–2272

36. Pedersen JS, Schurtenberger P (1996) Scattering functions of semiflexible polymers with and without excluded volume effects. Macromolecules 29:7602–7612

37. Glatter O, Kratky O (1982) Small angle X-ray scattering. Academic, London, New York, Paris, San Diego, San Francisco, Sao Paolo, Sydney, Tokyo, Toronto

38. Skrisovska L, Schubert M, Allain FH (2010) Recent advances in segmental isotope labeling of proteins: NMR applications to large proteins and glycoproteins. J Biomol NMR 46:51–56

39. Nemethy G, Scheraga HA (1962) Structure of water and hydrophobic bonding in proteins. I. A model for the thermodynamic properties in liquid water. J Chem Phys 36:3382–3417

Chapter 9

Dynamic and Static Light Scattering of Intrinsically Disordered Proteins

Klaus Gast and Christian Fiedler

Abstract

Molecular parameters such as size, molar mass, and intermolecular interactions, which are important to identify and characterize intrinsically disordered proteins (IDPs), can be obtained from light scattering measurements. In this chapter, we discuss the physical basis of light scattering, experimental techniques, sample treatment, and data evaluation with special emphasis on studies on proteins. Static light scattering (SLS) is capable of measuring molar masses within the range 10^3–10^8 g/mol and is therefore ideal for determining the state of association of proteins in solution. Since proteins are in general too small to obtain the geometric radius of gyration R_G from SLS, it is more useful to determine the hydrodynamic Stokes radius, R_S, which can be obtained easily and quickly from dynamic light scattering (DLS) experiments. Accordingly, DLS is an appropriate technique to monitor expansion or compaction of protein molecules. This is especially important for IDPs, which can be recognized and characterized by comparing the measured Stokes radii with those calculated for particular reference states, such as the compactly folded and the fully unfolded states. The combined application of DLS and SLS improves measurements of the molar mass and is essential when changes in the molecular dimensions and molecular association/dissociation take place simultaneously.

Key words: Dynamic light scattering, Static light scattering, Molecular mass, Stokes radius, Translational diffusion coefficient, Proteins, Compactness, Size distribution, Folding

1. Introduction

The experimental discovery of intrinsically disordered proteins (IDPs) was based mainly on two observations: lack of ordered secondary structure and atypically large molecular dimensions of particular proteins under conditions, where the proteins were found to be fully active. For an exploration of the molecular dimensions of IDPs, coupled with thorough control of their monomeric state, the combination of dynamic light scattering (DLS) and static light scattering (SLS) is a method of choice.

Vladimir N. Uversky and A. Keith Dunker (eds.), *Intrinsically Disordered Protein Analysis:*
Volume 2, Methods and Experimental Tools, Methods in Molecular Biology, vol. 896,
DOI 10.1007/978-1-4614-3704-8_9, © Springer Science+Business Media New York 2012

It is worth mentioning that the unexpected results of size exclusion chromatography (SEC) led to much confusion before the protein community became aware of the extraordinary properties of IDPs. Since a compact globular structure was expected, the apparently large molecular masses were interpreted assuming an oligomeric structure for the protein under study. Despite some earlier work pointing to the unfolded nature of those proteins (1–5), the beginning of extensive investigations leading to the realization of the importance of this unusual behavior can be dated roughly to the middle of the 1990s (6–8).

DLS and SLS measurements yield completely different macromolecular quantities. From DLS experiments, hydrodynamic parameters of macromolecules in solution, for example the Stokes radius R_S, can be obtained. An advantage over other hydrodynamic techniques is that DLS is fast, entirely noninvasive, and studies can be done under a wide variety of solvent conditions. The curiosity of being a hydrodynamic rather than an optical method results from the fact that DLS analyzes the temporal fluctuations of the light scattering intensity caused by hydrodynamic motions in solution. The connection to the light scattering process itself, which had been studied for a long time before the advent of DLS by measuring time-integrated or static light scattering (SLS), is important for several reasons. First, SLS enables the mass of the scattering particles to be determined, which is a useful complement to measurements of the hydrodynamic dimensions. Secondly, since the scattering effect is something like a weighting factor in a DLS experiment, its knowledge is essential for proper application of the method. Therefore, we shall discuss the basics of light scattering before turning to the application of DLS to studying the dimensions and conformational transitions of proteins.

As a consequence of recent technical developments, light scattering can now be considered as a standard laboratory method. Several compact instruments are on the market that are valuable tools for protein analysis. Measuring SLS and DLS in one and the same experiment is particularly useful for studying the molecular dimensions of IDPs. Further improvements have been achieved by combining light scattering and particle separation techniques.

1.1. Light Scattering: Basic Principles and Instrumentation

The following short introduction to both methods is an attempt to supply readers who are not familiar with these techniques with an intuitive understanding of the basic principles and the applicability to protein folding and aggregation problems. We will therefore only recall some of the more fundamental equations. A clearer and more detailed understanding can be achieved on the basis of numerous excellent textbooks (9–14).

The theory of light scattering can be approached from either the so-called single-particle analysis or the density fluctuation viewpoint. The single-particle analysis approach, which will be used here, is both simpler to visualize and adequate for studying the structure and dynamics of macromolecules in dilute solutions. The concepts of light scattering from small particles in vacuum can be applied to the scattering of macromolecules in solution by considering the excess of various quantities (e.g., light scattering, polarizability, refractive index) of macromolecules over that of the solvent.

The strength of the scattering effect depends primarily on the polarizability α of a small scattering element, which might be the molecule itself if it is sufficiently small. Larger molecules are considered as consisting of several scattering elements. The oscillating electric field vector $\vec{E}_0(t, \vec{r}) = \vec{E}_0 e^{-i(\omega t - \vec{k}_0 \vec{r})}$ of the incident light beam induces a small oscillating dipole with the dipole moment $\vec{p}(t) = \alpha \vec{E}(t)$. ω is the angular frequency and $\vec{k}_0$ is the wave vector with the magnitude $|\vec{k}_0| = 2\pi n_0/\lambda$. This oscillating dipole re-emits electromagnetic radiation, which has the same wavelength in the case of elastic scattering. The intensity of scattered light at distance r is

$$I_S = \frac{16\pi^4}{\lambda^4} \frac{\alpha^2}{r^2} I_0, \tag{1}$$

where λ is the wavelength in vacuum, n_0 is the refractive index of the surrounding medium, and $I_0 = |\vec{E}_0|^2$ is the intensity of the incident vertically polarized laser beam. Equation (1) differs only by a constant factor for the case of unpolarized light. In general, the scattered light is detected at an angle θ with respect to the incident beam in the plane perpendicular to the polarization of the beam (Fig. 1). The wave vector pointing in this direction is $\vec{k}_S$, where $\vec{k}_0$ and $\vec{k}_S$ have the same magnitude (Fig. 1).

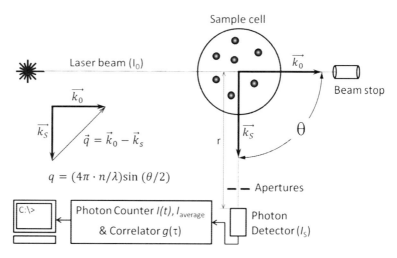

Fig. 1. Schematic diagram of a laser light scattering instrument.

An important quantity is the vector difference $\vec{q} = \vec{k}_0 - \vec{k}_S$, the so-called scattering vector, which determines the spatial distribution of the phases $\phi_i = \vec{q} \cdot \vec{r}_i$ of the scattered light wave emitted by individual scattering elements i (Eq. 1 and Fig.1). The magnitude of $\vec{q}$ is $q = (4\pi n/\lambda)\sin(\theta/2)$. The phases play an essential role for the total instantaneous intensity, which results from the superposition of light waves emitted by all scattering elements within the scattering volume v defined by the primary beam and the aperture of the detector. The instantaneous intensity fluctuates in time for non-fixed particles, like macromolecules in solution, due to phase fluctuations $\phi_i(t) = \vec{q} \cdot \vec{r}_i(t)$ caused by changes in their location $\vec{r}_i(t)$.

SLS measures the time average of the intensity, thus the term "time-averaged light scattering" would be more appropriate, although "static light scattering" appears the accepted convention now. SLS becomes q-dependent for large particles, when light waves emitted from scattering elements within an individual particle have distinct phase differences.

DLS analyzes the above-mentioned temporal fluctuations of the instantaneous intensity of scattered light. Accordingly, DLS can measure several dynamic processes in solution. It is evident that only those changes in the location of scattering elements lead to intensity fluctuations that produce a sufficiently large phase shift. This is always the case for the translational diffusion of macromolecules. The motion of segments of chain molecules and rotational motion can be studied for large structures. Rotational motion of monomeric proteins and chain dynamics of unfolded proteins are, in practice, not accessible by DLS.

1.2. Static Light Scattering

The light scattering intensity from a macromolecular solution can be calculated using (Eq. 1) and summing over the contributions of all macromolecules in the scattering volume v. It is useful to substitute the polarizability α by appropriate physical parameters. This can be done by applying the Clausius–Mosotti equation to macromolecular solutions, which leads to $n^2 - n_0^2 = 4\pi N'\alpha$, where n and n_0 are the refractive indices of the solution and the solvent, respectively. N' is the number of particles (molecules) per unit volume and can be expressed by $N' = N_A c/M$. N_A is Avogadro's number, c and M are the weight concentration and the molecular mass (molar mass) of the macromolecules, respectively. Using the approximation $n + n_0 \sim 2n_0$ we obtain $n^2 - n_0^2 \sim 2n_0(\partial n/\partial c)c$. $\partial n/\partial c$ is the specific refractive index increment of the macromolecules in the particular solvent.

The excess scattering of the solution over that of the solvent $I_{ex} = I_{solution} - I_{solvent}$ of noninteracting small molecules is then

$$I_{ex} = \frac{4\pi^2 n_0^2 (\partial n/\partial c)^2}{\lambda^4 N_A} \frac{v}{r^2} cMI_0 = H\frac{v}{r^2} cMI_0. \tag{2}$$

The optical constant $H = 4\pi^2 n_0^2 (\partial n/\partial c)^2 / \lambda^4 N_A$ depends only on experimental parameters and the scattering properties of the molecules in the particular solvent, which is reflected by $\partial n/\partial c$. The exact knowledge of $\partial n/\partial c$, the dependence of n on protein concentration in the present case, is very important for absolute measurements of the molecular mass. For proteins in aqueous solvents of low ionic strength $\partial n/\partial c$ is about 0.19 cm^3/g and does not depend significantly on the amino acid sequence. However, it is markedly different in solvents containing high concentrations of denaturants.

One can eliminate the instrument parameters in (Eq. 2) by using the Rayleigh (excess) ratio $R_q = (I_{ex}/I_0)(r^2/v)$. In practice, R_q of an unknown sample is calculated from the scattering intensities of the sample and that of a reference sample of known Rayleigh ratio R_{ref} by

$$R_q = R_{ref} f_{corr} (I_{ex}/I_{ref}), \qquad (3)$$

f_{corr} is an experimental correction factor accounting for differences in the refractive indices of sample and reference sample (12). In the more general case, including intermolecular interactions and large molecules that have a refractive index n_p and satisfy the condition $4\pi(n_p - n_0)d/\lambda \ll 1$ of the Rayleigh–Debye approximation, static light scattering data are conveniently presented by the relation

$$\frac{Hc}{R_q} = \frac{1}{MP(q)} + 2A_2 c. \qquad (4)$$

$P(q)$ is the particle scattering function, which is mainly expressed in terms of the product $q\overline{R_G}$, where $\overline{R_G} = <R^2>^{1/2}$ is the root mean-square radius of gyration of the particles. Analytical expressions for $P(q)$ are known for different particle shapes, whereas other characteristic size-dependent parameters are used instead of R_G (e.g., length L for rods or cylinders). In the limit $q\overline{R_G} \ll 1$, the approximation $P(q)^{-1} = 1 + (q\overline{R_G})^2/3$ can be used to estimate $\overline{R_G}$ from the angular dependence of R_q. A perceptible angular dependence of the scattering intensity can only be expected for particles with $R_G > 10$ nm. Thus, light scattering is not an appropriate method for monitoring changes of R_G during unfolding and refolding of small monomeric proteins. However a substantial angular dependence is observed for large protein aggregates.

The concentration dependence of the right-hand side of (Eq. 4) yields the second virial coefficient A_2, which reflects the strength and type of intermolecular interactions. The usefulness of measuring A_2 will be discussed below. A_2 is positive for predominantly repulsive (covolume and electrolyte effects) and negative for predominantly attractive intermolecular interactions.

In general, both extrapolation to zero concentration and to zero scattering angle ($q = 0$) are done in a single diagram

(Zimm plot) for calculations of M from (Eq. 4). The extrapolation to zero concentration at a fixed scattering angle is termed Debye plot.

The primary mass "moment" or "average" obtained is the weight-average molar mass

$$M = \sum_i c_i M_i / \sum_i c_i,$$

in the case of polydisperse systems. This has to be taken into consideration for proteins when both monomers and oligomers are present. Molar masses of proteins used in folding studies (i.e., $M < 50,000$ g/mol) can be determined at 90° scattering angle because the angular dependence of R_q is negligible. Measurements at different concentrations are mandatory, however, because remarkable electrostatic and hydrophobic interactions may exist under particular environmental conditions. In the following, the values of parameters measured at finite concentration are termed apparent values, e.g., M_{app}.

1.3. Dynamic Light Scattering

Information about dynamic processes in solution is contained primarily in the temporal fluctuations of the scattered electric field $\vec{E}_s(t)$. The time characteristics of these fluctuations can be described by the first-order time autocorrelation function (acf)

$$g^{(1)}(\tau) = <\vec{E}_s(t) \cdot \vec{E}_s(t + \tau)>.$$

The brackets denote an average over many products of $\vec{E}_s(t)$, with its value after a delay time τ. $g^{(1)}(\tau)$ is only accessible in the heterodyne detection mode, where the scattered light is mixed with a small fraction of the incident beam on the optical detector. This experimentally complicated detection method must be used if particle motion relative to the laboratory frame, e.g., the electrophoretic mobility in an external electric field, is to be measured. The less complicated homodyne mode, where only the scattered light intensity is measured, is normally the preferred optical scheme. As a consequence, only the second-order intensity correlation function

$$g^{(2)}(\tau) = <I(t)I(t + \tau)>$$

is directly available. Under particular conditions, which are met in the case of light scattering from dilute solutions of macromolecules, the Siegert relation

$$g^{(2)}(\tau) = 1 + \left| g^{(1)}(\tau) \right|^2$$

can be used to obtain the normalized first-order correlation function $g^{(1)}(\tau)$ from the measured $g^{(2)}(\tau)$. Analytical forms of $g^{(1)}(\tau)$ have been derived for different dynamic processes in solution. As we have already indicated above, essentially only translational diffusion motion contributes to the fluctuations of the scattered light in the case of monomeric proteins and we can reasonably neglect rotational effects. $g^{(1)}(\tau)$ for identical particles with a translational diffusion coefficient D has the form of an exponential

$$g^{(1)}(\tau) = e^{-q^2 D \tau}. \qquad (5)$$

D is related to the hydrodynamic Stokes radius R_S by the Stokes–Einstein equation

$$R_S = \frac{kT}{6\pi\eta D}, \qquad (6)$$

where k is the Boltzmann's constant, T is the temperature in K, and η is the solvent viscosity. $g^{(1)}(\tau)$ for a polydisperse solution containing L different macromolecular species (or aggregates) with masses M_i, diffusion coefficients D_i and weight concentrations c_i is

$$g^{(1)}(\tau) = \frac{1}{S} \sum_{i=1}^{L} a_i e^{-q^2 D_i \tau}, \qquad (7a)$$

where $S = \sum_{i=1}^{L} a_i$ is the normalization factor. The weights $a_i = c_i M_i = n_i M_i^2$ reflect the cM dependence of the scattered intensity (see Eq. 2). n_i is the number concentration (molar concentration) of the macromolecular species. Accordingly, even small amounts of large particles are considerably represented in the measured $g^{(1)}(\tau)$. The general case of an arbitrary size distribution, which results in a distribution of D, can be treated by an integral

$$g(\tau) = \frac{1}{S} \int a(D) e^{-q^2 D \tau} dD. \qquad (7b)$$

Equation (7b) has the mathematical form of a Laplace transformation of the distribution function $a(D)$. Thus, an inverse Laplace transformation is needed to reconstruct $a(D)$ or the related distribution functions $c(D)$ and $n(D)$. This is an ill-conditioned problem from the mathematical point of view because of the experimental noise in the measured correlation function. However, numerical procedures exist, termed "regularization," which allow stabilized, "smoothed" solutions to be obtained. A widely used program package for this purpose is "CONTIN" (15). Nevertheless, the distributions obtained can depend sensitively on the experimental noise and parameters used in the data evaluation procedure in special cases. Extraction of a reasonable amount of information from a DLS experiment is a critical step of the data evaluation procedure. For this purpose, a visual inspection of the pattern and the noise of the measured autocorrelation function is strongly recommended before applying the inverse Laplace transformation to obtain a size distribution. Figure 2 illustrates how the size distributions are related to the measured correlation functions for the cases of unimodal and bimodal distributions. More detailed instructions will be given in Subheading 3.

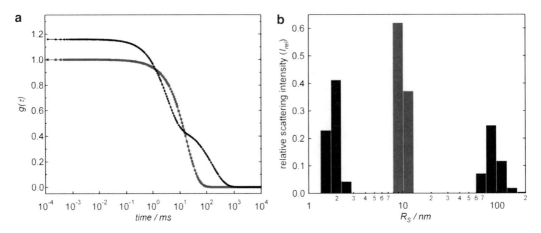

Fig. 2. Unimodal and bimodal size distributions and the related time-correlation functions (experimental conditions: $\lambda = 532$ nm, $\theta = 90$ °, $T = 20$ °, water).

In some cases, it might be more appropriate to use simpler but more stable data evaluation schemes like the method of cumulants (16), which yields the z-averaged diffusion coefficient $\bar{D}$ and higher moments reflecting the width and asymmetry of the distribution. $\bar{D}$ can be obtained simply from the limiting slope of the logarithm of $g^{(1)}(\tau)$, viz.

$$\bar{D} = -q^{-2} \frac{\mathrm{d}}{\mathrm{d}\tau} \left(\ln \left| g^{(1)}(\tau) \right| \right)_{\tau \to 0}.$$

This approach is very useful for rather narrow distributions.

D, like M, is concentration dependent, usually written in the form

$$D(c) = D_0(1 + k_D c), \qquad (8)$$

where k_D is the diffusive concentration dependence coefficient, which can be used to characterize intermolecular interactions. k_D can vary considerably in magnitude and sign, and in dependence on solvent, conformational state and net charge of the protein. However, k_D differs from A_2. The concentration dependence of D and other macromolecular parameters has been discussed in more detail by Harding and Johnson (17). Extrapolation to zero protein concentration, yielding D_0, is essential in order to calculate the hydrodynamic dimensions in terms of R_S for individual protein molecules.

1.4. Experimental Setup: Considerations for Studies on Proteins

The basic experimental setup for light scattering measurements in macromolecular solutions is relatively simple (Fig. 1). A beam from a continuous-wave (cw) laser is focused into a temperature-stabilized cuvet. Temperature stability better than ± 0.1 °C is required for DLS experiments because of the temperature dependence of the solvent viscosity. The scattered light intensity is detected at one or various scattering angles using either photomultiplier tubes or avalanche photodiode detectors (APD). The complexity of a particular light scattering apparatus can vary considerably. This depends on the type of the laser and the detection scheme. Expensive devices allow simultaneous multi-angle measurements of both SLS and DLS. For DLS or combined SLS/DLS experiments photon counting is the preferred mode of operation allowing a convenient use of digital autocorrelation techniques. Modern autocorrelators deliver time-correlation functions and average intensities as well. Analog signal detection with photodiodes is used in special simultaneous multi-angle measurements. While multi-angle experiments with either simultaneous or sequential detection are mandatory for studies on large particles ($> \lambda/20$), measurements at a fixed angle of 90 degrees are sufficient in the case of monomeric proteins. This simplifies the experiments considerably allowing the use of easily available (rectangular) measurement cells and minimizes the risk of distortions due to undesired stray light from the cell walls. Furthermore, measurements of concentration and light scattering can be done with the same cuvet (see Note 1).

It is very useful to measure the mass and the Stokes radius in one and the same experiment. However, the combination of the two experimental procedures involves some problems, since the optimum optical schemes for DLS and SLS are different. Briefly, DLS needs focused laser beams and only a much smaller detection aperture can be employed because of the required spatial coherence of the scattered light. This is unfavorable for SLS because it reduces the light intensity and demands higher beam stability.

Combined DLS/SLS offers two main advantages for studies of the conformational states of proteins. The first concerns the reliability of measurements of the hydrodynamic dimensions and plays an important role during folding or unfolding investigations and, particularly, for studies of IDPs. The observed changes or unexpected large values of R_S measured by DLS could partly or entirely result from an accompanying aggregation reaction. Such effects can clearly be recognized when the molecular mass is determined in the same experiment by SLS. The second advantage of combined DLS/SLS is its ability to measure correct molar masses of proteins in imperfectly clarified solutions that contain protein monomers and an unavoidable small amount of aggregates. This procedure is illustrated in Subheading 3.

A detailed evaluation of the advantages/disadvantages of different commercially available devices exceeds the scope of this chapter. The decision concerning the selection of an instrument should be based on the general requirements for SLS and DLS measurements outlined in Subheadings 2 and 3.

1.5. Supplementary Measurements and Accessory Devices

Additional physical quantities are required for estimating the molecular mass from SLS, and the diffusion coefficient and the Stokes radius from DLS. To calculate the optical constant in (Eq. 2), the refractive index of the solvent and the refractive index increment of the protein in the particular solvent must be known.

The former can be measured easily with an Abbe refractometer, whereas a differential refractometer is needed for precise measurements of $\partial n/\partial c$. Measurements of $\partial n/\partial c$ are often more expensive than the light scattering experiment itself. A comprehensive collection of $\partial n/\partial c$ values can be found in (18). Since the protein concentration c directly enters the equation for the calculation of the mass (Eq. 4), precise estimation of c is necessary. UV-VIS absorption measurements using the light scattering cuvet are the best choice.

For calculating R_S from DLS data by (Eq. 6), the dynamic viscosity η of the solvent must be known. It can be obtained from the kinematic viscosity v (e.g., measured by an Ubbelohde-type viscometer) and the density ρ (measured by a digital density meter) by $\eta = v\rho$.

A very useful option for light scattering instruments is online coupling with FPLC, HPLC, or field-flow fractionation (FFF) and a concentration detector, which can be a UV absorption monitor or a refractive index detector. This allows direct measurements of M of the eluting particles. In the case of known $\partial n/\partial c$ for the particular solvent conditions, the molecular mass can be obtained from the output signals of the SLS and refractive index detectors by $M = k_e(\partial n/\partial c)^{-1}(\text{output})_{\text{SLS}}/(\text{output})_{\text{RI}}$, where k_e is the instrument calibration constant. A parallel estimation of R_S is possible when the scattering is strong enough, allowing a sufficiently precise DLS experiment within a few minutes.

1.6. Use of Scaling Laws to Analyze the Dimensions of IDPs

A useful procedure to analyze the basic conformational type of a protein is to compare the measured Stokes radius with the expected hydrodynamic dimensions in particular reference states. Such states are the compactly folded globular state on the one hand and a highly unfolded state with random coil dimensions on the other (a highly unfolded state is obtained for proteins lacking disulfide bonds in the presence of high concentrations of GdmCl or urea). Hydrodynamic dimensions for particular states can be estimated on the basis of empirical relationships (scaling laws) derived from the analysis of large sets of experimental data. A linear dependence concerning the relation between the hydrodynamic dimensions and the number of amino acids N or the relative molecular mass M can be obtained by drawing R_S and N or M in a double logarithmic diagram (Fig. 3). The scaling laws can be written in two different forms:

$$R_S = R_{0,N}\,N^v \quad \text{or} \quad R_S = R_{0,M}\,M^v,$$

where the scaling exponent v and the pre-factors can be obtained from linear fits to the log–log plots. The results obtained by different authors (19–24) are shown in Table 1. For the analysis of Stokes radii we prefer the results from data sets used by Uversky (21) and Damaschun et al. (20) for the native and unfolded conformations,

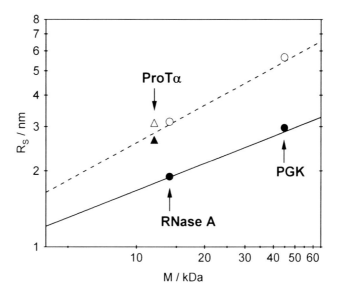

Fig. 3. Graphic illustration for the application of scaling laws of the type $R_S = aM^b$. The *straight line* corresponds to the scaling law for globular proteins and the *dashed line* for unfolded proteins in the presence of high concentrations of GdmCl. The Stokes radii for prothymosin α at pH 7.4 (*open triangle*) and pH 2.45 (*filled triangles*) are shown as an example for the hydrodynamic dimensions of a typical intrinsically unfolded protein.

Table 1
Scaling laws for proteins in natively folded and chemically unfolded states

State, conditions	Exponent ν	$R_{0,N}$ (nm)	$R_{0,M}$ (nm)	References
Natively folded				
	0.29	0.475		(22)
Figure 3	0.369		0.0557	(21)
	0.357		0.0625	(23)
	0.33	0.362		(19)
Unfolded				
GdmCl, urea	0.57	0.221		(22)
GdmCl	0.502		0.0286	(21)
Urea	0.524		0.022	(21)
GdmCl	0.543		0.0189	(23)
Urea	0.521		0.0224	(23)
GdmCl	0.50	0.280		(20)
GdmCl (Fig. 3)	0.495		0.0271	Rescaled (20)
	0.522	0.2518		(24)

The scaling laws are given by either $R_S = R_{0,N} N^\nu$ or $R_S = R_{0,M} M^\nu$, where N is the number of amino acid residues, M is the relative molecular mass in Da and R_S is obtained in nm. Errors are omitted respectively (Fig. 3). The original data from (20) have been rescaled for plotting R_S versus M. The hydrodynamic dimensions of two typical globular proteins, RNase A and phosphoglycerate kinase (PGK), in the native and unfolded states are also shown in Fig. 3. The Stokes radii of the IDP prothymosin α shown in Fig. 3 will be discussed in Chapter 3.

2. Materials

2.1. Light Scattering Equipment

The primary goal of light scattering investigations on IDPs is the determination of the molecular dimensions in terms of the hydro-dynamic Stokes radius. Nevertheless, the instrument should be able to measure SLS in addition to DLS with high precision and sensitivity to verify the monomeric state of the protein. Nowadays, most of the commercially available instruments fulfill this condition. These SLS/DLS systems can handle small sample volumes and are sensitive enough to study monomeric proteins at low concentrations. Some companies and the corresponding Web sites are listed in alphabetic order:

ALV-Laservertriebsgesellschaft Germany: http://www.alvgmbh.de.

Brookhaven Instruments Corporation, USA: http://www.bic.com.

Malvern Instruments Ltd., UK: http://www.malvern.co.uk.

Wyatt Technology Corporation, USA: http://www.wyatt.com or http://www.wyatt.de.

Several companies produce not only light scattering instruments but also important optional units.

2.2. Sample Cells

Most of the commercially available instruments can handle different sample cells. Exceptions are special built-in sample cells in multi-angle devices mostly working in the flow-through mode. As the laser beam waist in DLS instruments is very small, standard micro or ultra-micro fluorescence cells can be used. This reduces the necessary sample volume considerably, e.g., a sample volume of only 12 μl is needed for an ultra-micro fluorescence cell with a light path of 1.5 mm (Hellma, 105.252-QS). However, ultra-micro fluorescence cells with a pathlength of 3 mm (volume 45 μl, Hellma 105.251-QS) are very convenient to work with as long as sufficient protein is available. If extremely pure sample solutions are required for your light scattering experiment (see Subheading 3.3), flow cells (e.g., Hellma, 176.152-QS, pathlength 1.5 mm) preceded by an in-line filter with pore size 0.1 μm should be applied. Keep in mind that measuring both the light scattering signal and the concentration in one and the same cuvet is of great advantage.

Therefore, all cuvets used for light scattering should be compatible with standard UV-VIS photometers.

Cylindrical light scattering cells, if needed, can also be obtained from Hellma or from manufacturers of light scattering instruments.

2.3. Buffers and Solvents

Prepare all buffers and aqueous solvents using ultrapure deionized water. Buffers and solvents should be filtered (pore size 0.2 μm) and degassed directly before use.

2.4. Reference Samples

The standard reference sample for calibration of an SLS instrument is toluene (ultrapure, e.g., Merck Uvasol). Toluene should be filled into a carefully cleaned sample cell. Seal the reference cell tightly to avoid evaporation and contamination. Note that the Rayleigh ratio of toluene is strongly dependent on both wavelength and temperature (12). Most providers of commercial instruments incorporate the Rayleigh ratios of the reference sample for particular experimental conditions into the instrument software. A well-defined protein solution can also be used as a reference sample. However, such a reference solution is difficult to prepare and is in general not stable for long times.

DLS instruments do not require further calibration as long as all experimental parameters used for data treatment (e.g., λ, θ, n, T, η) are properly determined. However, test experiments with solutions of monodisperse latex spheres could be used to verify the proper function of an instrument.

2.5. Equipment for Sample Preparation and Purification

2.5.1. Filtration

Filters with pore sizes of 0.1 or 0.02 μm (e.g., Whatman Anotop or Anodisc filters) and inner diameters of 10 mm are very useful for sample purification. However, a disadvantage of filtration is the relatively large amount of sample volume needed to equilibrate the filter unit (see Subheading 3). While Anotop filters are ready to use, the Anodisc filter discs have to be mounted carefully into a special holder (see Note 2). Such filter holders are provided by some manufacturers of light scattering instruments (e.g., Wyatt technology corporation) and enable very small dead volumes (~10 μl). The Anotop filters (diameter of 10 mm) have dead volumes of about 0.1 ml.

2.5.2. Centrifugation / Ultracentrifugation

Centrifugation at about $10,000 \times g$ for at least 10 min is only sufficient to remove bubbles, dust particles, or visible aggregates. This procedure is only recommended for strongly scattering solutions or for proteins sensitive to strong mechanical stress conditions. Ultracentrifugation (accelerations up to about $75,000 \times g$) for 1–2 h permits the use of Eppendorf cups and is thus both useful and convenient to remove large aggregates completely. Elimination of species with rather small sedimentation coefficients, e.g., smaller aggregates ($R_S < 50$ nm) and, particularly, fibrous structures, requires high-speed ultracentrifugation with up to $300,000 \times g$. In this case, the use of special centrifugation tubes is obligatory.

2.5.3. Size Exclusion Chromatography and Other Particle Separation Techniques

Serious problems in light scattering experiments may arise from small aggregates or oligomers, because those structures can hardly be removed by filtration or centrifugation. Such particles can be separated from monomeric proteins by size exclusion chromatography (SEC). Besides conventional SEC-based purification schemes field-flow fractionation (FFF) is successfully utilized for separation. Avoiding the employment of stationary phases FFF is applicable to proteins larger than 10 kDa. A combination of FFF and SLS/DLS has already been offered for more than a decade by some companies (e.g., Wyatt technology corporation, Postnova Analytics GmbH).

2.5.4. Batch and Flow Experiments

The above-mentioned separation techniques (see Subheading 2.5.3) can be applied either in batch or in flow mode. In flow-SLS/DLS experiments, the light scattering flow-through cell is directly coupled to the separation device. This has the advantage that perfectly purified samples can be obtained provided that the column (or another separation device) is carefully flushed with filtered solvent. In the batch mode, collected fractions (e.g. from SEC) are transferred to the scattering cell, whereas an additional filtration step can be necessary to remove scattering contaminants.

2.6. Devices and Materials for Supplementary Measurements

2.6.1. Protein Concentration

Precise protein concentrations are especially important for evaluation of SLS data. The most convenient way is a spectrophotometric determination in the near-UV region (25). If the same cuvet is used for both the concentration determination and the light scattering experiment, the measured absorption may be comparatively small due to the disadvantageous combination of low protein concentration and short pathlength of the cell. Therefore, the use of a sensitive and fast instrument is recommended. Modern diode array spectrophotometers fulfill this condition very well.

2.6.2. Refractive Index n and Specific Refractive Index Increment $\partial n/\partial c$

The refractive index can be determined easily and with sufficient precision with an Abbe refractometer. Measurements of the refractive index increment $\partial n/\partial c$ are much more time-consuming and require the use of differential refractometers, which are also frequently used as concentration detectors in chromatography techniques. Therefore, instruments for measuring $\partial n/\partial c$ can be obtained from manufacturers of light scattering devices and chromatography systems as well (e.g., Optilab T-rEX from Wyatt technology). However, it is recommended to try to obtain appropriate values of $\partial n/\partial c$ from the literature (18) for particular buffer and solvents before starting tedious experiments.

2.6.3. Solvent Viscosities
and Densities

For calculations of Stokes radii the dynamic viscosity η of the solvent must be known, which requires measurements of the kinematic viscosity v and the solvent density ρ. Kinematic viscosities are measured using an Ubbelohde-type capillary viscometer (e.g., Schott AG, Germany). Precise density meters can be purchased from Anton Paar GmbH (Austria). Like calculations of $\partial n / \partial c$ viscosity measurements may become more time-consuming than a typical SLS/DLS experiment itself. Therefore, it is useful to check the literature for appropriate data (see Note 6).

3. Methods

In this section we describe light scattering experiments which are typical of intrinsically disordered proteins. Some results obtained with Prothymosin α (Protα) are shown for illustration. The subsequent description of all procedures refers to a hypothetic SLS/DLS setup comprising the following components and parameters:

- A continuous-wave laser ($\lambda = 532$ nm), laser power adjustable between 0.2 and 1 W.

- Detection at 90 degrees using a temperature-regulated (0–80 °C) sample holder for rectangular cells and a high quantum yield avalanche photo diode (APD) detector with maximum count rate of about 10^7 s^{-1} (10 MHz, an optimum count rate is adjustable by regulation of the laser power and changeable apertures).

- A multi-bit digital correlator with at least 200 pseudo-logarithmic spaced delay times that allows to handle exponential correlation functions with decay times from 10^{-6} s up to about 1 s without the need for adjustment of the correlator sample time $\Delta \tau$.

- An online computer for sophisticated instrument control, data storage, and data evaluation.

3.1. Control Experiments

1. Prepare a reference sample by filling toluene into a rectangular fluorescence cell ($d = 1$ cm; e.g., Hellma 111-QS) or follow the calibration instructions of the instrument provider. Avoid any contamination of the cell surfaces that could produce stray light!

2. Insert the reference cell into the sample holder at a defined temperature (typically, 20 °C or 25 °C) and check the light scattering intensity. For this purpose, standard conditions for the instrument settings (laser power, beam attenuation, etc.) should be used. Take into account the warm-up times

recommended by the instrument manufacturer. If the system setup is correct the measured "standard" intensity should be stable over long periods of time.

3. Thoroughly check both the sample preparation procedure and the quality of the sample cell in order to reduce stray light as far as possible (steps 4–6, see also Note 5).

4. Flush a 0.1-μm Anotop filter with at least 5 ml of ultrapure deionized water. Purge a carefully cleaned sample cell (Subheading 3.5) with nitrogen or clean air in order to remove dust (an improved procedure for filling sample cells is described in Subheading 3.3).

5. Filter an appropriate amount of water (e.g., 100 μl in the case of micro fluorescence cells with pathlength of 3 mm, Hellma 105.254) into the cell.

6. Observe the scattered intensity of the water sample for at least 2 min. The basic preconditions for light scattering experiments are fulfilled if (a) the scattering intensity is about ten times smaller than that of toluene (the exact ratio depends on the wavelength) and (b) the scattering signal is devoid of strong intensity fluctuations (signal spikes). It is particularly useful to carry out a short DLS experiment. The measured autocorrelation function (acf) should consist of a flat baseline with statistical noise.

The sample purification guidelines (steps 4–6) should be applied both to the buffer and protein solution. Repeat the sample preparation until all quality conditions are met. If the abovementioned protocols are not sufficient to meet the purification requirements, browse Subheading 3.3 for specialized sample preparation procedures.

3.2. Standard SLS/DLS Experiment with Proteins

The aim of the experiment is the determination of the diffusion coefficient D_0, Stokes radius R_S, and the molar mass M. This requires the extrapolation of M_{app} and D_{app} to zero protein concentration. If the concentration dependence turns out to be linear, 5 concentrations are sufficient. The concentration range depends on the protein under study and should be 0.5–2.5 mg/ml for a typical experiment. Each measurement requires a volume of about 0.5 ml. The entire experiment can be done in two ways regarding the order of concentrations. In the first case, 5 solutions with concentrations between 0.5 and 2.5 mg/ml are prepared separately and the experiment is started with the lowest concentration. In the second case, the experiment is started with the sample of the highest concentration. All subsequent samples are derived by dilution. In the latter case, less material is needed. The following protocol is based on the second method.

1. Prepare a stock protein solution with a concentration of 2.5 mg/ml in an adequate buffer, which was filtered and degassed immediately before use.

2. Flush a filter (0.1 μm pore size) with water, filter 100 μl water into the sample cell (here we use 45 μl-microcells, see above), check the cell for purity and measure the reference intensity as described (Subheading 3.1).

3. Empty and dry the sample cell.

4. Flush the same filter with 2 ml buffer, filter 100 μl buffer into the sample cell and measure the light scattering intensity I_B. This can be combined with a short DLS experiment. The acf must not show any correlations. However, buffers containing high salt concentrations may show a small, fast-decaying acf.

5. Measure the buffer baseline of the same cuvet in a UV-VIS spectrophotometer.

6. Empty the cuvet, purge it with ultrapure water and dry it.

7. Fill a separate 1-ml syringe with ~0.5 ml protein solution. Rinse about 0.3 ml very slowly through the filter previously used for buffer and collect the outlet in an Eppendorf tube. A volume of 0.3 ml should be enough to remove the buffer or the previous protein solution nearly completely from the filter unit. Filter 0.1 ml of the remaining protein solution into the sample cell.

8. Measure the near-UV absorption spectrum (or other available absorption bands) of the protein sample and determine the protein concentration.

9. Put the sample into the SLS/DLS instrument and watch the scattered intensity I_P or perform a short DLS experiment. The sample can be considered as suitable for measurements if no or only rare intensity spikes are observed. A new sample should be prepared, if the scattered intensity is strongly fluctuating.

10. Adjust beam attenuation. The mean pulse rate should not exceed 10^6 cps (see Note 4).

11. Start data acquisition. It is useful to accumulate data from a large number of short (5–10 s) time intervals instead of averaging over a single or a few long intervals.

    Whether or how an instrument can handle contaminated solutions depends on the quality of the data accumulation software. Some instruments come with special software, a so-called electronic dust-filter, to reject distorted measurement intervals or to stop data accumulation temporarily. The total duration of an experiment (typically, 1–30 min) that is needed to obtain an acf with a sufficiently high signal-to-noise ratio depends on the attainable pulse count rate and the complexity of the acf.

12. Empty the sample cell and collect all filtered material to prepare the next solution of lower concentration. Clean the sample cell either with water or according to Subheading 3.5 and dry it.

13. Repeat steps 7–12 until the lowest concentration is reached. Note, that longer data acquisition times are needed at lower concentrations if the decrease in scattering power cannot be compensated, e.g., by higher laser power. Normally, a single filter can be used for all concentrations. This simplifies the experiment and helps to save material.

14. Measure the reference intensity and turn to data evaluation (Subheading 3.4).

3.3. SLS/DLS Experiments with Special Precautions

In this section, some experimental situations are described, which may be encountered during SLS/DLS studies on protein solutions.

1. Small proteins or peptides, which scatter light only weakly.
For very weakly scattering systems it is crucial to *avoid any contamination of the sample*. Thus, working with open sample cells may be inappropriate. Instead, you should change to a flow experiment, in which a filter is permanently attached to the inlet of a flow-through cell. Without separating filter and cell, both elements should be successively purged with water, buffer and protein solution. This ensures perfectly cleaned samples allowing the measurement of rather small excess scattering intensities. However, the disadvantage of this method is the higher amount of sample volume that is needed due to the larger dead volume of the flow-through device.

2. Only a small amount of protein is available.
In this case you may be interested in reducing the sample volume as much as possible. There are two ways to work with sample volumes smaller than 50 µl. First, you can use a micro-filter device (see Subheading 2.5.1) and an ultra-micro fluorescence cell (see Subheading 2.2). The second possibility is ultra-centrifugation of the sample (see Subheading 2.5.2) and subsequent transfer of a small volume of the supernatant into an ultra-micro fluorescence cell. Try to estimate the protein concentration as good as possible with an appropriate UV-VIS spectrophotometer.

3. Data acquisition in the case of strongly contaminated protein solutions.
Even the purest sample solution can turn into an experimental nightmare if either the protein molecules have a tendency to form aggregates or the buffer solution facilitates the development of bubbles. The latter is frequently observed in mixed solvents and can hardly be prevented. Measurements under such conditions require special data acquisition schemes and the corresponding software. Briefly, instead of the usual data

accumulation scheme all short-time acfs and average intensities measured during the entire experiment need to be stored separately. After that, the individual acfs and intensities have to be inspected. Only undistorted members of the data set are then accumulated and used for data evaluation.

3.4. Data Evaluation

In this section we consider the exceptionally rare case of perfectly purified monomeric proteins first. This case allows an independent data evaluation for both SLS and DLS. A more realistic and thus more complicated situation is regarded in the last section.

1. Static light scattering.

    For the determination of the molar mass M and the second virial coefficient A_2 equations 3 and 4 have to be used. Even under unfolding conditions the size of proteins with $M < 100$ kDa is small compared to the laser wavelength (see Fig. 3). Therefore, at a scattering angle of 90 degree $P(q)$ is ~1 in (Eq. 4). The primary quantity for all further calculations is the relative excess scattering I_{ex}/I_{ref}, where I_{ex} is the difference between the scattering intensities of the protein solution, I_P and the buffer (or solvent), I_B. For proteins at concentrations below 1 mg/ml, I_{ex} can be small or even smaller than I_B. Therefore, the quality of primary data should be inspected before subjecting them to further calculations. The final data representation is plotting $1/M_{app} = Hc/R_q$ versus c (Debye plot). The results for the intrinsically disordered protein Protα are shown in Fig. 4. The slope of a linear fit yields the second virial coefficient A2, which is a measure of intermolecular

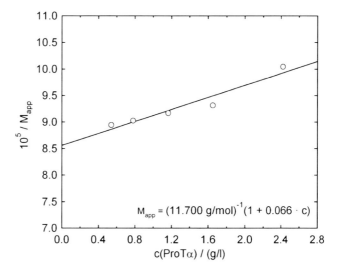

Fig. 4. Concentration dependence of the reciprocal apparent molecular mass (Debye plot) of prothymosin α in PBS, pH 7.4 at 20 °C.

interactions. The software of most of the commercially available instruments is able to generate Zimm or Debye plots.

2. Dynamic light scattering.

   The autocorrelation functions measured by DLS are subjected to mathematical procedures in order to obtain apparent diffusion coefficients D_{app}. These, in turn, can be converted into apparent Stokes radii by means of the Stokes–Einstein equation. In the ideal case of a monodisperse solution of protein monomers the acf is properly described by a single exponential decay and vice versa (Fig. 2) yielding D_{app} according to (Eq. 5). In the case of polydisperse solutions distributions of diffusion coefficients or Stokes radii can be calculated by means of an inverse Laplace transform. The application of this procedure is generally recommended, because it can be applied to polydisperse and monodisperse systems as well. The mathematical procedure of an inverse Laplace transform for obtaining the distribution is, however, the most crucial step of DLS data evaluation. Though the software supplied with many DLS instruments is able to calculate distributions automatically the following remarks should be considered:

   – Check the measured acf for distortions (oscillations, steps, spikes) before executing the inverse Laplace algorithm.

   – Check the stability of the obtained result both by varying the range of D_{app} or R_{app} and the strength of the regularization parameter which is important for obtaining reliable solutions (15).

   The mathematical procedure can be instructed to calculate different types of distributions, namely intensity, mass concentration or number distributions of D_{app} or $R_{S,app}$. Only the first one can be calculated without making additional assumptions on the particle structure. This may be difficult, however, particularly in case of IDPs. Normally, the strongest or sole peak in the calculated distribution can be attributed to the monomeric protein.

   Keep in mind that the diffusion coefficient of a protein D_0 can only be obtained by linearly extrapolating the experimentally determined apparent diffusion coefficients D_{app} to zero protein concentration. Illustrating the concentration dependence of D_{app}, experimental data obtained for prothymosin α are shown in Fig. 5. It becomes evident from these data that both the hydrodynamic dimensions and the intermolecular interactions remarkably change with the solvent conditions. The results are discussed in more detail in (7).

3. Advantage of combined SLS/DLS experiments in the presence of aggregates.

   For the calculation of the molar mass we have assumed so far that the total light scattering intensity, I_{ex}, results exclusively

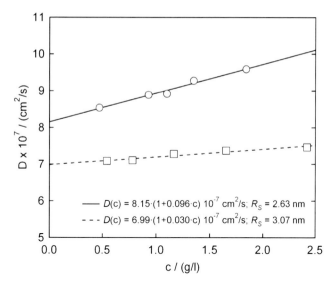

Fig. 5. Concentration dependence of the translational diffusion coefficient D of prothymosin α in PBS, pH 7.4 (*squares*) and in 10 mM glycine-HCl, pH 2.45 (*circles*) at 20 °C.

from the monomeric protein. However, if aggregates contribute significantly to I_{ex}, static light scattering experiments yield meaningless molar masses. Since aggregates are much larger than monomers their contribution to the total light scattering intensity may be notable even at low concentrations. If the aggregate fraction is small enough, so that the monomer concentration is essentially identical to the total concentration, proper values of M can be obtained simply by considering the size distribution obtained from DLS. SLS/DLS experiments with the Syrian hamster prion protein (ShPrP) may serve as an example (26). No corrections to I_{ex} are needed at pH 7, since the size distribution consists of only one peak (Fig. 6). By contrast, analysis of the size distribution at pH 4 tells us that aggregates with Stokes radii in the region of 20 nm contribute with 14 % to I_{ex}. Once the aggregate-related scattering intensity is subtracted from I_{ex} evaluation of the remaining scattering intensity yielded a molar mass of ShPrP that was close to that calculated from the amino acid sequence (26). Underlining the strength of DLS, the data evaluation procedure works still when the contributions of aggregates and monomers to I_{ex} are comparable. This opportunity simplifies measurements of molar masses of proteins by light scattering considerably.

3.5. Cleaning of Sample Cells

After finishing the experiment sample cells should be thoroughly cleaned by special cleaning solutions provided by manufacturers of glass or quartz cells (e.g., Hellmanex from Hellma GmbH). In general, it is sufficient to leave the cells in dilute solutions (1–2 %) of the concentrate for some hours at room temperature. After that

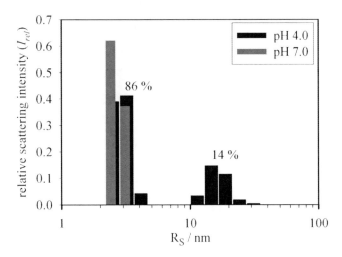

Fig. 6. Size distributions in terms of the relative scattered intensity versus R_S for the recombinant Syrian hamster prion protein, $c = 1.1$ mg/ml at pH 7 (unimodal distribution) and pH 4 (bimodal distribution).

the cells have to be rinsed carefully with ultrapure water. Since cleaning solutions as well as ultrapure water may harm the surface of your cells, avoid very long contact time.

3.6. Remarks Concerning Supplemental Experiments

Supplementary measurements of the solvent viscosity, the solvent density and the refractive increment of proteins should be done with great care. Since a detailed description of these procedures exceeds the scope of this chapter, the reader should follow the instructions of the manufacturers of the corresponding instruments.

4. Notes

1. Cells.
   - Use quartz cells only.
   - Check the cell concerning stray light by measuring the scattering intensity of water. Change the cell position checking for signal fluctuations that may be related to contaminations at the cell surface.

2. Filtration.
   - Filter your solutions directly into the quartz cell.
   - Purge the filter with water or buffer solution and repeatedly check the scattering intensity of the filtrate.
   - Avoid bubble formation during filtration. This may be observed most frequently with mixed solvents and can be reduced by optimizing pressure and flow rate during

filtration. Low and stable flow rates can be obtained by using syringe pumps.

- Keep in mind that some organic solvents or strong acid/ alkaline solutions may harm your filter material.

- Filtration through a 0.02 μm filter membrane can cause shear stress that promotes protein aggregation. Increase the pore size to 0.1 μm to avoid or reduce this effect.

- Some proteins tend to irreversibly bind to the filter membrane. This may reduce the protein concentration in the filtrate. Therefore, determining the protein concentration after each filtration is crucial.

- Assemble micro-filter units very carefully to avoid fracture of the Anodisc filter plates.

3. Centrifugation.

The choice of an optimum relative centrifugal force depends on the mass and the structure of the aggregates. For example, sedimentation of long fibrillar aggregates is rather slow. Therefore, protein aggregates do not necessarily form "pellets" at the bottom of the centrifugation vessel. Care should be taken when transferring the supernatant from the vessel into the quartz cell.

4. Excess scattering intensities.

- As a rule of thumb, the scattering of an aqueous buffer solution roughly accounts for 1/10 of the toluene standard. Proper correction of the measured scattering intensity of a protein solution for the solvent scattering is especially important in the case of weakly scattering solutions of small proteins at low concentrations. For example, the excess scattering of a 0.4 mg/ml protein solution (M = 10,000 Da) equals that of pure water.

- High signal intensities (e.g., in strongly scattering solutions) can lead to detector overload. This, in turn, leads to nonlinear signal detection causing not only incorrect average scattering intensities but also serious distortions of the recorded acf. In order to prevent detector overload, the average intensity of scattered light (I_S) should not exceed about 1/10 of the specified maximum intensity (photon count rate) suggested in the manufacturer's documentation.

5. Turbid and other absorbing solutions.

- Solutions with visible turbidities are not applicable to SLS/ DLS experiments.

- Difficulties with turbid solutions arise from two main aspects. First, turbidity leads to attenuation of the incident and the scattered light. Second, multiple scattering, particularly at large scattering angles, becomes far more likely. This leads to alterations of the average intensity and to distortions of the measured autocorrelation functions.

– It is very important that the sample does not absorb light at the wavelength of the laser light. Absorption of the intense laser beam can lead to an uncontrolled heating of the sample in addition to the attenuation of incident and scattered light.

6. Viscosities.

– The individual components of a sample do not additively contribute to the viscosity.

– Working with multicomponent systems it is generally recommended to determine the correct viscosity experimentally rather than adding up theoretical values.

7. Concentration dependence of $1/M_{app}$ and D_{app}.
The measured dependences of $1/M_{app}$ and the apparent diffusion coefficient D_{app} on the protein concentration can be described by the second virial coefficient A_2 (or additional higher-order virial coefficients in particular cases) or the diffusive concentration dependence coefficient k_D, respectively. A precise determination of M_0 and D_0 requires measurement at many different protein concentrations. If the observed concentration dependence is linear, it is usually sufficient to measure M_{app} or D_{app} at five concentrations. However, a non-linear dependence requires an increase in both concentration range and number of concentrations.

References

1. Nimmo GA, Cohen P (1978) The regulation of glycogen metabolism. Purification and characterisation of protein phosphatase inhibitor-1 from rabbit skeletal muscle. Eur J Biochem 87:341–351

2. Hemmings HC et al (1984) DARPP-32, a dopamine- and adenosine 3':5'-monophosphate-regulated phosphoprotein enriched in dopamine-innervated brain regions. II. Purification and characterization of the phosphoprotein from bovine caudate nucleus. J Neurosci 4:99–110

3. Hernandez MA, Avila J, Andreu JM (1986) Physicochemical characterization of the heat-stable microtubule-associated protein MAP2. Eur J Biochem 154:41–48

4. Lynch WP, Riseman VM, Bretscher A (1987) Smooth muscle caldesmon is an extended flexible monomeric protein in solution that can readily undergo reversible intra- and intermolecular sulfhydryl cross-linking. A mechanism for caldesmon's F-actin bundling activity. J Biol Chem 262:7429–7437

5. Watts JD et al (1990) Thymosins: both nuclear and cytoplasmic proteins. Eur J Biochem 192:643–651

6. Schweers O et al (1994) Structural studies of tau protein and Alzheimer paired helical filaments show no evidence for beta-structure. J Biol Chem 269:24290–24297

7. Gast K et al (1995) Prothymosin alpha: a biologically active protein with random coil conformation. Biochemistry 34:13211–13218

8. Weinreb PH et al (1996) NACP, a protein implicated in Alzheimers disease and learning, is natively unfolded. Biochemistry 35:13709–13715

9. Huglin M (1972) Light scattering from polymer solutions. Academic, New York

10. Kratochvil P (1987) Classical light scattering from polymer solutions. Elsevier, Amsterdam

11. Schmitz KS (1990) An introduction to dynamic light scattering by macromolecules. Academic, New York

12. Chu B (1991) Laser light scattering. Academic, New York

13. Brown W (1993) Dynamic light scattering. Claredon, Oxford

14. Berne BJ, Pecora R (2000) Dynamic light scattering with applications to chemistry, biology, and physics. Dover Publications, Mineola, New York

15. Provencher SW (1982) CONTIN—a general-purpose constrained regularization program for inverting noisy linear algebraic and integral-equations. Comp Phys Commun 27:229–242

16. Koppel DE (1972) Analysis of macromolecular polydispersity in intensity correlation spectroscopy: the method of cumulants. J Chem Phys 57:4814–4820

17. Harding SE, Johnson P (1985) The concentration dependence of macromolecular parameters. Biochem J 231:543–547

18. Theisen A et al (2000) Refractive increment data-book. Nottingham University Press, Nottingham

19. Damaschun G, Damaschun H, Gast K, Misselwitz R, Zirwer D, Guhrs KH, Hartmann M, Schlott B, Triebel H, Behnke D (1993) Physical and conformational properties of staphylokinase in solution. Biochim Biophys Acta 1161:244–248

20. Damaschun G et al (1998) Denatured states of yeast phosphoglycerate kinase. Biochemistry (Moscow) 63:259–275

21. Uversky VN (1993) Use of fast protein size-exclusion liquid-chromatography to study the unfolding of proteins which denature through the molten globule. Biochemistry 32:13288–13298

22. Wilkins DK et al (1999) Hydrodynamic radii of native and denatured proteins measured by pulse field gradient NMR techniques. Biochemistry 38:16424–16431

23. Uversky VN (2002) What does it mean to be natively unfolded? Eur J Biochem 269:2–12

24. Zhou HX (2002) Dimensions of denatured protein chains from hydrodynamic data. J Phys Chem 106:5769–5775

25. Pace CN et al (1995) How to measure and predict the molar absorption-coefficient of a protein. Protein Sci 4:2411–2423

26. Sokolowski F et al (2003) Formation of critical oligomers is a key event during conformational transition of recombinant Syrian hamster prion protein. J Biol Chem 278:40481–40492

Chapter 10

Estimation of Intrinsically Disordered Protein Shape and Time-Averaged Apparent Hydration in Native Conditions by a Combination of Hydrodynamic Methods

Johanna C. Karst, Ana Cristina Sotomayor-Pérez, Daniel Ladant, and Alexandre Chenal

Abstract

Size exclusion chromatography coupled online to a Tetra Detector Array in combination with analytical ultracentrifugation (or with quasi-elastic light scattering) is a useful methodology to characterize hydrodynamic properties of macromolecules, including intrinsically disordered proteins. The time-averaged apparent hydration and the shape factor of proteins can be estimated from the measured parameters (molecular mass, intrinsic viscosity, hydrodynamic radius) by these techniques. Here we describe in detail this methodology and its application to characterize hydrodynamic and conformational changes in proteins.

Key words: Molecular mass, Intrinsic viscosity, Protein shape, Time-averaged apparent hydration, Protein hydration, Size exclusion chromatography, Static light scattering, Dynamic light scattering, Analytical ultracentrifugation, Online viscometer

Abbreviations

M	Molecular mass, g/mol
T	Absolute temperature, K
C	Protein concentration, mol/L (M)
dn/dc	Refractive index increment, mL/g
dA/dc	Absorbance increment, L/g cm
$[\eta]$	Intrinsic viscosity, mL/g
v	Hydrodynamic shape function, viscosity increment, Simha–Saito shape factor, unitless
δ	Time-averaged apparent hydration, $g_{H_2O}/g_{protein}$
f/f_0	Translational frictional ratio of the protein, including shape and hydration parameters
f	Frictional coefficient of the protein, g/s
f_0	Frictional coefficient of an anhydrous sphere of the mass of the protein, g/s
R_H	Hydrodynamic radius of the protein, cm
R_0	Radius of an anhydrous sphere of the mass of the protein, cm
V_H	Hydrodynamic volume calculated from the R_H, cm^3

Vladimir N. Uversky and A. Keith Dunker (eds.), *Intrinsically Disordered Protein Analysis: Volume 2, Methods and Experimental Tools*, Methods in Molecular Biology, vol. 896, DOI 10.1007/978-1-4614-3704-8_10, © Springer Science+Business Media New York 2012

D_t Translational diffusion coefficient, cm^2/s

s Sedimentation coefficient obtained at the temperature of the experiment, Svedberg, 10^{-13} s

η Viscosity of the solvent, Poise: g/cm s

ρ Density of the solvent, g/mL

k_B Boltzmann's constant, erg/K (K_B: 1.38065 $\times$ 10^{-16} erg/K, with erg: g cm^2/s^2 = 10^{-7} J 1.38065 $\times$ 10^{-23} J/K)

N_A Avogadro's number, molecules/mol

$\bar{v}$ Partial specific volume, mL/g

a/b Axial ratio of ellipsoid

$RALS$ Right Angle Light-Scattering

$LALS$ Low Angle Light-Scattering

IP Internal Pressure

DP Differential Pressure

UV Ultraviolet absorption

RI   Refractive Index

1. Introduction

Many methods are available for the characterization of folded proteins, allowing the determination of structural and hydrodynamic parameters. In contrast, fewer methods can be used to study intrinsically disordered proteins, because the particular behavior of such proteins makes their study difficult. Unfolded proteins are hydrated and flexible with limited residual structure. These biophysical characteristics preclude the use of X-ray crystallography and limit the use of NMR. In this context, experimental approaches providing information on the shape and the hydration of intrinsically disordered proteins are valuable. As opposed to large-scale instruments required for small-angle X-ray and neutron solution scattering (SAXS and SANS), SEC-TDA is a benchmark for in-lab molecular mass, protein shape, and hydration determination.

Here, we describe a methodology to characterize the hydrodynamic properties consisting in the combination of several experimental biophysical approaches: analytical ultracentrifugation (AUC), quasi-elastic light scattering (QELS), and size exclusion chromatography coupled online to a Tetra Detector Array (SEC-TDA) (Fig. 1). This latter technique, which is described in detail here, combines right and low angles static light-scattering (RALS and LALS) detectors, a spectrophotometer (UV), a refractometer (RI), and pressure transducers (differential pressure (DP) and internal pressure (IP)) (Fig. 2). Importantly, this methodology allows the characterization of intrinsically disordered proteins in solution and in native conditions and provides an estimation of their hydrodynamic parameters, such as shape and hydration.

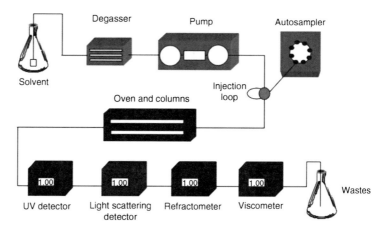

Fig. 1. Scheme of size exclusion chromatography system connected online to a Tetra Detector Array and controlled by a GPCmax module (Viscotek Ltd., a Malvern Company). The GPCmax module provides an integrated solvent pump, an autosampler, and a degasser. The Tetra Detector Array contains a UV detector, a static light-scattering cell, a differential refractive index (RI) detector, and a four-capillary differential viscometer. All the detectors reside within a temperature-controlled compartment. It is noteworthy that the linkup of an online QELS instrument would allow the acquisition of all hydrodynamic parameters (M, R_{H}, $[\eta]$) with a unique sample injection.

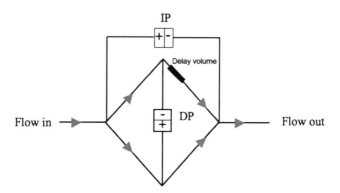

Fig. 2. Schematic diagram of differential viscometer (Viscotek Ltd., a Malvern Company). It consists of a four-capillary bridge design, developed by Dr. Max Haney. The four capillaries are arranged in a balanced bridge configuration, analogous to the Wheatstone bridge commonly present in electrical circuits. Differential pressure transducers measure the pressure difference (DP) across the midpoint of the bridge and the pressure difference IP from inlet to outlet. A delay column is inserted in the circuit to create the differential pressure (providing a reference flow of solvent during elution of the sample).

Molecular mass and intrinsic viscosity are measured by SEC-TDA. The protein concentration is determined using the photometer or the deflection refractometer. The static light-scattering signals combined with the protein concentration provide online measurements of the molecular mass (M) of each eluting species, whereas differential viscometer measurements in combination with protein concentration provide their intrinsic viscosity ($[\eta]$) values. The hydrodynamic radius of the macromolecule, R_{H}, is obtained either by the sedimentation coefficient (s) value determined by AUC, or by using the translational diffusion coefficient (Dt) determined by QELS. From the hydrodynamic parameters (M, $[\eta]$, R_H) measured by these techniques, we can ascribe the respective contributions of hydration and shape factor (viscosity increment) to the intrinsic viscosity value.

The methodology described here has been shown to be useful for the studies of intrinsically disordered proteins, protein/ligand interaction (1, 2), protein/protein interaction (3, 4) and proteins exhibiting anomalous behavior (5, 6).

2. Materials

1. A gel filtration column of appropriate fractionation range for the samples of interest.

2. A gel filtration buffer that is compatible with both the column and the sample (see Note 1).

3. Calibration standards (see Note 2):

   (a) NIST-1923 Polyethylene Oxide 22 KDa (PEO from Viscotek PolyCal^{TM} TDS-PEO-N at 4 g/L).

   (b) BSA (SIGMA A0281 at 2 g/L).

4. A GPCmax module or a similar instrument that provides an integrated solvent pump, an autosampler and a degasser (Fig. 1). It can be controlled manually or using the OmniSEC software.

5. A tetra detector array model 302 (Viscotek Ltd., a Malvern Company) or a similar instrument that consists of a UV detector, a differential refractive index (RI) detector, a four-capillary differential viscometer (Fig. 2) and a static light-scattering cell with two photodiode detectors at 7 ° for low angle (LALS) and at 90 ° for right angle laser light scattering (RALS) (Fig. 1). All the detectors reside within a temperature-controlled compartment (see Note 3).

3. Methods

3.1. Data Acquisition

1. Filter all solutions using a 0.22-μm filter to degas and remove any small particles that may clog the column frit (see Note 4).

2. Change the RALS filter to avoid injection of column particle in the static light-scattering cell.

3. Purge the pulse dampener.

4. Switch on the TDA electronic part, the laser source and the UV detector (see Notes 5 and 6).

5. Open the OmniSEC software (Viscotek Ltd., a Malvern Company) and follow the general procedures described in the user manual.

6. In the panel "tool/options," choose a folder to save your files. Enter parameters such as time and volume of an experimental run, file name format, and detectors sensitivity in the panel "acquire/configuration."

7. Indicate the flow rate (see Note 7).

8. Select the minimum and maximum pressure (see Note 8).

9. Wash out ethanol 20 % (in water) from the GPCmax-TDA with water (0.1 mL/min) and purge the Refractometer and Viscometer detectors (see Note 9).

10. Wash extensively the Hamilton syringe and its needle as well as the injection loop of the autosampler with water and then with buffer.

11. Equilibrate the GPCmax-TDA with at least 60 mL of buffer (0.5 mL/min) and examine the baselines (see Note 10).

12. Start a quick run.

13. Purge the Refractometer and Viscometer detectors for 5 min with buffer and repeat the operation until the displayed values become stable (see Note 9).

14. Connect the column online; the column should have been previously equilibrated with at least two column volumes of buffer (see Note 11).

15. When the system is equilibrated, flat all baselines until an acceptable signal/noise value is reached (see Note 10). Then, turn off the flow, wait until the IP baseline is flat and reset the Inlet Pressure Transducer by pushing the zero button on the instrument (see Note 12).

16. Measure the viscometer bridge balance by using the following formula:

    Specific viscosity: 4DP/(IP-2DP)

    The value should be below 0.03. The bridge is very well balanced for a specific viscosity value below 0.01 (i.e., 1 % of difference across the viscometer bridge).

17. Clarify and degas the samples using a 0.22-μm filter and/or by centrifugation ($10,000 \times g$ for 10 min). Vials, containing the sample, can be inserted at defined numbered positions in the autosampler (see Note 13).

    To determine M and $[\eta]$ (and subsequently shape and hydration), a concentration of 2 g/L is required for a folded protein, whereas $\approx 0.5–1$ g/L is enough for an intrinsically disordered protein, as $[\eta]$ of such proteins is higher. To determine the molecular mass only, a concentration around 0.5 g/L is enough.

18. Create an analytical sequence in the software by listing the injections to be realized: standards (BSA, PEO), sample injections, and finally standards again. Indicate the sample names, concentrations, volume, number, and the order of sample injections (see Note 14).

19. Start the sequence.

20. At the end of a sequence, replace the buffer from all the system by water. Make purges of the Refractometer and Viscometer as described in (11). Then store the system in filtered and degassed 20 % ethanol (in water).

3.2. Data Processing

Standards are used for TDA internal constant calibrations. BSA is employed to calibrate the conversion from static light scattering to molecular mass, as this protein is commonly used as a reference in numerous biochemical applications. However, as its intrinsic viscosity is low, PEO can be used to calibrate both M and $[\eta]$ parameters (see Note 2). Then, BSA injections are used to validate the method.

Polypeptide concentrations can be determined using the photometer (UV) and/or the deflection refractometer (RI). RALS and/or LALS data combined with the protein concentrations (UV or RI) provide the molecular mass, while the differential viscometer measurements (Differential Pressure, DP), in combination with the protein concentrations provide the intrinsic viscosity.

3.2.1. Calibration

1. Open a file of standard for calibration.

2. For each detector, add the baselines automatically with OmniSEC. If necessary, modify their positions manually and note the noise (in Volt).

3. Surround the monomeric species by the integration limits on all channels.

4. Create a method, in which you indicate all parameters concerning the standard (dn/dc, dA/dc, MM, $[\eta]$...) (see Note 2). Click the box "*Calculate Concentration from Detectors*." Two different methods can be created, one using the refractometer (RI) and the other using the photometer (UV) as detector to determine the sample concentration.

5. Press the button "*calibration*" and save the method.

6. Check the calibration constant values (see Note 15).

7. Test the calibration with the same standard file as sample and with other injections of standard. If the standard changes, do not forget to change the parameters (dn/dc, dA/dc...) in the "sample parameters" window.

1. Open a sample file.

2. Determine the baselines automatically on all channels with OmniSEC. If necessary, modify their position manually.

3. Place the integration limits on all channels by surrounding the entire elution profile, from the void volume to the salt elution, or restrict the window to the area of interest.

4. Open the appropriate method depending on the selected detector used to determine the concentration (UV or RI). Enter dn/dc and dA/dc of the protein in the "sample parameters" window.

5. Start analysis by pressing the button *execution*: the software computes molecular mass and intrinsic viscosity of the macromolecule (see example given in Fig. 3).

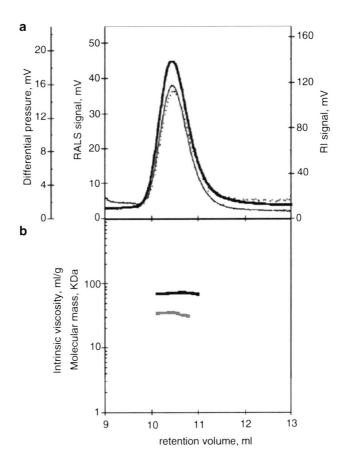

Fig. 3. Typical traces of a protein analyzed by size exclusion chromatography connected to a tetra detector array. (**a**) Deflection refractometer (*bold continuous line*), right angle light scattering (*thin continuous line*) and differential pressure (*dotted line*) chromatograms. (**b**) The molecular mass (*black line*) and intrinsic viscosity (*grey line*) are computed with the OmniSEC software (Viscotek Ltd., a Malvern Company). This figure shows typical chromatograms from the apo-state of the RD protein (1), an intrinsically disordered protein (13).

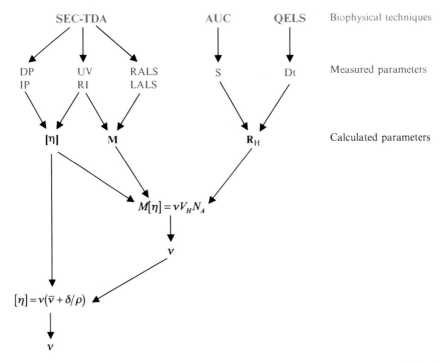

Fig. 4. Determination of protein shape (v) and hydration (δ) parameters. The molecular mass (M), the intrinsic viscosity $[\eta]$, and the hydrodynamic radius R_{H} of the macromolecule determined by SEC-TDA, AUC, and/or QELS are required to estimate protein shape and hydration.

6. Check the robustness of the results by testing the two different methods (RI- and UV-based) if applicable, and different injections (volume or concentration) of the same protein sample.

3.3. Protein Shape and Hydration Estimation

For the determination of protein shape and hydration, the molecular mass (M), the intrinsic viscosity $[\eta]$, the hydrodynamic radius R_{H} and the partial specific volume $\bar{v}$ of the protein are required (Fig. 4). The hydrodynamic radius R_{H} is calculated from the translational diffusion coefficient measured by QELS using the Stock Einstein relation $R_{\mathrm{H}} = (k_{\mathrm{B}} T)/(6\pi\eta Dt)$ or more accurately, using the Svedberg equation $R_{\mathrm{H}} = M(1 - \rho\bar{v})/(6\pi\eta N_{\mathrm{A}}s)$ with molecular mass determined by SEC-TDA (see above) and sedimentation coefficient (s) determined by AUC (see Note 16). The partial specific volume (which usually ranges between 0.69 and 0.75 mL/g) can be either experimentally determined by AUC equilibrium measurement or estimated from the amino acids sequence with, for instance, the SEDNTERP software (http://www.rasmb.bbri.org). The solvent viscosity η and density ρ can also be computed with the SEDNTERP software.

The intrinsic viscosity of a protein, $[\eta]$ (7, 8), is expressed according to the relation $[\eta] = v V_s = v(\bar{v} + \delta/\rho)$ (see Note 17). The intrinsic viscosity is the product of (a) a hydrodynamic function, the viscosity increment v, and (b) the swollen volume V_S. A v value of 2.5 suggests that the protein adopts a spherical shape, while increasing values of v are indicative of ellipsoidal shapes. The swollen volume is the sum of two volumic factors, i.e., the partial specific volume, $\bar{v}$, and the time-averaged apparent hydration of the protein (δ in $g_{H_2O}/g_{protein}$). The hydration parameter includes the water molecules bound to the protein and the water molecules entrained by the diffusion of the protein. Hydration corresponds to the water molecules within the hydrodynamic volume, i.e., to the water included in protein cavities and between the protein surface and the plan of shearing (slipping plan).

Three approaches can be used to estimate protein shape and hydration (see Fig. 4 and Note 18 for an example)

1. The shape factor v can be calculated using the Einstein viscosity relation $M[\eta] = v V_H N_A$ (see Note 19) inverted to $v = M[\eta]/V_H N_A$, where V_H is the hydrodynamic volume defined by $V_H = 4\pi R_H^3/3$. The viscosity increment is then used to estimate the ratio ($a/b = \vartheta$) of the semi-axes a and b (with $a > b$) for an equivalent prolate or oblate ellipsoid. Harding and Colfen described polynomial equations that can be used to convert the viscosity increment into the axial ratio value (9). Solutions for triaxial ellipsoid have also been described (10). The solution for a prolate ellipsoid ($R_H^3 = a \times b \times b$ with $a > b$ and $a/b = \vartheta$) is given by $a = R_H \times \vartheta^{2/3}$ and $b = a/\vartheta$ (or $b = R_H \times \vartheta^{-1/3}$), while the solution for an oblate ellipsoid ($R_H^3 = a \times a \times b$ with $a > b$ and $a/b = \vartheta$) is given by $b = R_H \times \vartheta^{-2/3}$ and $a = \vartheta \times b$ (or $a = R_H \times \vartheta^{1/3}$). Then, from the intrinsic viscosity relation $[\eta] = v(\bar{v} + \delta/\rho)$, the hydration δ is calculated from the parameters $[\eta]$, v and $\bar{v}$ according to $\delta = (([\eta]/v) - \bar{v})\rho$ (see Note 20).

2. A second way to calculate protein shape and hydration is first to determine the hydration (δ) from the experimental values of M, $\bar{v}$ and R_H using the relation $M(\bar{v} + \delta) = V_H N_A$. Then, the viscosity increment v is calculated with the intrinsic viscosity relation $v = [\eta]/(\bar{v} + \delta/\rho)$, providing the a/b ratio as described above.

3. As an alternative estimation of the semi-axial ratio ($a/b = \vartheta$), the Perrin hydrodynamic function P (11) can be calculated according to the equation $P = (f/f_0)\big/\left[((\delta/\bar{v}\rho) + 1)^{1/3}\right]$ (12). For this purpose, the frictional ratio f/f_0 and the previously determined hydration values δ are combined. R_H values are used to calculate the frictional ratio f/f_0 from the relation $f/f_0 = R_H/R_0$. The anhydrous radius R_0 (by definition with $\delta = 0$), is defined by $M\bar{v} = V_0 N_A = (4\pi R_0^3/3) N_A$, which gives $R_0 = (3M\bar{v}/4\pi N_A)^{1/3}$. An estimation of the semi-axial value can be computed according to the polynomial equations provided by Harding and Colfen (9). The shape factor estimated from the Perrin Hydrodynamic function is generally lower than that obtained from the viscosity increment.

4. Notes

1. It is recommended to work at neutral pH. Acidic buffers can damage pressure transducers, whereas basic buffers can damage quartz of the static light-scattering cell.

2. PEO and BSA parameters are the following:

   PEO (Viscotek PolyCal^{TM} TDS-PEO-N): its intrinsic viscosity is 38.8 ± 0.4 mL/g, its molecular mass is 22.411 g/mol and its refractive index increment, dn/dc, is 0.132. It is noteworthy that PEO cannot be used for a UV-method as it does not absorb in the UV region.

   BSA (SIGMA A0281): its intrinsic viscosity is 4 mL/g, its molecular mass is 66,430 g/mol, its molar extinction coefficient (ε) is 45,000 M^{-1} cm^{-1}, dn/dc is 0.185 and dA/dc is 0.667 L/g cm

   The refractive index increment, dn/dc, is the slope of the plot of the total scattered light (dn) as a function of sample concentrations (dc). dA/dc is easily computed as $dA/dc = \varepsilon/M$.

3. The instrument described in the present review was purchased from Malvern (Malvern, UK). Other companies, e.g., Wyatt Technology, sell similar apparatus.

4. Buffers must be degassed before use to prevent air bubbles from becoming trapped in the pump, column, or detectors.

5. TDA power supply should not be switched off; it controls the temperature of the oven that must be comprised between 15 and 80 °C and be at least 3 °C higher than the room (or refrigerated cabinet) temperature to ensure thermal stability.

   The GPCmax system must be switched on in order to let the degasser pump functioning. Degasser pressure must be comprised between 0.7 and 0.8 mbar.

6. With flow, the pressure on DP and IP detectors should not exceed 2.5 V (2.5 kPa) and 100 mV (100 kPa), respectively.

7. The flow rate must be 0.1 mL/min for the transition from 20 % ethanol to water. Flow rate during an injection is comprised between 0.2 and 0.5 mL/min. The flow of the pump can be adjusted at any time during operation by pressing the Pump ON key and type the desired flow in the box. In any case, it should not exceed 1.5 mL/min when the viscometer is connected online (the IP detector limit pressure is 100 mV, which corresponds to 100 kPa).

8. Pressure must be limited to that of the column, including the backpressure of the system. If the pressure drops under or exceeds the programmed pressure values, the pump stops automatically.

9. Purges are performed in order to obtain optimum, flat baselines. Purge time can be settled from the software. Do not make them too long (from 2 to 5 min) to avoid damage of purge valves.

10. Baseline detector responses are approximately: RI (variable), UV: 20 mV; RALS: 30–50 mV; LALS: 300–500 mV; IP: 5–6 mV (at 0.1 mL/min) to 25–30 mV (at 0.5 mL/min). It is noteworthy that a lower value can indicate a leakage; DP: 20–100 mV, as a function of the flow rate. Baseline noise should be around: IP: 0.02–0.04; DP, RALS, RI, and UV: 0.1; and LALS: 0.5 mV.

11. It is better to equilibrate the column with water and then with buffer before connecting it to the GPCmax system in order to avoid injection in the detectors of particles that may come out from the column.

12. The IP measurement is treated as an "absolute" measurement by the data system, unlike the others signals, which are baseline-corrected. For this reason, the IP must never be zeroed with the flow on.

13. Each vial is accessible to a needle, which can take a defined volume of sample and inject it onto the column for analysis. Put in the vial a higher volume than the one required because you have to take into account the volume (1) lost in the capillar during loop loading (typically, 50 µL) and (2) required to fill the space between the bottom of the vial and the extremity of the needle (this volume is dependent on the shape of the vial used).

14. At least three injections of different concentrations (0.5; 1; 2 g/L) or different volumes (100; 150; 200 µL) must be performed for each sample. Usually, two or three injections of each standard are performed at the same volume (150 µL) and concentration (2 and 4 g/L for BSA and PEO, respectively) to be sure of the signal and baseline stability, and the reproducibility of the data. If required, injections of variables volumes (50–200 µL) can be performed in order to determine the dn/dc of the samples (see Note 2).

15. Calibration constants values should range as following: RI and UV: 1×10^6 to 20×10^6; RALS and LALS: 1×10^8 to 20×10^8; DP: 0.7–1.3.

16. The Svedberg equation is given by $s = M(1 - \rho\bar{v})/(6\pi\eta N_A R_H)$, where s is the sedimentation coefficient (S), M the molecular mass (g/mol), $\bar{v}$ the partial specific volume (mL/g), and R_H the hydrodynamic radius of the macromolecule (cm). ρ and η are the solvent density (g/mL) and viscosity (poise: g/cm s), respectively.

17. The intrinsic viscosity relation is given by $[\eta] = v(\bar{v} + \delta/\rho)$, where $[\eta]$ is the intrinsic viscosity (mL/g), v is the viscosity increment (unitless), $\bar{v}$ the partial specific volume (mL/g),

and δ the hydration of the macromolecule (g/g). ρ is the solvent density (g/mL).

18. Here is a theoretical example describing how to access to hydrodynamic parameters from M, $[\eta]$ and R_H (Fig. 4). We consider a protein with a molecular mass of 50,000 g/mol, an intrinsic viscosity of 12 mL/g, a partial specific volume of 0.73 mL/g, a sedimentation coefficient of 3 S (3×10^{-13} s) and a translational diffusion coefficient of 5.6×10^{-7} cm^2/s. The macromolecule is studied at 25 °C in Hepes 20 mM, NaCl 150 mM pH 7.4 of density 1.00382 g/mL and viscosity 0.00908 poise (or 0.00908 g/cm s).

We first calculate the hydrodynamic radius (R_H) using the Svedberg equation:

$$R_H = M(1 - \rho\bar{v})/(6\pi\eta N_A s)$$

with M in g/mol, ρ in g/mL, $\bar{v}$ in mL/g, η in poise and s in seconds.

$$R_H = 50,000(1 - 1.00382 \times 0.73)/(6\pi \times 0.00908 \times N_A \times 3 \times 10^{-13})$$

$$R_H = 4.32 \times 10^{-7} \text{ cm}$$

Alternatively, the hydrodynamic radius R_H can be calculated from the translational diffusion coefficient measured by QELS using the Stock Einstein relation

$$R_H = (k_B T)/(6\pi\eta D t)$$

with $k_B = 1.38 \times 10^{-16}$ g cm^2/s^{2}K, T in Kelvin, η in poise and $D t$ in cm^2/s.

$$R_H = (1.38 \times 10^{-16} \times 298)/(6\pi \times 0.00908 \times 5.6 \times 10^{-7})$$

$$R_H = 4.32 \times 10^{-7} \text{ cm}$$

From the R_H, we calculate the hydrodynamic volume (V_H) of the macromolecule according to

$$V_H = 4\pi R_H^3/3$$

with R_H in cm.

$$V_H = 4\pi(4.32 \times 10^{-7})^3/3$$
$$V_H = 3.38 \times 10^{-19} \text{ cm}^3$$

- According to the first approach, the shape factor (the viscosity increment, v) is calculated using the Einstein viscosity relation:

$$v = M[\eta]/V_H N_A$$

with M in g/mol, $[\eta]$ in mL/g, and V_H in cm^3.

$$v = 50,000 \times 12/3.38 \times 10^{-19} N_A$$

$$v = 2.95$$

The viscosity increment provides an estimation of the ratio of the lengths of the semi-axes a and b of an ellipsoid of revolution according to the polynomial equations described by Harding and Colfen (9). A viscosity increment of 2.95 gives an axial ratio of 2.07. We can then calculate the semi-axes a and b for an equivalent prolate or oblate spheroid.

For a prolate ellipsoid (defined by a, b, b semi-axes):

$a = R_H \times \vartheta^{2/3}$	$b = a/\vartheta$
$a = 4.32 \times 2^{2/3}$	$b = 7/2$
$a = 7$ nm	$b = 3.5$ nm

For an oblate ellipsoid (defined by a, a, b semi-axes):

$a = R_H \times \vartheta^{1/3}$	$b = a/\vartheta$
$a = 4.32 \times 2^{1/3}$	$b = 5.4/2$
$a = 5.4$ nm	$b = 2.7$ nm

Finally, from the intrinsic viscosity relation, the hydration parameter is extracted:

$$\delta = (([\eta]/v) - \bar{v})\rho$$

with $[\eta]$ in mL/g, $\bar{v}$ in mL/g, and ρ in g/mL. v is unitless.

$$\delta = ((12/2.95) - 0.73)1.00382$$

$$\delta = 3.35 \text{ g/g}$$

Altogether, these data indicate that the protein is elongated with an axial ratio of 2 and displays a high hydration.

• The second approach is used to calculate first the hydration parameter according to

$$\delta = ((V_H N_A)/M) - \bar{v}$$

with V_H in cm^3, M in g/mol, and $\bar{v}$ in mL/g.

$$\delta = ((3.38 \times 10^{-19} \times N_A)/50000) - 0.73$$

$$\delta = 3.34 \text{ g/g}$$

The viscosity increment is then determined using

$$v = [\eta]/(\bar{v} + \delta/\rho)$$

with $[\eta]$ in mL/g, $\bar{v}$ in mL/g, δ in g/g, and ρ in g/mL.

$$v = 12/(0.73 + 3.34/1.00382)$$

$$v = 2.96$$

Then, the semi-axes a and b of an ellipsoid of revolution are estimated as described above.

- Finally, the third approach is used to calculate the Perrin hydrodynamic function P ($P = (f/f_0)/\left[((\delta/\bar{v}\rho) + 1)^{1/3}\right]$) as an alternative estimation of the semi-axial ratio using the hydration previously calculated by the first and second procedures and the frictional ratio determined, according to

$$f/f_0 = R_H/R_0$$

R_0 is the hydrodynamic radius of an anhydrous and spherical molecule with equivalent M and $\bar{v}$ and is defined as

$$R_0 = \left(\frac{3M\bar{v}}{4\pi N_A}\right)^{1/3}$$

with M in g/mol and $\bar{v}$ in mL/g.

19. The Einstein viscosity relation is given by $M[\eta] = v V_H N_A$, where M is the molecular mass (g/mol), $[\eta]$ the intrinsic viscosity (mL/g), v the viscosity increment, and V_H the hydrodynamic volume of the macromolecule (cm^3).

20. The partial specific volume can be either calculated by AUC equilibrium measurement or estimated with the SEDNTERP software. An approximation of the $\bar{v}$ is enough for the calculation of the hydration parameter. Indeed, $\bar{v}$ commonly ranges between 0.69 and 0.75 mL/g, which makes little changes to hydration that can varies between 0.2 and 0.5 g/g for folded proteins, and from 1 to 10 g/g for partially and intrinsically disordered proteins. These values are given as rough indications.

Acknowledgements

This work was supported by the Institut Pasteur (Grant PTR374), the Centre National de la Recherche Scientifique (CNRS UMR 3528), and the Agence Nationale de la Recherche, programme Jeunes Chercheurs (ANR, grant ANR-09-JCJC-0012).

References

1. Chenal A, Guijarro JI, Raynal B, Delepierre M, Ladant D (2009) RTX calcium binding motifs are intrinsically disordered in the absence of calcium: implication for protein secretion. J Biol Chem 284:1781–1789

2. Sotomayor Perez AC, Karst JC, Davi M, Guijarro JI, Ladant D, Chenal A (2010) Characterization of the regions involved in the calcium-induced folding of the intrinsically disordered RTX motifs from the bordetella

pertussis adenylate cyclase toxin. J Mol Biol 397:534–549

3. Sotomayor-Pérez AC, Ladant D, Chenal A (2011) Calcium-induced folding of intrinsically disordered repeat-in-toxin (RTX) motifs via changes of protein charges and oligomerization states. J Biol Chem 286:16997–17004.

4. Chenal A, Vendrely C, Vitrac H, Karst JC, Gonneaud A, Blanchet CE, Pichard S, Garcia E, Salin B, Catty P, Gillet D, Hussy N, Marquette C, Almunia C, Forge V (2011) Amyloid fibrils formed by the programmed cell death regulator Bcl-xL. J Mol Biol 415:584–599.

5. Chenal A, Vendrely C, Vitrac H, Karst JC, Gonneaud A, Blanchet CE, Pichard S, Garcia E, Salin B, Catty P, Gilltet D, Hussy N, Marquette C, Almunia C, Gorge V (2011) Amyloid fibrils formed by the programmed cell death regulator Bcl-xL. J Mol Biol 415:584–599

6. Bourdeau RW, Malito E, Chenal A, Bishop BL, Musch MW, Villereal ML, Chang EB, Mosser EM, Rest RF, Tang WJ (2009) Cellular functions and X-ray structure of anthrolysin O, a cholesterol-dependent cytolysin secreted by Bacillus anthracis. J Biol Chem 284:14645–14656

7. Simha R (1940) The influence of Brownian movement on the viscosity of solutions. J Phys Chem 44:25–34

8. Harding SE (1997) The intrinsic viscosity of biological macromolecules. Progress in measurement, interpretation and application to structure in dilute solution. Prog Biophys Mol Biol 68:207–262

9. Harding SE, Colfen H (1995) Inversion formulae for ellipsoid of revolution macromolecular shape functions. Anal Biochem 228:131–142

10. Harding SE, Horton JC, Colfen H (1997) The ELLIPS suite of macromolecular conformation algorithms. Eur Biophys J 25:347–359

11. Perrin F (1936) Mouvement brownien d'un ellipsoïde (II). Rotation libre et dépolarisation des fluorescences. Translation et diffusion de molécules ellipsoïdales. J Phys Rad 7:1–11

12. Squire PG, Himmel ME (1979) Hydrodynamics and protein hydration. Arch Biochem Biophys 196:165–177

13. Chenal A, Karst JC, Perez AC, Wozniak AK, Baron B, England P, Ladant D (2010) Calcium-induced folding and stabilization of the intrinsically disordered RTX domain of the CyaA toxin. Biophys J 99:3744–3753

Chapter 11

Size-Exclusion Chromatography in Structural Analysis of Intrinsically Disordered Proteins

Vladimir N. Uversky

Abstract

Gel-filtration chromatography, also known as size-exclusion chromatography (SEC) or gel-permeation chromatography, is a useful tool for structural and conformational analyses of intrinsically disordered proteins (IDPs). SEC can be utilized for the estimation of the hydrodynamic dimensions of a given IDP, for evaluation of the association state, for the analysis of IDP interactions with binding partners, and for the induced folding studies. It also can be used to physically separate IDP conformers based on their hydrodynamic dimensions, thus providing a unique possibility for the independent analysis of their physicochemical properties.

Key words: Gel-filtration chromatography, Size-exclusion chromatography, Gel-permeation chromatography, Hydrodynamic dimensions, Molecular mass, Compaction, Induced folding, Intrinsically disordered protein, Coil-like, Pre-molten globule, Molten globule

1. Introduction

Due to a wide range of physical methods utilized for separation and analysis of complex mixtures, chromatography deserves a unique position among various analytical techniques of modern biochemistry, biophysics, and molecular biology. In chromatography, the components to be separated are partitioned between a stationary phase (which is typically packed into a column) and a mobile phase (that usually contains dissolved sample and percolates through the stationary phase in a definite direction). These phases are chosen based on the different capabilities of components of the sample to interact with the stationary and mobile phases. The ability of the stationary phase to differently interact with the components of the mobile phase determines their retention and separation, since a

Vladimir N. Uversky and A. Keith Dunker (eds.), *Intrinsically Disordered Protein Analysis: Volume 2, Methods and Experimental Tools*, Methods in Molecular Biology, vol. 896, DOI 10.1007/978-1-4614-3704-8_11, © Springer Science+Business Media New York 2012

component with higher affinity to the stationary phase will travel longer through the column than a component with lower affinity. Importantly, various retention mechanisms based on reversible physical interactions can be utilized (e.g., adsorption at a surface, absorption in an immobilized solvent layer, electrostatic interactions, etc.). Furthermore, more than one type of interaction may contribute simultaneously to the separation mechanism, and various means may be employed to achieve the reversibility of the component interaction with the stationary phase. All this defines the uniqueness of chromatography as an exceptionally useful and pliable analytical tool with almost endless applications.

One of the most commonly used chromatography types is the size-exclusion chromatography (SEC), also known as molecular exclusion chromatography, gel-filtration or gel-permeation chromatography (GFC or GPC), which represents a unique laboratory tool for separation of biomolecules, including proteins, based on their hydrodynamic dimensions. Separation in SEC is achieved via the use of the porous beads with a well-defined range of pore sizes as the stationary phase. Therefore, the separation mechanism of gel-filtration is very gentle, nonadsorptive, and is typically independent of the eluent system used.

In SEC, molecules in the mobile phase pass by a number of these porous beads while flowing through the column. Molecules whose hydrodynamic dimensions are smaller than a particular limit can fit inside all the pores in the beads. They will be drawn in pores by the force of diffusion, where they will stay for a short time and then will move out. These molecules are totally included as they have total access to all the mobile phase inside and between the beads. They have the largest retention and therefore will elute last during the gel filtration separation. On the other hand, large molecules that are too massive to fit inside any pore will have access only to the mobile phase between the beads. These molecules are excluded as they just follow the solvent flow and reach the end of the column before molecules with smaller size. Finally, molecules of intermediate size are partially included as they can fit inside some but not all of the pores in the beads and therefore possess an intermediate retention and elute between the large ("excluded") and small ("totally included") molecules. Within the fractionation range chosen, molecules are eluted in order of decreasing size. It is important to remember that all the molecules larger than all the pores in a matrix will elute together regardless of their size. Likewise, any molecules that can fit into all the pores in the beads will elute at the same time.

In the protein field, the most frequent uses of SEC are separation of proteins based on their size and estimation of their molecular masses. Formally, SEC is a separation technique based on hydrodynamic radius (see below); however, for similarly shaped molecules hydrodynamic radius is proportional to molecular mass. Therefore, we can talk about SEC as a mass-based separation, even though this is not strictly true.

The hydrodynamic volume is one of the most important and fundamental structural parameters of a protein molecule. Hydrodynamic volume changes dramatically during the denaturation and unfolding of a globular protein (1–4), and the evaluation of the protein hydrodynamic dimensions (compact, extended, or partially swollen) is an absolute prerequisite for an accurate classification of a protein conformation. Many experimental techniques were elaborated to estimate the protein hydrodynamic dimensions, including viscometry, sedimentation, dynamic light scattering, small angle X-ray scattering, small angle neutron scattering, and so on.

Although many of the listed above approaches are based on the well-developed theories, all of them have some difficulties and pitfalls. Some hydrodynamic techniques require large protein quantities, while others use complex and expensive equipment and sophisticated approaches for data processing. All techniques based on scattering (dynamic light scattering, small angle X-ray scattering, small angle neutron scattering) require very homogeneous samples, as presence of even small fraction of the aggregated material is known to dramatically affect the scattering profile, making interpretation of data difficult. Furthermore, practically all these methods consume a lot of time for sample preparation and precise measurements and have some limitations in the variation of experimental conditions. Application of SEC allows researchers to overcome many of these experimental difficulties. For example, the protein concentration can be decreased up to 0.001 mg/ml by using the 226 nm filter in the optical registration system or even to the nanogram level if the radiolabeled proteins are studied.

In comparison with the classical hydrodynamic methods such as viscometry and sedimentation, the use of SEC as a technique for the macromolecular dimension evaluation is a relatively novel approach. In 1959, it was recognized that the SEC-based fractionation of macromolecules is determined by their molecular sizes (5). This brought the molecular sieve hypothesis of the gel-forming polymer action to the existence. SEC now is considered as a general separation technique where size and shape of molecules are the prime separation parameters (6). Therefore, the elution behavior of proteins on SEC column is determined by their Stokes radii rather than by molecular masses (2, 3, 7–14).

Currently, SEC-HPLC (FPLC) is commonly used as a convenient tool for the estimation of molecular dimensions and analysis of their changes under the variety of conditions. For example, the processes of globular protein denaturation and unfolding are often analyzed by SEC either in terms of changes in the retention time which correlate with the changes of Stokes radii of the protein conformers (2, 3, 13, 15–18) or by following the appearance of a new elution peak (2, 3, 13, 15, 16, 19–23). Hydrodynamic dimensions of intrinsically disordered proteins are also studied by SEC (24–28). This chapter describes the peculiarities of the SEC application for evaluation of protein hydrodynamic dimensions, for conformational classification of IDPs, for separation of different conformational states of a protein by their dimensions, and for the independent structural characterization of these separated conformers.

2. Materials

1. A gel filtration column of appropriate fractionation range for the samples of interest.

2. A gel filtration buffer that is compatible with both the column and the sample (see Note 1).

3. Calibration standards, i.e., a set of proteins with known hydrodynamic dimensions. Possible standards are thyroglobulin (Pharmacia AB, gel filtration calibration kit); ferritin, catalase, aldolase, bovine serum albumin, ovalbumin from hen egg, chymotrypsinogen A, and cytochrome *c* (calibration proteins 1 for gel chromatography, Combithek, Boehringer Mannheim GmbH), carbonic anhydrase II, myoglobin, β-lactamaze, α-lactalbumin, lysozyme, etc.

4. Proteins with known hydrodynamic dimensions in their folded and completely unfolded states can be used as standards for the SEC column calibration (see Note 2).

5. Blue dextran (Sigma) and acetone should also used in column calibration.

6. The protein solution, in volume and concentration required for SEC measurements (see Note 3).

7. Concentrated urea and guanidinium hydrochloride (GdmHCl) solutions. The denatured concentration in solution is measured by the refractive index.

8. Any chromatographic equipment, e.g., FPLC equipment (Pharmacia, Uppsala, Sweden). To analyze the diluted protein solutions absorption detector (e.g., 2158 Uvicord SD (LKB)) should be equipped with the 226-nm filter or set to measure absorption in the peptide bond region ~220 nm.

3. Methods

3.1. Estimation of the Hydrodynamic Dimensions by SEC

3.1.1. Calibration of the SEC Column: Theoretical Background

All the chromatographic materials suitable for gel-filtration are characterized by specific "exclusion limits." This parameter defines an approximate upper limit for the size of molecules that can be separated using a given column matrix. Gel-filtration columns are characterized by two parameters, the void volume (V_O) and the total volume (V_T). V_O is essentially the volume of the mobile phase between the beads of the chromatographic medium. Molecules larger than the exclusion limit, i.e., excluded molecules, elute in the V_O. V_T is the volume of all of the liquid within the column (i.e., both within the porous beads, as well as between them). The smallest, or included, molecules appear in the V_T.

Calibration of a gel-filtration column represents a crucial primary step in obtaining the quantitative information on the protein molecular dimensions by SEC. Column calibration implies the determination of a correlation between the parameters characterizing the column permeation properties (or the retention capability) and the protein hydrodynamic dimensions. In SEC, the retention of solute molecules by the column depends on their continuous exchange between the mobile phase and the stagnant mobile phase within the pores of the column matrix. This exchange is an equilibrium entropy-controlled process, as enthalpic processes such as adsorption are undesirable in SEC. Thus, the SEC retention volume (V_r) is expressed by the following equation (29):

$$V_r = V_O + V_P K_{SEC} + V_S K_{LC},\qquad(1)$$

where V_O is the void volume, V_P is the pore volume, whereas V_S is the stationary phase volume, K_{SEC} corresponds to the SEC solute distribution coefficient, and K_{LC} is the coefficient characterizing the liquid chromatography solute distribution. The ideal SEC retention has to be governed only by entropic contributions (i.e., it has to exclude both specific and nonspecific interactions of solute molecules with the column matrix). Therefore, the column packing material–eluent combination should be chosen such that K_{LC} is minimized to be as close to zero as possible (29).

The value of K_{SEC} for peaks eluting in the region resolvable by the SEC column is $0 < K_{SEC} < 1$ (see Note 4). The retention of a given molecule by a SEC column can be described by the column partition coefficient, K_d, which is determined from the elution profiles by the following equation:

$$K_d = \frac{V_{el} - V_O}{V_T - V_O},\qquad(2)$$

where V_O and V_T are void and total solvent-accessible column volumes, respectively, whereas V_{el} is the elution (or retention) volume of a given molecule under given conditions.

3.1.2. Calibration of the SEC Column: Experimental Steps

The dependence of the retention volume of a solute on its hydrodynamic dimension represents the SEC calibration curve.

1. First, one has to make sure that the void volume (V_O) and the total volume (V_T) of the column are independent of solution pH, denaturants in wide concentration intervals, and temperature. Since in our experiments, the value of the column void volume is based on the elution volume of blue dextran, and the value of the total solvent-accessible column volume is based on the elution volume of acetone, blue dextran and acetone are injected into the column under the variety of solvent/environment conditions to be used in the IDP analysis.

2. A starting point of column calibration is an injection of a series of well-characterized SEC standards, proteins with known hydrodynamic dimensions, followed by the determination of the corresponding retention volumes (see Note 5).

3. The information from the step 2 is then used for the conversion of the retention volume axis in SEC to a hydrodynamic dimension axis (that is, calibration), which can be accomplished in a number of ways. A simple column calibration curve is a direct R_S versus K_d plot (10, 30). Alternatively, when the column permeation properties are independent of the experimental conditions (i.e., when $V_T - V_O = \textit{const}$ for all conditions and buffers used), the simplified calibration procedure, plotting the migration rate ($1,000/V_{el}$) versus R_S, can be used (13) (see Note 6).

3.1.3. Measuring Hydrodynamic Dimensions of Proteins by SEC: Experimental Steps

Thorough SEC-based analysis of several proteins whose hydrodynamic dimensions in different conformational states were estimated by other hydrodynamic techniques revealed that the SEC-determined R_S values were in good agreement with those obtained by traditional hydrodynamic methods such as viscometry, sedimentation, and dynamic light scattering (2, 3, 13). The accuracy of the SEC measurements is typically high enough to obtain the reliable information on the hydrodynamic dimensions of a protein in different conformational states. In fact, even molten globules, whose hydrodynamic dimensions are very close to the respective values of the globular ordered proteins, were reliably discriminated form the corresponding folded proteins (2, 3).

SEC represents a very useful tool to follow changes in the hydrodynamic dimensions accompanying denaturation and unfolding of globular proteins (2, 3, 13, 15). Importantly, it has been established that the unfolding curve retrieved for a given protein by SEC coincides with the unfolding curves measured for this protein by other techniques. This clearly indicated that the reliable R_S measurements can be done not only under the conditions preceding and following the conformational transition but also within the transition region (2, 3, 13, 15).

1. Inject the protein sample to the calibrated column and determine the corresponding retention volume.

2. Using calibration curve from steps in Subheading 3.1.2 determine the Stokes radius of a given protein.

3. In order to obtain more accurate data, the above steps 1 and 2 should be repeated three to five times.

3.2. Structural Classification of IDPs Based on the Results of SEC Analysis

3.2.1. Theoretical Considerations

Molecular density, and hence hydrodynamic dimensions, is one of the most unambiguous characteristics of a polymer molecule. Additional knowledge can be gained via the analysis of the molecular mass dependence of the molecular density for a polymer in different conformational states. In fact, the density of a globule is expected to be independent of the chain length, whereas the density of a partially collapsed or swelled macromolecules depends on both the chain length, and therefore on its molecular weight M, and on the non-specific interactions of the monomer units with the solvent (31). Keeping this in mind, data retrieved by SEC for several proteins in different conformational states can be utilized for finding a potential correlation between the hydrodynamic dimensions of a protein molecule in a variety of conformational states and the length of polypeptide chain (4, 32–37). The analyzed proteins were grouped in the following classes: native globular proteins with nearly spherical shapes; equilibrium molten globules and equilibrium pre-molten globule states in the presence of strong denaturants; denaturant-unfolded proteins without cross-links; and natively unfolded proteins. Figure 1 represents the results of this analysis and clearly shows that in all cases studied an excellent correlation between the apparent molecular density (determined as $\rho = M/(4\pi R_S{}^3/3)$, where M is a molecular mass and R_S is a hydrodynamic radius of a given protein) and molecular mass was observed. Thus, regardless of the differences in the amino acid sequences and biological functions, protein molecules behave as polymer homologues in a number of conformational states (4, 32–37).

This analysis gave rise to a set of the standard equations for a polypeptide chain in a number of conformational states (37):

$$\log(R_S^N) = -(0.204 \pm 0.023) + (0.357 \pm 0.005) \cdot \log(M), \quad (3)$$

$$\log(R_S^{MG}) = -(0.053 \pm 0.094) + (0.334 \pm 0.021) \cdot \log(M), \quad (4)$$

$$\log(R_S^{PMG}) = -(0.21 \pm 0.18) + (0.392 \pm 0.041) \cdot \log(M), \quad (5)$$

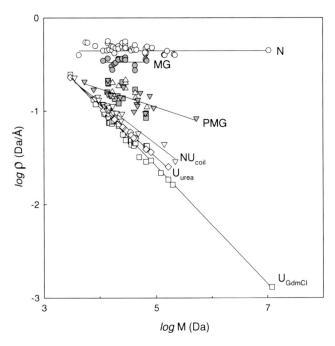

Fig. 1. Variation of the density of protein molecules, ρ, with protein molecular weight, M, for a number of thermodynamically stable conformational states: N, ordered globular protein; MG, molten globule; PMG, partially folded and partially collapsed conformations (native pre-molten globules are shown as *reversed triangles*; proteins with intact disulfate bridges in 8 M urea or 6 M GdmHCl are shown as *squares*; intermediates accumulated during the unfolding by urea or GdmCl are shown by *circles*); NU$_{coil}$, native coil-like proteins under the physiological conditions; U$_{urea}$, unfolded in 8 M urea (proteins without cross-links or with reduced cross-links); and U$_{GdmCl}$, unfolded in 6 M GdmHCl (proteins without cross-links or with reduced cross-links). *Lines* represent the best fits. Modified from ref. 33.

$$\log(R_S^{U(\text{urea})}) = -(0.649 \pm 0.016) + (0.521 \pm 0.004) \cdot \log(M),$$
$$(6)$$

$$\log(R_S^{U(\text{GdmCl})}) = -(0.723 \pm 0.033) + (0.543 \pm 0.007) \cdot \log(M),$$
$$(7)$$

$$\log(R_S^{NU(\text{coil})}) = -(0.551 \pm 0.032) + (0.493 \pm 0.008) \cdot \log(M),$$
$$(8)$$

$$\log(R_S^{NU(\text{PMG})}) = -(0.239 \pm 0.055) + (0.403 \pm 0.012) \cdot \log(M),$$
$$(9)$$

where N, MG, PMG, U(urea), and U(GdmCl) correspond to the native, molten globule, pre-molten globule, urea-, and GdmCl-unfolded globular proteins, respectively, whereas NU(coil) and

NU(PMG) correspond to native coil-like and native pre-molten globule-like proteins, respectively (see Note 7).

3.2.2. Knowing IDP Conformation from SEC Measurements

As it follows from discussion above, SEC is very useful in ascertaining the degree of compactness of a protein, and can distinguish between partially and fully unfolded states, since an increase in the hydrodynamic volume is associated with unfolding. Transformation of a typical globular protein into a molten globule state results in a ~15–20% increase in its hydrodynamic radius (2–4, 35, 38, 39). The relative increase in hydrodynamic volume of less folded intermediates is even larger (4, 21, 22, 33, 35, 39). Figure 1 shows that different protein conformations possess very different molecular mass dependencies of their hydrodynamic radii, R_S (4, 33–36). Therefore, equilibrium conformation of any given IDP (coil-like, PMG-like, or molten globule-like) can easily be discriminated by the degree of compactness of the polypeptide chain. Thus, based on its hydrodynamic dimensions evaluated by SEC and the unique molecular mass dependencies of the Stokes radii described by a set of equations above, an IDP can be assigned to one of the structural classes.

1. To this end, using a set of Eqs. 3–9 (see above) one should calculate the expected Stokes radii for a set of standard conformations (folded, molten globular, pre-molten globular, urea-, and GdmCl-unfolded globular proteins, and native coil-like and native pre-molten globule-like IDP) of a hypothetical protein with the appropriate molecular mass.

2. Compare the calculated values with the Stokes radius of a given IDP evaluated by SEC.

3.2.3. Illustrative Examples of Structural Classification of IDPs Based on the SEC Experiments

Examples below illustrate the usefulness of this approach for structural classification of several IDPs. The hydrodynamic properties of the members of the synuclein family, α-, β-, and γ-synucleins, were studied under the variety of experimental conditions by SEC (26). This analysis revealed that under the conditions of neutral pH β-synuclein was slightly more extended and possessed the hydrodynamic dimensions typical of a completely unfolded polypeptide chain, whereas α- and γ-synucleins were more compact than expected for a random coil (26). This conclusion followed from the comparison of the measured Stokes radius values with those calculated for a completely unfolded polypeptide chain of the appropriate molecular mass. In the case of β-synuclein, the experimentally determined value (33.9 ± 0.4 Å) perfectly matched the calculated one (34.1 Å). However, the Stokes radii measured for α- (31.8 ± 0.4 Å) and γ-synucleins (30.4 ± 0.4 Å) were notably lower than the corresponding calculated values (34.3 and 32.8 Å, respectively) (24). The conclusion on the partially collapsed form in α-synuclein was further confirmed by measurement of its R_S in the presence of 8 M urea, where the protein behaved as a random coil ($R_S = 34.5 \pm 0.4$ Å) (24). Furthermore, SEC analysis revealed that the decrease in pH was accompanied by the formation of partially folded conformation in all three synucleins as evidenced by the substantial decrease in their hydrodynamic dimensions ($R_S = 27.9 \pm 0.4$, 27.5 ± 0.4, and 26.5 ± 0.4 Å for α-, β-, and γ-synuclein, respectively). As these data were in perfect agreement with the values calculated from Eq. 9, it has been concluded that at acidic pH all three proteins formed pre-molten globule-like conformation (26).

SEC analysis of another IDP, inhibitory γ subunit of the cGMP phosphodiesterase (PDEγ) revealed that the hydrodynamic dimensions of PDEγ at neutral pH in the absence of 8 M urea were close to those measured in the presence of 8 M urea (28). Furthermore, the R_S value determined for native PDEγ ($R_S = 24.8 \pm 0.8$ Å) was very close to R_S calculated for a native coil with a molecular mass of 9,669 Da ($R_S = 25.9$ Å), suggesting that PDEγ should be classified as a native coil (28).

Finally, the results of the SEC analysis of the C-terminal domain of chicken gizzard caldesmon (CaD136, residues 636–771) agreed well with the data of the far-UV CD, SAXS, and intrinsic fluorescence and showed that this domain was essentially unfolded under the conditions of neutral pH (40). Here, the hydrodynamic dimensions of CaD136 were relatively close to those measured in the presence of 6 M GdmCl ($R_S = 28.1 \pm 0.8$ and 35.3 ± 0.8 Å, respectively), confirming the fact that CaD136 is essentially unfolded even in the absence of denaturant. Comparison of these measured values with R_S calculated using Eqs. 4, 7–9 for a protein with a molecular mass of 14,514 Da (19.1, 21.7, 34.4, 31.7, and 27.4 Å for N, MG, U, NU(coil), and NU(PMG), respectively) suggested that CaD136 belonged to the native pre-molten globule class (40).

3.3. Assessing the Induced Folding and Association of IDPs by SEC

3.3.1. Analysis of Induces Folding

Presented above data for the members of the synuclein family illustrate that SEC is a useful tool for the evaluation of the IDP partial folding induced by changes in the environment. Similar results on the pH-induced gaining of partially folded conformation were obtained for another typical IDP, prothymosin α (27). The hydrodynamic dimensions of this protein at neutral pH were close to those measured in the presence of 8 M urea (31.4 ± 0.3 and 32.8 ± 0.3 Å, respectively). The small difference between the two values was explained by swelling of the unfolded polypeptide chain in a good solvent. Both values were virtually indistinguishable from R_S, calculated for the completely unfolded protein with a molecular mass of 12.21 kDa (31.3 Å). A decrease in pH led to a pronounced decrease of the prothymosin α hydrodynamic dimension ($R_S = 24.9 \pm 0.3$ Å) (27). It should be emphasized that this R_S value determined at pH 2.5 was still far from that expected for a globular protein of 12 kDa, but correlated well with the dimensions of the pre-molten globule

(25.5 Å). Although we considered here only the data on the pH-induced partial folding of IDPs, SEC can be utilized for the analysis of folding induced in IDPs by any other environmental factors.

The situation when IDP partially folds while oligomerizing is more complex, as here we are dealing with two opposite effects—decrease in the hydrodynamic dimensions induced by folding and increase in the molecular mass (and consequently hydrodynamic volume) caused by oligomerization. However, even in this case very useful information can be extracted both on the polypeptide conformation and on its oligomerization state. For a protein with a molecular mass of M, this information can be retrieved using Eqs. 3, 4, 7–9 a set of molecular masses $n \times M$, with $n = 1, 2, 3, \ldots$, N being the oligomerization stage.

Analysis of the heat-induced dimerization of α-synuclein represents an illustrative example of this approach. Incubation of α-synuclein at high temperature induced partially folded conformation (24). This structural transformation was completely reversible, when the heat treatment was transient. It has been hypothesized that if the partially folded conformation serves as an intermediate of the fibril assembly, then populating this structure for a longer period of time should induce the self-assembly and might trap the structure in oligomeric forms (41). To test this hypothesis and see if the sustained heat treatment, thereby sustained partially folded structure, can stabilize the structure and initiate the olig-omerization, purified wild type human recombinant α-synuclein was incubated at different temperatures for up to three days. This incubation resulted in a temperature-dependent, progressive aggregation followed by the gel electrophoretic analysis. The incubation at 65°C showed small oligomers (mostly dimers) at day 1, and at day 3, larger aggregates were detected with increased amount of small oligomers (dimers). Whereas the olig-omerization at 50°C was slower, but apparent at day 3, no oligomers of any size were detectable at 37°C or room tempera-ture for up to 3 days (41).

At the next stage, hydrodynamic properties of different asso-ciated forms of α-synuclein have been analyzed by SEC (41). This analysis revealed that the initial conformation of α-synuclein was essentially unfolded polypeptide chain with the Stokes radius R_S = 31.3 Å, whereas trapped dimeric form, being characterized by $R_S = 36.3$ Å, had to be comprised of more compact protein mole-cules. These conclusions followed from the comparison of the measured Stokes radius, R_S, values with those calculated for native coil or native pre-molten globule with a molecular mass of 14,460 kDa. In the case of initial α-synuclein conformation, the experimentally determined value perfectly matched the value calcu-lated for the native coil of 14,460 Da ($R_S = 31.6$ Å). However, R_S measured for the trapped conformation coincided with expected dimensions of the pre-molten globule protein with molecular mass of 28,920 Da ($R_S = 36.2$ Å). Therefore, incubation of α-synuclein at elevated temperatures for prolonged periods of time may trap the stable dimers comprised of partially folded pre-molten globule-like intermediates (41). These experiments suggested that partially folded PMG-like conformation of α-synuclein was stabilized as the protein underwent a highly selective self-assembly process during prolonged incubation at elevated temperatures (41).

3.4. SEC-Based Physical Separation of IDP Conformers and Their Independent Analysis

An exceptional advantage of SEC, in comparison with the vast majority of traditional structural methods, is its capability of physical separation of protein conformers, which are different in their hydrodynamic dimensions. Although such separation takes place only under the particular conditions (e.g., under the conditions which are favoring slow conformational exchange between these species or upon the formation of stable oligomeric forms), this property of SEC allows one to perform an independent investigations of different physical properties of compact and less compact or monomeric and oligomeric conformations. Various traditional spectroscopic techniques, being combined with the chromatographic facilities, can be used for such structural characterization.

This property of SEC was successfully applied for studying the formation of baicalein-stabilized oligomers of α-synuclein. Baicalein is the main component of a traditional Chinese herbal medicine *Scutellaria baicalensis* and has multiple biological activities including antiallergic, anticarcinogenic, and anti-HIV properties (42–45). Furthermore, baicalein was shown to markedly inhibit α-synuclein fibrillation in vitro (46, 47). This inhibition occurred via inducing the specific oligomerization. This ability of baicalein to effectively induce oligomerization of α-synuclein was shown using SEC-HPLC. After incubation for 2 days with 100 μM baicalein, the HPLC profile of α-synuclein showed a new peak with a retention time of 11.5 min, indicating formation of the stable oligomeric species (47). The peak corresponding to the monomeric protein was also observed in the elution profile. Purified samples eluting from the HPLC column were monitored by UV spectroscopy to confirm the baicalein binding (47). The baicalein has three characteristic maxima in the absorption spectrum, at 216, 277, and 324 nm. Zhu et al. (46) showed that when the baicalein was oxidized, the absorbance at 324 nm disappeared, whereas when it was bound to α-synuclein, a new peak at ~360 nm was observed. The UV absorption spectrum of α-synuclein oligomer showed an absorbance at around 360 nm, suggesting the effective baicalein binding. Interestingly, the peak in the HPLC profile corresponding to the monomeric α-synuclein coincubated with baicalein also showed an absorbance at 360 nm, indicating baicalein binding (47).

These two samples separated by SEC were used for the detailed biophysical analysis, including atomic force and electron microscopy, SAXS, FTIR and far-UV CD (47). Furthermore, thermodynamic stability of the baicalein-stabilized oligomers was evaluated via the analysis of their GdmCl-induced unfolding (47). The purified baicalein-stabilized oligomers were incubated at 37°C for 1 month. No fibrils were formed and no dissociation was observed after this prolonged incubation, suggesting the high stability of the oligomers. Inhibitory effects of these oligomers on α-synuclein fibrillation were also evaluated. Finally, the effect of the baicalein-stabilized oligomers on the integrity of lipid membranes was evaluated (47). All these very important studies became possible due to the ability of SEC to physically separate monomeric and oligomeric α-synuclein species.

4. Notes

1. It is recommended to work at neutral pH. Acidic buffers can damage pressure transducers, whereas basic buffers can damage quartz of the static light scattering cell.

2. The accuracy of the evaluation of protein's hydrodynamic dimensions (e.g., estimations of its Stokes radius, R_S) by SEC depends significantly on the number of proteins used for the column calibration. Furthermore, if the determination of hydrodynamic dimensions for denatured and unfolded globular proteins is planned, then a set of denatured and unfolded proteins with known R_S values should be used for column calibration (3). This requirement is also applicable for the evaluation of dimensions of intrinsically disordered proteins. However, in the case of IDPs, a set of globular proteins with known hydrodynamic dimensions in various denatured and unfolded conformations can be used for calibration.

3. To avoid the influence of protein association on the results of SEC-FPLC measurements, the usual protein concentration should be low, about 0.001 mg/ml.

4. As the largest species is entirely excluded from the pores in the column matrix, its retention volume is equal to the void volume, in which case K_{SEC} is zero. On the other hand, the smallest molecule permeates all of the pores within the SEC column, and its retention volume equals the sum of the void volume and the pore volume, i.e., the total volume or total permeation limit (V_T). The value of K_{SEC} for species eluting at the total volume is 1. Finally, the intermediate-size molecules can permeate the pores to some extent and thus can be separated according to their respective hydrodynamic volumes.

5. It is important to repeat all the measurements several times in order to produce more accurate calibration curve.

6. It is also useful to perform calibration procedure under the variety of experimental conditions and in various solvent systems to be utilized in the IDP analysis. The goal of this analysis is to make sure that the dependences of the migration rate $(1,000/V_{el})$ versus the logarithm of protein molecular weight obtained for different solvents have virtually the same slopes. These data would provide additional support to the conclusion that the permeation properties of the column do not change significantly under the conditions used in the IDP analysis.

7. Importantly, statistical analysis has revealed that the relative errors of the recovered approximations exhibit random distribution over the wide range of chain lengths and do not generally exceed 10% (33). This means that the effective protein dimensions in a variety of conformational states can be predicted based on the chain length with an accuracy of 10%. In other words, this set of equations can be used to estimate the R_S value for any protein with known molecular mass M in any conformational state. Another important point is that having the R_S measured by SEC and knowing the molecular mass of the protein, one can understand what conformational state the studied protein is in under the given conditions.

References

1. Tanford C (1968) Protein denaturation. Adv Protein Chem 23:121–282

2. Uversky VN (1993) Use of fast protein size-exclusion liquid chromatography to study the unfolding of proteins which denature through the molten globule. Biochemistry 32:13288–13298

3. Uversky VN (1994) Gel-permeation chromatography as a unique instrument for quantitative and qualitative analysis of protein denaturation and unfolding. Int J Bio-Chromatogr 1:103–114

4. Uversky VN (2003) Protein folding revisited. A polypeptide chain at the folding-misfolding-nonfolding cross-roads: which way to go? Cell Mol Life Sci 60:1852–1871

5. Porath J, Flodin P (1959) Gel filtration: a method for desalting and group separation. Nature 183:1657–1659

6. Porath J (1968) Molecular sieving and adsorption. Nature 218:834–838

7. Andrews P (1965) The gel-filtration behaviour of proteins related to their molecular weights over a wide range. Biochem J 96:595–606

8. Ackers GK (1967) Molecular sieve studies of interacting protein systems. I. Equations for transport of associating systems. J Biol Chem 242:3026–3034

9. Ackers GK (1970) Analytical gel chromatography of proteins. Adv Protein Chem 24:343–446

10. le Maire M, Ghazi A, Moller JV, Aggerbeck LP (1987) The use of gel chromatography for the determination of sizes and relative molecular masses of proteins. Interpretation of calibration curves in terms of gel-pore-size distribution. Biochem J 243:399–404

11. Le Maire M, Aggerbeck LP, Monteilhet C, Andersen JP, Moller JV (1986) The use of high-performance liquid chromatography for the determination of size and molecular weight of proteins: a caution and a list of membrane proteins suitable as standards. Anal Biochem 154:525–535

12. Potschka M (1987) Universal calibration of gel permeation chromatography and determination of molecular shape in solution. Anal Biochem 162:47–64

13. Corbett RJ, Roche RS (1984) Use of high-speed size-exclusion chromatography for the study of protein folding and stability. Biochemistry 23:1888–1894

14. Fish WW, Reynolds JA, Tanford C (1970) Gel chromatography of proteins in denaturing solvents. Comparison between sodium dodecyl sulfate and guanidine hydrochloride as denaturants. J Biol Chem 245:5166–5168

15. Corbett RJ, Roche RS (1983) The unfolding mechanism of thermolysin. Biopolymers 22:101–105

16. Endo S, Saito Y, Wada A (1983) Denaturant-gradient chromatography for the study of protein denaturation: principle and procedure. Anal Biochem 131:108–120

17. Lau SY, Taneja AK, Hodges RS (1984) Synthesis of a model protein of defined secondary and quaternary structure. Effect of chain length on the stabilization and formation of two-stranded alpha-helical coiled-coils. J Biol Chem 259:13253–13261

18. Brems DN, Plaisted SM, Havel HA, Kauffman EW, Stodola JD, Eaton LC, White RD (1985) Equilibrium denaturation of pituitary- and recombinant-derived bovine growth hormone. Biochemistry 24:7662–7668

19. Gupta BB (1983) Determination of native and denatured milk proteins by high-performance size exclusion chromatography. J Chromatogr 282:463–475

20. Uversky VN, Semisotnov GV, Pain RH, Ptitsyn OB (1992) 'All-or-none' mechanism of the molten globule unfolding. FEBS Lett 314:89–92

21. Uversky VN, Ptitsyn OB (1994) "Partly folded" state, a new equilibrium state of protein molecules: four-state guanidinium chloride-induced unfolding of beta-lactamase at low temperature. Biochemistry 33:2782–2791

22. Uversky VN, Ptitsyn OB (1996) Further evidence on the equilibrium "pre-molten globule state": four-state guanidinium chloride-induced unfolding of carbonic anhydrase B at low temperature. J Mol Biol 255:215–228

23. Withka J, Moncuse P, Baziotis A, Maskiewicz R (1987) Use of high-performance size-exclusion, ion-exchange, and hydrophobic interaction chromatography for the measurement of protein conformational change and stability. J Chromatogr 398:175–202

24. Uversky VN, Li J, Fink AL (2001) Evidence for a partially folded intermediate in alpha-synuclein fibril formation. J Biol Chem 276:10737–10744

25. Receveur-Brechot V, Bourhis JM, Uversky VN, Canard B, Longhi S (2006) Assessing protein disorder and induced folding. Proteins 62:24–45

26. Uversky VN, Li J, Souillac P, Millett IS, Doniach S, Jakes R, Goedert M, Fink AL (2002) Biophysical properties of the synucleins and their propensities to fibrillate: inhibition of alpha-synuclein assembly by beta- and gamma-synucleins. J Biol Chem 277:11970–11978

27. Uversky VN, Gillespie JR, Millett IS, Khodyakova AV, Vasiliev AM, Chernovskaya TV, Vasilenko RN, Kozlovskaya GD, Dolgikh DA, Fink AL, Doniach S, Abramov VM (1999) Natively unfolded human prothymosin alpha adopts partially folded collapsed conformation at acidic pH. Biochemistry 38:15009–15016

28. Uversky VN, Permyakov SE, Zagranichny VE, Rodionov IL, Fink AL, Cherskaya AM, Wasserman LA, Permyakov EA (2002) Effect of zinc and temperature on the conformation of the gamma subunit of retinal phosphodiesterase: a natively unfolded protein. J Proteome Res 1:149–159

29. Meunier DM (1997) Molecular weight determinations. In: Settle F (ed) Handbook of instrumental techniques for analytical chemistry. Prentice-Hall, Inc., Upper Saddle River, NJ, pp 853–866

30. Ui N (1979) Rapid estimation of the molecular weights of protein polypeptide chains using high-pressure liquid chromatography in 6 M guanidine hydrochloride. Anal Biochem 97:65–71

31. Grossberg AY, Khokhlov AR (1989) Statistical physics of macromolecules. Nauka, Moscow

32. Abramov VM, Vasiliev AM, Khlebnikov VS, Vasilenko RN, Kulikova NL, Kosarev IV, Ishchenko AT, Gillespie JR, Millett IS, Fink AL, Uversky VN (2002) Structural and functional properties of Yersinia pestis Caf1 capsular antigen and their possible role in fulminant development of primary pneumonic plague. J Proteome Res 1:307–315

33. Tcherkasskaya O, Davidson EA, Uversky VN (2003) Biophysical constraints for protein structure prediction. J Proteome Res 2:37–42

34. Tcherkasskaya O, Uversky VN (2001) Denatured collapsed states in protein folding: example of apomyoglobin. Proteins 44:244–254

35. Tcherkasskaya O, Uversky VN (2003) Polymeric aspects of protein folding: a brief overview. Protein Pept Lett 10:239–245

36. Uversky VN (2002) Natively unfolded proteins: a point where biology waits for physics. Protein Sci 11:739–756

37. Uversky VN (2002) What does it mean to be natively unfolded? Eur J Biochem 269:2–12

38. Ptitsyn OB (1995) Molten globule and protein folding. Adv Protein Chem 47:83–229

39. Ptitsyn OB, Bychkova VE, Uversky VN (1995) Kinetic and equilibrium folding intermediates. Philos Trans R Soc Lond B Biol Sci 348:35–41

40. Permyakov SE, Millett IS, Doniach S, Permyakov EA, Uversky VN (2003) Natively unfolded C-terminal domain of caldesmon remains substantially unstructured after the effective binding to calmodulin. Proteins 53:855–862

41. Uversky VN, Lee HJ, Li J, Fink AL, Lee SJ (2001) Stabilization of partially folded conformation during alpha-synuclein oligomerization in both purified and cytosolic preparations. J Biol Chem 276:43495–43498

42. Li BQ, Fu T, Gong WH, Dunlop N, Kung H, Yan Y, Kang J, Wang JM (2000) The flavonoid baicalin exhibits anti-inflammatory activity by binding to chemokines. Immunopharmacology 49:295–306

43. Ikezoe T, Chen SS, Heber D, Taguchi H, Koeffler HP (2001) Baicalin is a major component of PC-SPES which inhibits the proliferation of human cancer cells via apoptosis and cell cycle arrest. Prostate 49:285–292

44. Gao Z, Huang K, Xu H (2001) Protective effects of flavonoids in the roots of *Scutellaria baicalensis* Georgi against hydrogen peroxide-induced oxidative stress in HS-SY5Y cells. Pharmacol Res 43:173–178

45. Shieh DE, Liu LT, Lin CC (2000) Antioxidant and free radical scavenging effects of baicalein, baicalin and wogonin. Anticancer Res 20:2861–2865

46. Zhu M, Rajamani S, Kaylor J, Han S, Zhou F, Fink AL (2004) The flavonoid baicalein inhibits fibrillation of alpha-synuclein and disaggregates existing fibrils. J Biol Chem 279:26846–26857

47. Hong DP, Fink AL, Uversky VN (2008) Structural characteristics of alpha-synuclein oligomers stabilized by the flavonoid baicalein. J Mol Biol 383:214–223

Part III

Methods to Analyze Conformational Behavior

Chapter 12

Denaturant-Induced Conformational Transitions in Intrinsically Disordered Proteins

Paolo Neyroz, Stefano Ciurli, and Vladimir N. Uversky

Abstract

Intrinsically disordered proteins (IDPs) differ from ordered proteins at several levels: structural, functional, and conformational. Amino acid biases also drive atypical responses of IDPs to changes in their environment. Among several specific features, the conformational behavior of IDPs is characterized by the low cooperativity (or the complete lack thereof) of the denaturant-induced unfolding. In fact, the denaturant-induced unfolding of native molten globules can be described by shallow sigmoidal curves, whereas urea- or guanidinium hydrochloride-induced unfolding of native pre-molten globules or native coils is a noncooperative process and typically is seen as monotonous feature-less changes in the studied parameters. This chapter describes some of the most characteristic features of the IDP conformational behavior.

Key words: Conformational stability, Denaturant-induced unfolding, Molten globule, Pre-molten globule, Two-state transition, Three-state transition

1. Introduction

1.1. Introducing IDPs

Intrinsically disordered proteins (IDPs) exist as highly dynamic ensembles of interconverting structures, carry numerous vital biological functions, and are abundant in various proteomes. Recently, the structure–function paradigm stating that ordered 3D structures represent the indispensable prerequisite for the effective protein functioning has been redefined to include IDPs (1–12). According to this redefined paradigm, native proteins (or their functional regions) can exist in any of the known conformational states: ordered, molten globule, pre-molten globule, and coil. Function can arise from any of these conformations and transitions between them. Therefore, in addition to the "protein folding" problem, where the correct folding of a globular protein into the rigid biologically active conformation is determined by its amino

Vladimir N. Uversky and A. Keith Dunker (eds.), *Intrinsically Disordered Protein Analysis:*
Volume 2, Methods and Experimental Tools, Methods in Molecular Biology, vol. 896,
DOI 10.1007/978-1-4614-3704-8_12, © Springer Science+Business Media New York 2012

acid sequence (13), the "protein nonfolding problem" does exist too, where the lack of a rigid globular structure in a given IDP may be encoded in some specific features of its amino acid sequence. Since structural classification of IDPs frequently utilizes definitions developed for the description of the partially folded globular proteins, these conformations are briefly introduced below.

The unique 3D structure of a globular protein is stabilized by noncovalent interactions of different nature. These include hydrogen bonds, hydrophobic interactions, electrostatic interactions, van der Waals interactions, etc. Complete (or almost complete) disruption of all these interactions can be achieved in concentrated solutions of strong denaturants (such as urea or guanidinium hydrochloride (GdmHCl)). Here, an initially folded and highly ordered molecule of a globular protein unfolds, i.e., transforms into a highly disordered random coil-like conformation (14–17). However, environmental changes can decrease (or even completely eliminate) only some noncovalent interactions, whereas the remaining interactions could stay unchanged (or even could be intensified). Very often, a globular protein will lose its biological activity under these conditions, thus becoming denatured (17). It is important to remember that denaturation is not necessarily accompanied by the unfolding of a protein, but rather might result in the appearance of various partially folded conformations with properties intermediate between those of the folded (ordered) and the completely unfolded states. In fact, globular proteins exist in at least four different equilibrium conformations: folded (ordered), molten globule, pre-molten globule, and unfolded (6, 18–24). The ability of a globular protein to adopt different stable partially folded conformations is believed to be an intrinsic property of a polypeptide chain.

The molten globular protein is denatured and has no (or has only a trace of) rigid cooperatively melted tertiary structure. Small-angle X-ray scattering analysis reveals that it has a globular structure typical of folded globular proteins (22, 25–28). 2D-NMR, coupled with hydrogen–deuterium exchange, shows that the molten globule is characterized not only by the native-like secondary structure content but also by the native-like folding pattern (29–36). A considerable increase in the accessibility of a protein molecule to proteases is noted as a specific property of the molten globule (37, 38). The transformation into this intermediate state is accompanied by a considerable increase in the affinity of a protein molecule to hydrophobic fluorescence probes (such as 8-anilinonaphthalene-1-sulfonate, ANS), and this behavior is a characteristic property of the molten globules (39, 40). Finally, on the average, the hydrodynamic radius of the molten globule is increased by no more than 15% compared to that of the folded state, which corresponds to the volume increase of ~50% (41).

The globular protein in the pre-molten globule state is also denatured. It has a considerable amount of residual secondary structure, which is much less pronounced than that of the native or the molten globule protein. The pre-molten globule form is considerably less compact than the molten globule or folded states, being still noticeably more compact than the random coil. The pre-molten globule can effectively interact with ANS, though this interaction is weaker than the molten globule. The pre-molten globule has no globular structure (6, 22, 23), suggesting that the pre-molten globule probably represents a "squeezed," partially collapsed and partially ordered form of a coil. Finally, the pre-molten globule was shown to be separated from the molten globule by an all-or-none-transition (6, 18, 20, 21), suggesting that the molten globule and the pre-molten globule are different thermodynamic (phase) states of a globular protein.

By analogy with the mentioned above partially folded conformations of ordered globular proteins, IDPs can be grouped into three structurally different subclasses: native molten globules (so-called collapsed IDPs), native pre-molten globules, and native coils (both are known as extended IDPs).

1.2. Amino Acid Determinants of Intrinsic Disorder

In an attempt to understand the relationship between the amino acid sequence and protein intrinsic disorder, a set of experimentally characterized IDPs was systematically compared with a set of ordered globular proteins (2, 11, 42, 43). This analysis revealed that IDPs differ from structured globular proteins and domains with regard to many attributes, including amino acid composition, sequence complexity, hydrophobicity, charge, and flexibility. IDPs are significantly depleted in a number of so-called order-promoting residues (Ile, Leu, Val, Trp, Tyr, Phe, Cys, and Asn), being substantially enriched in so-called disorder-promoting amino acids (Ala, Arg, Gly, Gln, Ser, Pro, Glu, Asp, and Lys) (2, 11, 42–45).

Furthermore, the combination of low mean hydrophobicity and relatively high net charge was shown to constitute an important prerequisite for the lack of compact structure in extended (coil-like and pre-molten globule-like) IDPs (46). Overall, these analyses revealed that the amino acid sequences of IDPs and ordered proteins are very different, supporting a hypothesis that the propensity of a polypeptide chain to fold or stay disordered is encoded in its amino acid sequence. For example, extended IDPs are disordered under physiological conditions because of the strong electrostatic repulsion (due to their high net charges) and weak hydrophobic attraction (due to their low contents of hydrophobic residues). These data also suggest that IDPs (especially extended IDPs) with their highly biased amino acid sequences might possess unpredictable conformational responses to changes in their environment.

2. Materials

1. Prepare all solutions using ultrapure water (prepared by purifying deionized water to attain a sensitivity of 18 MΩ cm at 25°C).

2. Guanidine hydrochloride (GdmHCl) and urea are of analytical grade and stored at room temperature.

3. Prepare IDP for the analysis. Here, purification of the recombinant *Bacillus pasteurii* UreG (*Bp*UreG) is performed as reported previously (47).

4. Prepare protein samples (15 μM) in a buffer containing 50 mM Tris–HCl at pH 8.0.

5. Prepare stock solutions of GdmHCl (6 M), urea (8 M) (see Note 1), and NaCl (4 M) (see Note 2).

3. Methods

3.1. Use of Intrinsic Fluorescence for the Protein Unfolding Analysis

Experiments described below aim at the characterization of the folded state of a protein. In particular, different intrinsic fluorescence parameters are used to monitor the structural transition between a native state, N, and a denatured state, U, and to determine the thermodynamic parameters of the corresponding conformational transitions (48).

3.1.1. Sample Preparation

In a quartz fluorescence cuvette of 1 cm path length and 4 mL volume capacity, small aliquots (25–1,500 μL) of the denaturant (GdmHCl) from a 6 M stock solution are added using a microsyringe to a protein sample (0.7 mg/mL) of the initial volume of 2 mL. Then, using this procedure, the data must be corrected for dilution as, $F_{cor} = F_{obs} (V_t/V_i)$, where F_{cor} and F_{obs} indicate the dilution-corrected and the observed fluorescence intensities, V_i represents the initial sample volume in the absence of denaturant, and V_t is the total sample volume after the denaturant addition.

As an alternative, sample of the same volume can be prepared by adding 100 μL of a protein stock (final concentration of 15 μM) to denaturant solutions of the same volume (1.9 mL) and increasing denaturant concentration.

In both the preparation methods described, the protein folding structure can be evaluated over a range of denaturant concentrations from 0 M to 3 M GdmHCl. The samples are equilibrated at 25°C before the fluorescence measurements are performed.

3.1.2. Equilibrium Unfolding Fluorescence Measurements

To monitor the protein unfolding transition as a function of GdmHCl concentration, different fluorescence parameters are appropriate, such as (1) the change in the intensity of the fluorescence emission, (2) the change of the maximum emission wavelength (see Experiment 1), (3) the change of the steady-state emission anisotropy, $<r>$ (see Experiment 2), or (4) the change of the time-resolved fluorescence anisotropy (correlation time ϕ, usually associated with changes in the tumbling of the protein).

Experiment 1. This experiment describes this application for the case of *B. pasteurii* UreG (52). Using the excitation wavelength of 295 nm the entire emission spectra is recorded from 300 nm to 450 nm for each sample at increasing denaturant concentrations. In a typical PC-driven fluorometer apparatus the data are saved as an array of pairs of column for each denaturant concentration: a first column containing the incrementing wavelength (x) and a second column containing the measured fluorescence intensity (y). Then, to monitor the changes of the emission maximum (λ_{MAX}) as a function of the denaturant concentration it is a good practice to use the Center of Mass (COM) of the fluorescence emission. In this case, COM is calculated as follows: $\Sigma x_k y_k / \Sigma y_k$, where x and y indicate the kth wavelength and intensity coordinate of the emission spectrum, respectively (49).

Experiment 2. This experiment describes this application for the case of *B. pasteurii* UreG (52). To collect fluorescence steady-state anisotropy measurements, the fluorometer must be equipped with two polarizers placed in the excitation and the emission path, respectively (2). The best UV light transmittance is obtained using Glan–Thomson prisms, but also less expensive Polaroid HNP'B can be used, provided that an excitation wavelength longer than 290 nm is chosen. In addition, due to the polarizers' light transmission cut, the excitation and the emission band passes are increased to 4–5 nm each. Then, by the appropriate rotation of the polarizers of 0° and 90° with each other, the measure of the intensities of the parallel (I_{vv}) and the perpendicular (I_{vh}) components of the polarized fluorescence emission can be obtained. These fluorescence intensity components are used to calculate the steady-state anisotropy, $<r>$, from Eq. 1 (50):

$$\langle r \rangle = \frac{I_{vv} G - I_{vh}}{I_{vv} G + 2 I_{vh}} \tag{1}$$

where G, the "grating" correction factor, is calculated as $G = I_{hh} / I_{hv}$.

3.1.3. Equilibrium Unfolding Data Analysis

To describe a two-state transition, $N \leftrightarrow U$, between the native and the unfolded states with an unfolding equilibrium constant, $K_{un} = [U]/[N]$, and a free energy change given by,

$$\Delta G^0_{un} = -RT \ln K_{un} \tag{2}$$

the following linear relationship (linear extrapolation model) is used to describe the thermodynamics of the denaturant-induced unfolding of proteins (48, 51):

$$\Delta G_{un}^0 = \Delta G_{0,un}^0 - m\,[D] \qquad (3)$$

where $\Delta G_{0,un}^0$ is the free energy change of unfolding extrapolated to zero denaturant concentration at a reference temperature, and m, the denaturant concentration index, is a measure of the dependence of ΔG_{un}^0 on denaturant concentration. Equilibrium unfolding data analysis is then performed as previously described (52). The following equation is used to fit the experimental data by a two-state transition model (52):

$$X = \frac{X_{0N} + S_N[D] + (X_{0U} + S_U[D]).\,e^{\frac{-\Delta G_{0,un}^0 + m[D]}{RT}}}{1 + e^{\frac{-\Delta G_{0,un}^0 + m[D]}{RT}}} \qquad (4)$$

where X_{0N} and X_{0U} indicate either the wavelength λ_{MAX} (COM) or the steady-state anisotropy $<r>$ of the native and the unfolded states in the absence of the denaturant D, and the terms S_N and S_U represent the baseline slopes for the native and unfolded regions. Since the latter slope terms tend to be zero when λ_{MAX} or $<r>$ are used, they are not considered in these data analysis procedures.

Fitting equations can be obtained by nonlinear least-squares utilities included in several commercial software packages. The results that can be obtained by these experiments are presented in Figs. 1 and 2. The ΔG obtained in this way can be considered as a measure of the free energy involved in the noncooperative transition observed in these experiments.

3.2. Designing the Unfolding Curves Using Size Exclusion Chromatography

This section is dedicated to the design of the unfolding curves based on the results of the size exclusion chromatography (SEC) analysis. Although described approaches were originally developed for the analysis of the unfolding of globular proteins, they can also be used for the analysis of the unfolding behavior of IDPs.

3.2.1. Preliminary Considerations

To build an unfolding curve based on the SEC data, one has to analyze the changes induced in the elution profile by variations of the environmental conditions. Similar to the profiles retrieved by other separation techniques, e.g., sedimentation or urea-gradient electrophoresis (53), the shape of the elution profile within the transition region depends dramatically on two rates: the rate of the exchange between the protein conformers and the characteristic rate of the heterogeneous equilibrium between the bound and unbound state of the protein on the stationary phase of the chromatographic process. In the case of conformational exchange faster than the chromatographic process, a single peak corresponding to the hydrodynamic dimensions averaged over compact and less

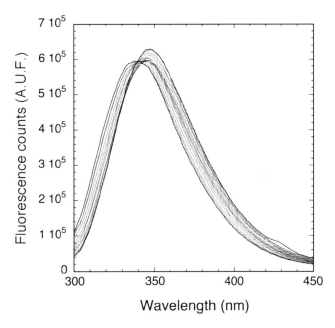

Fig. 1. Conformational transition of BpUreG as revealed by steady-state fluorescence signals. Steady-state emission spectra of BpUreG (0.7 mg/mL) in 50 mM Tris–HCl (pH 8.0) at 24°C at increasing concentrations of GdmHCl (from 0 to 3 M, incubation time of 10 min). The excitation wavelength is 295 nm, and the band pass is 2.5 nm on both the excitation and the emission side. (Modified from ref. 52).

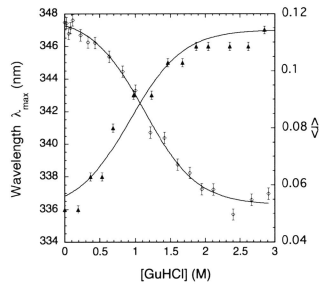

Fig. 2. Conformational transition of BpUreG as revealed by steady-state fluorescence signals. Changes in λ_{max} (*filled triangle*) and steady-state anisotropy (*open diamond*) as a function of denaturant concentration. The *solid lines* represent the fits by a nonlinear least-squares method of the experimental data. A two-state model in the form described by Eq. 4 is used (modified from ref. 52).

compact conformations is observed. The position of this peak is condition-dependent. In the case of conformational exchange slower than the chromatographic process, two different elution peaks appear, corresponding to the different conformers. The intensities of these peaks are redistributed as conditions are changed. Importantly, this behavior (the presence of two separate peaks within the transition region) illustrates the most promising application of size-exclusion chromatography for studying "all-or-none" transitions between two states of different compactness (18, 21, 41, 54–58). Furthermore, in this case, SEC opens unique possibilities for independent estimation of the hydrodynamic dimensions for these compact and less compact species, and, in principle, their physical separation and independent study of other structural properties.

SEC data on protein conformational changes can be represented as unfolding curves monitored either by relative areas of the two peaks or, for fast exchange, by the position of the average peak (see, for example, refs. 41, 54–56). When exchange is fast, the sigmoidal unfolding curve is obtained simply as a dependence of the elution peak position on the denaturant concentration, whereas when exchange is slow, the same dependence consists of two different curves, describing the individual behavior of compact and less compact species. However, even in the case of slow exchange, sigmoidal unfolding curves can be obtained (41, 54–56). The simplest approach to analyze the data is to plot a curve representing the denaturant-determined dependence of the relative area under one of the elution peaks. These dependencies can be determined as fractions of compact (f_C) and less compact (f_{LC}) conformations, being simply equal to relative areas of the corresponding elution peaks:

$$f_C = \frac{S_C}{S_C + S_{LC}}; \qquad f_{LC} = 1 - f_C = \frac{S_{LC}}{S_C + S_{LC}}, \qquad (5)$$

where S_C and S_{LC} are the areas under the elution peaks corresponding to compact and less compact species, respectively. Alternatively, one can reconstruct the denaturant-induced dependence of the position of an averaged elution peak ($<V_{el}>$) using the obvious equation (18, 21, 54):

$$<V_{el}> = f_C <V_{el}^C> + f_{LC} <V_{el}^{LC}>$$
$$= (1 - f_{LC}) <V_{el}^C> + f_{LC} <V_{el}^{LC}>, \qquad (6)$$

where $<V_{el}^C>$ and $<V_{el}^{LC}>$ are the elution volumes of the peaks corresponding to the compact and less compact molecules respectively. Then the denaturant dependence of this $<V_{el}>$ parameter should be plotted. This unfolding curve, being similar to that followed by the traditional methods, describes the changes in hydrodynamic dimensions averaged over all protein conformations existing within the transition region. Note that this equation also describes a case when compact and less compact conformations are in fast exchange.

Let us now consider two different unfolding scenarios: when a protein unfolds according to a simple two-state mechanism and when it is a more complex process. It is believed that the denaturation of a small protein is an all-or-none process (59), where each of the structural parameters (X) can be presented as superposition of contributions of native (N) and denatured (D) molecules:

$$X = f_N X_N + f_D X_D = f_N X_N + (1 - f_N) X_D, \qquad (7)$$

where X is the measured parameter and X_N and X_D are its values for the native and denatured molecules, respectively, whereas f_N and f_D are the fractions of the native and denatured molecules. Equation 7 leads to the well-known equations:

$$f_D = \frac{X - X_N}{X_D - X_N}; \qquad f_N = 1 - f_D = \frac{X_D - X}{X_D - X_N}, \qquad (8)$$

which are valid in all cases when protein denaturation as such is an "all-or-none" process, irrespectively of other transitions which may follow denaturation.

In the case of a simple two-state N $\rightarrow$ U transition between native and unfolded states, Eqs. 5 and 8 give identical information, as $f_C = f_N$ and $f_{LC} = f_D = f_U$ (with f_U being a fraction of unfolded molecules), $<V_{el}^{LC}> = V_{el}^{U}$ and Eq. 6 leads to (18, 21, 54):

$$f_U = \frac{<V_{el}> - V_{el}^{C}}{V_{el}^{U} - V_{el}^{C}}, \qquad (9)$$

where $<V_{el}^{C}>$ and V_{el}^{U} for any given denaturant concentration can be determined by extrapolation of the corresponding baselines. Importantly, Eq. 9 describes also a situation when protein unfolds according to the three-state mechanism, being accompanied by the accumulation of an intermediate state, the hydrodynamic dimensions of which are close to the molecular size of the native molecule (as observed for the molten globule state) (18).

Now, if a protein unfolds according to the three-state transition (i.e., through a compact molten globule-like intermediate), then the denaturation and unfolding become decoupled and Eqs. 5 and 8 permit one to determine the fraction of molecules in all three states, f_N, f_{MG}, and f_U, as $f_U = f_{LC}$ and

$$f_{MG} = f_D - f_{LC} = f_C - f_N, \qquad (10)$$

It is important to remember, that in this case the changes of any structural parameter of a protein are the results of averaging over all possible conformations:

$$X = f_N X_N + f_I X_I + f_U X_U, \qquad (11)$$

where f_I is a fraction of intermediate(s) and X_I is a value of parameter X in this intermediate (or its value averaged over all intermediates). Ignoring this fact and using a traditional approach to describe this complex process in terms of the unfolding degree derived from the Eq. 8

$$f_U = \frac{X - X_N}{X_U - X_N},$$ (12)

is generally incorrect, as, in fact, f_U depends not only on X_N and X_U but also on X_I. However, this approach might work if X_I is close either to X_N or to X_U.

Furthermore, even in the case of a four-state transition through a compact (molten globule) and a less compact (pre-molten globule) intermediates when the less compact intermediate exchanges slowly with the compact intermediate but quickly with unfolded molecules, SEC permits evaluation to be done separately for fractions of molecules in both less compact states (pre-molten globule and unfolded). In this case, elution volume averaged over the both less compact states is equal to:

$$<V_{el}^{LC}> = \frac{f_{PMG} V_{el}^{PMG} + f_U V_{el}^U}{f_{PMG} + f_U}$$
$$= V_{el}^{PMG} + \frac{f_U}{f_{LC}} (V_{el}^U - V_{el}^{PMG}),$$ (13)

where f_{PMG} is a fraction of the less compact (pre-molten globule-like) intermediate and V_{el}^{PMG} is its elution volume. Therefore

$$f_U = f_{LC} \frac{<V_{el}^{LC}> - V_{el}^{PMG}}{V_{el}^U - V_{el}^{PMG}},$$ (14)

and f_U (as well as $f_{PMG} = f_{LC} - f_U$) can be calculated from f_{LC} and $<V_{el}^{LC}>$ if the values of V_{el}^U and V_{el}^{PMG} can be estimated (18). The elution volume of the unfolded state, V_{el}^U, can be extrapolated from the corresponding baseline, whereas for V_{el}^{PMG} one can have only the upper estimate, as the $<V_{el}^{LC}>$ value obtained in the point of appearance of the less compact species. Thus, even in the case of very complex unfolding scenario (when protein unfolds through two intermediate states (compact and less compact) separated by the all-or-none transition and when less compact intermediate is in the fast conformational exchange with the unfolded conformation) the application of Eq. 14 to the SEC data allows one to extract unique information on the independent behavior of less compact intermediate and unfolded species.

3.2.2. Illustrative Examples of SEC-Based Unfolding Curves

SEC-based analysis of the GdmHCl-induced unfolding of lysozyme and carbonic anhydrase revealed dramatic difference in the conformational behavior of these two proteins. For example, there was a pronounced difference in the magnitude of the second branch of the plot presenting $<V_{el}>$ versus GdmHCl concentration, i.e., the part describing the GdmHCl dependence of $<V_{el}^{LC}>$. In the case of the lysozyme unfolding, this curve, being less pronounced, reflects the swelling of the unfolded polypeptide chain, whereas in the case of carbonic anydrase (21), as well as β-lactamase (18) and DNA ligase (58), it reflects the existence of the less compact intermediate (pre-molten globule) rapidly interconverting with the unfolded state. Furthermore, the unfolding of lysozyme represents a simple two-state transition, which describes the GdmHCl-induced changes of all studied parameters and coincides with the all-or-none transition monitored by the bimodality of SEC profile (see below for further discussion of this phenomenon). In the case of β-lactamase (18) and carbonic anhydrase (21), at least three decoupled conformational transitions were detected: (a) the denaturation curve reflecting the destruction of the rigid tertiary structure and monitored by the disappearance of near-UV CD signal and biological activity; (b) the all-or-none transition followed by SEC; (c) the real unfolding curve, monitored by changes in the protein hydrodynamic dimensions. This decoupled unfolding mechanism with several discrete transitions monitored by different experimental techniques reflects the accumulation of the intermediate states. Another important point is that the all-or-none transition monitored by the bimodality of SEC profiles characterizes the transition between two denatured conformations, the molten globule and its precursor (18, 21).

3.3. Using Slopes of the Unfolding Curves in the Conformational Analysis of Proteins

3.3.1. Some Theoretical Considerations

IDPs, being highly dynamic, are characterized by low conformational stability, which is reflected in low steepness of the transition curves describing their unfolding induced by strong denaturants or even in the complete lack of the sigmoidal shape of the unfolding curves. This is in strict contrast to the solvent-induced unfolding of ordered globular proteins, which is known to be highly cooperative process. In fact, we can find here an extreme case of the cooperative transition, which is an all-or-none transition where a cooperative unit includes the whole molecule, i.e., no intermediate states can be observed in the transition region. Based on the analysis of the unfolding transitions in ordered globular proteins it has been concluded that the steepness of urea- or GdmHCl-induced unfolding curves depends strongly on whether a given protein has a rigid tertiary structure (i.e., it is ordered) or is already denatured and exists as a molten globule (60, 61). In fact, urea-induced or GdmHCl-induced protein unfolding of globular proteins often involves at least two steps: the ordered (native) state to molten globule (N ↔ MG) and the molten globule to unfolded state (MG ↔ U) transitions (19, 20, 62–65). Both transitions are rather cooperative (as they follow a sigmoidal curve). However, for a long time the dimensions of the cooperative units for these transitions were unknown and, as a consequence, it was unclear whether or not these transitions have a real all-or-none nature.

The usual method for estimation of the cooperativity of transition is the measurement of the slope of the transition curve at its middle point. This slope is proportional to the change of the thermodynamic quantity conjugated with the variable provoking the transition, i.e., to the difference in the numbers of denaturant molecules "bound" to the initial and final states in the urea-induced or GdmHCl-induced transitions, Δv_{eff}. To understand the thermodynamic nature of solvent-induced transitions in globular proteins, the dependence of their transition slopes on the protein molecular mass (M) was analyzed (60, 61). This analysis was based on the hypothesis that the slope of a phase transition in small systems depends on the dimensions of this system (66, 67). In the case of first-order phase transition, the slope increases proportionally to the number of units in a system (66), whereas the slope for second-order phase transition is proportional to the square root of this number (67). Therefore, it is possible to distinguish between phase and nonphase intramolecular transitions by measuring whether their slopes depend on molecular weight.

Taking these observations into account, the available at that moment experimental data on urea-induced and GdmHCl-induced N↔U, N↔MG and MG↔U transitions in small globular proteins were analyzed (60, 61). Here, the cooperativity was measured based on the corresponding Δv_{eff} values, which were obtained from the equilibrium constants K_{eff} as (60, 61):

$$\Delta v_{eff} = \left(\frac{\partial \ln K_{eff}}{\partial \ln a}\right)_{a=a_t} = 4a_t \left(\frac{\partial \Theta}{\partial a}\right)_{a=a_t} \qquad (15)$$

where Θ is the fraction of molecules in one of the states separated by the all-or-none transition, a is the activity of a denaturing agent, and a_t is its activity at the middle-transition point. The activities of urea and GdmHCl as functions of their molar concentrations (m) can be calculated by the empirical equations (51):

$$a_{urea} = 0.9815(m) - 0.02978(m)^2 + 0.00308(m)^3 \qquad (16)$$

and

$$a_{GdmHCl} = 0.6761(m) - 0.1468(m)^2 + 0.02475(m)^3$$
$$+ 0.00132(m)^4 \qquad (17)$$

The analysis revealed that the $\Delta v_{eff}(M)$ curve comprised of two parts: for small globular proteins (with $M < 25$–30 kDa), cooperativity of unfolding transition increased with M, whereas for large proteins (with $M > 25$–30 kDa), cooperativity did not depend on M. The existence of pronounced molecular mass dependence of the degree of cooperativity showed that denaturant-induced unfolding of small globular proteins exhibited the characteristics of phase transition. On the other hand, the independence of the cooperativity of unfolding from M for large proteins was related to their multidomain organization.

The analysis of the $\log(\Delta v_{\mathrm{eff}}^{N \leftrightarrow U})$ versus $\log(M)$ dependence for solvent-induced $N \leftrightarrow U$ transitions in small globular proteins revealed that it can be described as:

$$\log(\Delta v_{\mathrm{eff}}^{N \leftrightarrow U}) = 0.97 \log(M) - 0.07 \qquad (18)$$

with the root mean square deviation (rmsd) of 0.112 and the correlation coefficient (r) of 0.87 (60, 61). As the proportionality coefficient in the above equation is equal to 0.97 ± 0.15, we can conclude that $\Delta v_{\mathrm{eff}}^{N \leftrightarrow U}$ is proportional to M, clearly showing that the urea- and GdmHCl-induced unfolding of small globular proteins is an all-or-none transition, i.e., an intramolecular analog of first-order phase transition in macroscopic systems.

The analogous analyses for the denaturant-induced $N \leftrightarrow MG$ and $MG \leftrightarrow U$ transitions revealed that the slopes of the corresponding transition curves clearly increased with the molecular mass of a protein. These increases were approximated by the following equations (60, 61):

$$\log(\Delta v_{\mathrm{eff}}^{N \leftrightarrow MG}) = 1.02 \log(M) - 0.49 \qquad (19)$$

with rmsd $= 0.090$ and $r = 0.82$; and

$$\log(\Delta v_{\mathrm{eff}}^{MG \leftrightarrow U}) = 0.89 \log(M) - 0.40 \qquad (20)$$

with rmsd $= 0.092$ and $r = 0.84$.

These data suggested that all denaturant-induced transitions in small globular proteins can be described in terms of all-or-none transitions (60, 61).

3.3.2. Using Slopes of the Unfolding Curves for Prediction of the Conformational Nature of the Protein Before Unfolding: Experimental Approach

1. Measure denaturant-induced changes in a given structural parameter for a given protein (e.g., as described in sections above).

2. Calculate the unfolding degrees, θ, for a given transition as $\theta = (X - X_N)/(X_U - X_N)$, where X is the parameter by which unfolding is monitored, while X_N, and X_U are the values of this parameter in the native, and unfolded states, respectively.

3. Build the unfolding curve as θ versus denaturant concentration dependence.

4. Measure the Δv_{eff} value for a given transition using the equation $\Delta v_{\mathrm{eff}} = 4/(\ln a_1 - \ln a_2)$ which is practically equivalent to Eq. 15. Here a_1 and a_2 are the activities of urea or GdmHCl at the beginning and at the end of the transition, defined as the

intersection of the tangent to the transition curve at the middle of the transition with the lines $\theta = 0$ and $\theta = 1$.

5. Compare measured Δv_{eff} value with $\Delta v_{\text{eff}}^{N \leftrightarrow U}$ and $\Delta v_{\text{eff}}^{MG \leftrightarrow U}$ values calculated from Eqs. 18 and 20 for a protein with a given molecular mass.

3.3.3. Using Slopes of the Unfolding Curves for Prediction of the Conformational Nature of the Protein Before Unfolding: Illustrative Example

In application to IDPs, it has been proposed that this type of analysis can be used to differentiate whether a given protein has ordered (rigid) structure or exists as a native molten globule (7). In fact, the comparison of Eqs. 18 and 20 suggests that the slope of the N $\leftrightarrow$ U transition for a protein with a molecular mass of M is more than twice as steep as the slope of the MG$\leftrightarrow$U transition. For example, for a protein with the molecular mass of 30 kDa, $\Delta v_{\text{eff}}^{N \leftrightarrow U} = 23.1$, whereas $\Delta v_{\text{eff}}^{MG \leftrightarrow U} = 8.2$. Therefore, to extend this type of analysis to IDP, the corresponding Δv_{eff} value should be determined from the denaturant-induced unfolding experiments. Then, this quantity should be compared to the $\Delta v_{\text{eff}}^{N \leftrightarrow U}$ and $\Delta v_{\text{eff}}^{MG \leftrightarrow U}$ values corresponding to the N$\leftrightarrow$U and MG$\leftrightarrow$U transitions in globular protein of a given molecular mass, evaluated by Eqs. 18 and 20, respectively (see Note 3).

4. Notes

1. It is a good practice to measure GdmHCl and urea concentrations in all the solutions refractometrically.

2. GdmHCl stock solutions are stored at 4°C and are used for no more than 2–3 days. The NaCl stock solution is stored at 4°C.

3. Although the denaturant-induced unfolding of a native molten globule can be described by a shallow sigmoidal curve (e.g., see ref. 52), urea- or GdmHCl-induced structural changes in native pre-molten globules or native coils are noncooperative and typically seen as monotonous feature-less changes in the studied parameters. This is due to the low content of the residual structure in these species.

References

1. Wright PE, Dyson HJ (1999) Intrinsically unstructured proteins: re-assessing the protein structure-function paradigm. J Mol Biol 293:321–331

2. Dunker AK, Lawson JD, Brown CJ, Williams RM, Romero P, Oh JS, Oldfield CJ, Campen AM, Ratliff CM, Hipps KW, Ausio J, Nissen MS, Reeves R, Kang C, Kissinger CR, Bailey RW, Griswold MD, Chiu W, Garner EC, Obradovic Z (2001) Intrinsically disordered protein. J Mol Graph Model 19:26–59

3. Dunker AK, Obradovic Z (2001) The protein trinity – linking function and disorder. Nat Biotechnol 19:805–806

4. Dunker AK, Brown CJ, Lawson JD, Iakoucheva LM, Obradovic Z (2002) Intrinsic disorder and protein function. Biochemistry 41:6573–6582

5. Dunker AK, Brown CJ, Obradovic Z (2002) Identification and functions of usefully disordered proteins. Adv Protein Chem 62:25–49

6. Uversky VN (2003) Protein folding revisited. A polypeptide chain at the folding-misfolding-nonfolding cross-roads: which way to go? Cell Mol Life Sci 60:1852–1871

7. Uversky VN (2002) Natively unfolded proteins: a point where biology waits for physics. Protein Sci 11:739–756

8. Uversky VN (2002) What does it mean to be natively unfolded? Eur J Biochem 269:2–12

9. Dunker AK, Cortese MS, Romero P, Iakoucheva LM, Uversky VN (2005) Flexible nets. The roles of intrinsic disorder in protein interaction networks. FEBS J 272:5129–5148

10. Uversky VN, Oldfield CJ, Dunker AK (2005) Showing your ID: intrinsic disorder as an ID for recognition, regulation and cell signaling. J Mol Recognit 18:343–384

11. Radivojac P, Iakoucheva LM, Oldfield CJ, Obradovic Z, Uversky VN, Dunker AK (2007) Intrinsic disorder and functional proteomics. Biophys J 92:1439–1456

12. Dunker AK, Oldfield CJ, Meng J, Romero P, Yang JY, Chen JW, Vacic V, Obradovic Z, Uversky VN (2008) The unfoldomics decade: an update on intrinsically disordered proteins. BMC Genomics 9 Suppl 2: S1

13. Anfinsen CB, Haber E, Sela M, White FH Jr (1961) The kinetics of formation of native ribonuclease during oxidation of the reduced polypeptide chain. Proc Natl Acad Sci USA 47:1309–1314

14. Anson ML, Mirsky AE (1932) The effect of denaturation on the viscosity of protein systems. J Gen Physiol 15:341–350

15. Mirsky AE, Pauling L (1936) On the structure of native, denatured and coagulated proteins. Proc Natl Acad Sci USA 22:439–447

16. Neurath H, Greenstein JP, Putnam FW, Erickson JO (1944) The chemistry of protein denaturation. Chem Rev 34:157–265

17. Tanford C (1968) Protein denaturation. Adv Protein Chem 23:121–282

18. Uversky VN, Ptitsyn OB (1994) "Partly folded" state, a new equilibrium state of protein molecules: four-state guanidinium chloride-induced unfolding of beta-lactamase at low temperature. Biochemistry 33:2782–2791

19. Ptitsyn OB (1995) Structures of folding intermediates. Curr Opin Struct Biol 5:74–78

20. Ptitsyn OB (1995) Molten globule and protein folding. Adv Protein Chem 47:83–229

21. Uversky VN, Ptitsyn OB (1996) Further evidence on the equilibrium "pre-molten globule state": four-state guanidinium chloride-induced unfolding of carbonic anhydrase B at low temperature. J Mol Biol 255:215–228

22. Uversky VN, Karnoup AS, Segel DJ, Seshadri S, Doniach S, Fink AL (1998) Anion-induced folding of Staphylococcal nuclease: characterization of multiple equilibrium partially folded intermediates. J Mol Biol 278:879–894

23. Uversky VN, Fink AL (1999) Do protein molecules have a native-like topology in the pre-molten globule state? Biochemistry (Mosc) 64:552–555

24. Tcherkasskaya O, Uversky VN (2001) Denatured collapsed states in protein folding: example of apomyoglobin. Proteins 44:244–254

25. Eliezer D, Chiba K, Tsuruta H, Doniach S, Hodgson KO, Kihara H (1993) Evidence of an associative intermediate on the myoglobin refolding pathway. Biophys J 65:912–917

26. Kataoka M, Hagihara Y, Mihara K, Goto Y (1993) Molten globule of cytochrome c studied by small angle X-ray scattering. J Mol Biol 229:591–596

27. Kataoka M, Kuwajima K, Tokunaga F, Goto Y (1997) Structural characterization of the molten globule of alpha-lactalbumin by solution X-ray scattering. Protein Sci 6:422–430

28. Semisotnov GV, Kihara H, Kotova NV, Kimura K, Amemiya Y, Wakabayashi K, Serdyuk IN, Timchenko AA, Chiba K, Nikaido K, Ikura T, Kuwajima K (1996) Protein globularization during folding. A study by synchrotron small-angle X-ray scattering. J Mol Biol 262:559–574

29. Baum J, Dobson CM, Evans PA, Hanley C (1989) Characterization of a partly folded protein by NMR methods: studies on the molten globule state of guinea pig alpha-lactalbumin. Biochemistry 28:7–13

30. Bushnell GW, Louie GV, Brayer GD (1990) High-resolution three-dimensional structure of horse heart cytochrome c. J Mol Biol 214:585–595

31. Chyan CL, Wormald C, Dobson CM, Evans PA, Baum J (1993) Structure and stability of the molten globule state of guinea-pig alpha-lactalbumin: a hydrogen exchange study. Biochemistry 32:5681–5691

32. Jeng MF, Englander SW, Elove GA, Wand AJ, Roder H (1990) Structural description of acid-denatured cytochrome c by hydrogen exchange and 2D NMR. Biochemistry 29:10433–10437

33. Bose HS, Whittal RM, Baldwin MA, Miller WL (1999) The active form of the steroidogenic acute regulatory protein, StAR, appears to be a molten globule. Proc Natl Acad Sci USA 96:7250–7255

34. Bracken C (2001) NMR spin relaxation methods for characterization of disorder and folding in proteins. J Mol Graph Model 19:3–12

35. Eliezer D, Yao J, Dyson HJ, Wright PE (1998) Structural and dynamic characterization of partially folded states of apomyoglobin and implications for protein folding. Nat Struct Biol 5:148–155

36. Wu LC, Laub PB, Elove GA, Carey J, Roder H (1993) A noncovalent peptide complex as a model for an early folding intermediate of cytochrome c. Biochemistry 32:10271–10276

37. Merrill AR, Cohen FS, Cramer WA (1990) On the nature of the structural change of the colicin E1 channel peptide necessary for its translocation-competent state. Biochemistry 29:5829–5836

38. Fontana A, Polverino de Laureto P, De Philipps V (1993) Molecular aspects of proteolysis of globular proteins. In: van den Tweel W, Harder A, Buitelear M (eds) Protein stability and stabilization. Elsevier, Amsterdam, pp 101–110

39. Semisotnov GV, Rodionova NA, Razgulyaev OI, Uversky VN, Gripas AF, Gilmanshin RI (1991) Study of the "molten globule" intermediate state in protein folding by a hydrophobic fluorescent probe. Biopolymers 31:119–128

40. Uversky VN, Winter S, Lober G (1996) Use of fluorescence decay times of 8-ANS-protein complexes to study the conformational transitions in proteins which unfold through the molten globule state. Biophys Chem 60:79–88

41. Uversky VN (1993) Use of fast protein size-exclusion liquid chromatography to study the unfolding of proteins which denature through the molten globule. Biochemistry 32:13288–13298

42. Romero P, Obradovic Z, Li X, Garner EC, Brown CJ, Dunker AK (2001) Sequence complexity of disordered protein. Proteins 42:38–48

43. Williams RM, Obradovi Z, Mathura V, Braun W, Garner EC, Young J, Takayama S, Brown CJ, Dunker AK (2001) The protein non-folding problem: amino acid determinants of intrinsic order and disorder. Pac Symp Biocomput, 89–100

44. Daughdrill GW, Pielak GJ, Uversky VN, Cortese MS, Dunker AK (2005) Natively disordered proteins. In: Buchner J, Kiefhaber T (eds) Handbook of protein folding. Wiley-VCH, Weinheim, Germany, pp 271–353

45. Vacic V, Uversky VN, Dunker AK, Lonardi S (2007) Composition profiler: a tool for discovery and visualization of amino acid composition differences. BMC Bioinformatics 8:211

46. Uversky VN, Gillespie JR, Fink AL (2000) Why are "natively unfolded" proteins unstructured under physiologic conditions? Proteins 41:415–427

47. Zambelli B, Stola M, Musiani F, De Vriendt K, Samyn B, Devreese B, Van Beeumen J, Turano P, Dikiy A, Bryant DA, Ciurli S (2005) UreG, a chaperone in the urease assembly process, is an intrinsically unstructured GTPase that specifically binds Zn++. J Biol Chem 280:4684–4695

48. Eftink MR (1994) The use of fluorescence methods to monitor unfolding transitions in proteins. Biophys J 66:482–501

49. Simeoni F, Masotti L, Neyroz P (2001) Structural role of proline residues of the β-hinge region of p13suc1 as revealed by site-directed mutagenesis and fluorescence studies. Biochemistry 40:8030–8042

50. Lakowicz JR (1999) Principle of fluorescence spectroscopy. Plenum Publishing, New York

51. Pace CN (1986) Determination and analysis of urea and guanidine hydrochloride denaturation curves. Methods Enzymol 131:266–280

52. Neyroz P, Zambelli B, Ciurli S (2006) Intrinsically disordered structure of Bacillus pasteurii UreG as revealed by steady-state and time-resolved fluorescence spectroscopy. Biochemistry 45:8918–8930

53. Goldenberg DP, Creighton TE (1984) Gel electrophoresis in studies of protein conformation and folding. Anal Biochem 138:1–18

54. Uversky VN (1994) Gel-permeation chromatography as a unique instrument for quantitative and qualitative analysis of protein denaturation and unfolding. Int J Bio-Chromatogr 1:103–114

55. Corbett RJ, Roche RS (1983) The unfolding mechanism of thermolysin. Biopolymers 22:101–105

56. Corbett RJ, Roche RS (1984) Use of high-speed size-exclusion chromatography for the study of protein folding and stability. Biochemistry 23:1888–1894

57. Uversky VN, Semisotnov GV, Pain RH, Ptitsyn OB (1992) 'All-or-none' mechanism of the molten globule unfolding. FEBS Lett 314:89–92

58. Georlette D, Blaise V, Dohmen C, Bouillenne F, Damien B, Depiereux E, Gerday C, Uversky VN, Feller G (2003) Cofactor binding modulates the conformational stabilities and unfolding patterns of NAD(+)-dependent DNA ligases from Escherichia coli and Thermus scotoductus. J Biol Chem 278:49945–49953

59. Privalov PL (1979) Stability of proteins: small globular proteins. Adv Protein Chem 33:167–241

60. Ptitsyn OB, Uversky VN (1994) The molten globule is a third thermodynamical state of protein molecules. FEBS Lett 341:15–18

61. Uversky VN, Ptitsyn OB (1996) All-or-none solvent-induced transitions between native, molten globule and unfolded states in globular proteins. Fold Des 1:117–122

62. Ptitsyn OB (1987) Protein folding: hypotheses and experiments. J Prot Chem 6:273–293

63. Kuwajima K (1989) The molten globule state as a clue for understanding the folding and cooperativity of globular-protein structure. Proteins 6:87–103

64. Christensen H, Pain RH (1991) Molten globule intermediates and protein folding. Eur Biophys J 19:221–229

65. Rodionova NA, Semisotnov GV, Kutyshenko VP, Uverskii VN, Bolotina IA (1989) Staged equilibrium of carbonic anhydrase unfolding in strong denaturants. Mol Biol (Mosk) 23:683–692

66. Hill TL (1963-1964) Thermodynamics of the small systems. Wiley, New York

67. Grosberg AY, Khokhlov AR (1989) Statistical physics of macromolecules. Nauka, Moscow

Chapter 13

Identification of Intrinsically Disordered Proteins by a Special 2D Electrophoresis

Agnes Tantos and Peter Tompa

Abstract

Intrinsically disordered proteins (IDPs) lack a well-defined three-dimensional structure under physiological conditions. They constitute a significant fraction of various proteomes and have significant roles in key cellular processes. Here we report the development of a two-dimensional electrophoresis technique for their de novo recognition and characterization. This technique consists of the combination of native and 8 M urea electrophoresis of heat-treated proteins where IDPs are expected to run into the diagonal of the gel, whereas globular proteins either precipitate upon heat treatment or unfold and run off the diagonal in the second dimension.

Key words: Intrinsically unstructured protein, Large-scale separation, Proteomics, 2D electrophoresis, Heat-stable proteins, Diagonal electrophoresis

1. Introduction

The long-standing dogma that tied protein function to a well-defined three-dimensional structure has been increasingly challenged by the recognition that for many proteins/protein domains the native, functional state is intrinsically unstructured/disordered (1–4). Such proteins constitute a significant fraction of various proteomes: from studies based on their sequence attributes (3, 5) and heat stability (6) it has been ascertained that as much as 25% of all residues may fall into disordered regions in the proteomes of different species. IDPs have so far been identified by the chance observation of the structural anomaly of proteins studied for their functional interest. Systematic studies aimed at identifying novel IDPs are all the more compelling as IDPs play essential physiological and pathological roles (2–4, 7). We reasoned that a straightforward technique to separate IDPs from globular proteins in a cellular extract could be established by the

Vladimir N. Uversky and A. Keith Dunker (eds.), *Intrinsically Disordered Protein Analysis:*
Volume 2, Methods and Experimental Tools, Methods in Molecular Biology, vol. 896,
DOI 10.1007/978-1-4614-3704-8_13, © Springer Science+Business Media New York 2012

combination of a native gel electrophoresis of heat-treated proteins followed by a second, denaturing gel containing 8 M urea. The rationale for the first dimension is that IDPs are very often heat stable, thus heat treatment results in a good initial separation from globular proteins, most of which aggregate and precipitate. In the native gel, IDPs and rare heat-stable globular proteins will then be separated according to their charge/mass ratios. Combining this first dimension with an 8 M urea second step is rationalized by the usual structural indifference of IDPs to chemical denaturation by trichloroacetic acid, guanidine HCl, or urea as reported for CsD1 (8), NACP (9), stathmin (10), and p21Cip1 (11), for example. As urea is uncharged and IDPs are just as "denatured" in 8 M urea as under native conditions, they are expected to run the same distance in the second dimension and end up along the diagonal. Heat-stable globular proteins, on the other hand, will unfold in urea, slow down in the second gel, and arrive above the diagonal. Because of this effective separation, IDPs are amenable to subsequent identification by mass spectrometry. The technique has been set up by a variety of controls and tested in terms of its performance in identifying novel IDPs from cellular extracts and characterizing proteins for structural disorder (12).

The technique is reproducible, is easy to perform, and is readily adaptable to a high-throughput format. Its comparison with other experimental and bioinformatic techniques showed that it provides dependable assessment of global structural disorder even in contradictory cases. Although its resolving power does not match that of the conventional 2D technique, it enables specific applications in two directions. Its first practical application is the rapid characterization of a single protein in terms of its disorder status. A simple run of the native/8 M-urea 2D electrophoresis can tell with high certainty if the protein is ordered or disordered in solution (Fig. 1). Given that the technique can provide information on a protein of very small quantity and limited purity, it will be a useful complement to other techniques that are more demanding on protein quantity and quality. The other application is in the analysis of cellular extracts. Using this technique we could identify a range of novel, mostly disordered proteins from *Escherichia coli* and *Saccharomyces cerevisiae* (Fig. 1) (13) and *Drosophila melanogaster* (12) extracts. Whereas the 2D technique offers unique applications, it has some limitations as well. Because of the application of a native gel in the first dimension, its resolving power does not match that of the conventional 2D electrophoresis. This can be partially overcome by applying mild, non-charged detergents. An important further limitation is that under the standard conditions of native electrophoresis, proteins of net positive charge are lost. As roughly half of the IDPs have a basic isoelectric point (14), they are lost in this type of analysis.

Notwithstanding these limitations, this technique is capable of yielding significant information on the identity of novel intrinsically disordered proteins.

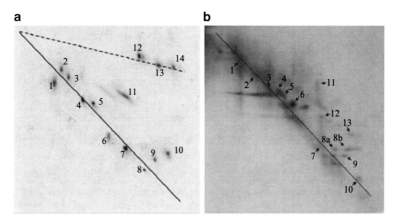

Fig. 1. The native/8 M urea 2D electrophoresis of IUPs and globular proteins. (**a**) A mixture of proteins (1 μg of each) was run on a 7.5% native gel in the first dimension (without heat treatment) and on a 7.5% gel containing 8 M urea in the second dimension. Individual proteins marked are as follows: IUPs: (1) stathmin; (2) MAP2c; (3) Mypt1-(304–511); (4) ERD10; (5) -casein; (6) NACP; (7) Csd1; (8) Bob-1; (9) DARPP32; and (10) -casein; globular proteins: (11) fetuin; (12) IPMDH; (13) BSA; and (14) ovalbumin. The continuous line marks the diagonal of the gel to where IUPs run. The *dashed line* marks the position of globular proteins. Picture taken from ref. 13. (**b**) Separation and identification of IUPs from *S. cerevisiae*. A heat-treated extract of *S. cerevisiae*. *Dots* marked were cut out and sent for MS identification. Picture taken from ref. 13.

2. Materials

Prepare all solutions using deionized water and analytical grade reagents. Prepare and store all reagents at room temperature (unless indicated otherwise).

1. Native polyacrylamide gel components.

   (a) Stacking gel buffer: 500 mM Tris–HCl; pH 6.8. Dissolve 60.6 g Tris–HCl in about 900 ml of deionized water and adjust pH to 6.8 with HCl. Using a graduated 1-L glass cylinder adjust the volume with water to 1,000 ml. Store at 4°C.

   (b) Resolving gel buffer: 1.5 M Tris–HCl; pH 8.8. Dissolve 181.7 g Tris–HCl in about 900 ml of deionized water and adjust pH to 6.8 with HCl. Using a graduated 1-L glass cylinder adjust the volume with water to 1,000 mL. Store at 4°C.

   (c) 30% acrylamide/bisacrylamide solution: dissolve 146 g acrylamide monomer and 4 g Bis in about 100 ml water and mix for about 30 min (see Note 1). Adjust the volume to 500 ml with water and filter the solution with a 45-μm filter. Store in a bottle covered with aluminum foil at 4°C (see Note 2).

2. 8 M urea polyacrylamide gel components.

(a) Gel buffer: the same buffer as the separation gel buffer for the native gel, supplemented with 8 M urea. Dissolve 480.48 g urea in around 300 ml of gel buffer (see Note 3). Heat the solution to 60°C in order to make the solvation of urea easier. Adjust the volume to 1,000 ml with gel buffer. Store at room temperature.

(b) 30% acrylamide/bis solution with 8 M urea: dissolve 146 g acrylamide monomer and 4 g Bis in about 100 ml water and mix for about 30 min. Dissolve 240.24 g urea by gentle heating. Adjust the volume to 500 ml with water and filter the solution with a 45-μm filter. Store in a bottle covered with aluminum foil at room temperature (see Note 4).

3. Ammonium persulphate (APS) solution: 10% solution in water.

4. N,N,N,N'-Tetramethyl-ethylenediamine   (TEMED)   purchased from a commercial supplier. Store at 4°C.

5. Native PAGE running buffer: 0.025 M Tris–HCl, pH 8.3, 0.192 M glycine (see Note 5).

6. Sample buffer for the native PAGE (4×): 0.12 M Tris–HCl (pH 6.8), 0.1% bromophenol blue (BPB), 45% glycerol. Prepare 1-ml aliquots and store at −20°C. Immediately before use, add a few milligrams of DTE (see Note 6).

7. Coomassie brilliant blue stain: dissolve 0.1 g Coomassie Brilliant Blue 250 (CBB250 G) in 100 ml ethanol. Add 800 ml water and 3 ml concentrated HCl. Stir overnight. Next day adjust the volume to 1 L (see Note 7).

8. Proteins: prepare a protein mixture that contains both globular and disordered proteins, at approximately 1 mg/ml concentration each. Add your protein of unknown nature to this mixture (see Note 8).

3. Methods

All procedures should be carried out at room temperature unless otherwise specified.

1. First dimension

(a) Prepare the resolving gel by mixing 1.5 mL of resolving buffer, 1.5 mL of acrylamide mixture, and 3 mL water in a 50-mL conical flask. Add 45 μL of ammonium persulfate, and 6 μL of TEMED, and cast the gel within a 7.25 cm × 10 cm × 0.7 mm gel cassette. Use a separator cut short enough that the comb will fit on top of it to create a gel strip according to Fig. 2. Pour the solution on both sides of

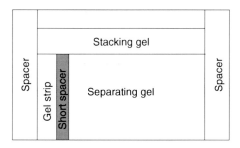

Fig. 2. Casting of the first-dimension gel strip.

the separator in order to avoid leakage from the strip. Allow space for the stacking gel and overlay gently with water.

(b) Prepare the stacking gel by mixing 1.0 mL of stacking gel buffer, 0.54 mL of acrylamide mixture, and 2.5 mL water in a 50-mL conical flask. Add 20 μL of ammonium persulfate, and 3 μL of TEMED. After cleaning away the water from the separating gel, pour the mixture on top of the separating gel and immediately insert a 10-well comb without introducing any bubbles (see Note 9).

(c) Mix the protein sample with the sample buffer. Do not boil the sample, but centrifuge it briefly at $3,000 \times g$ prior to loading on the gel. Load 20–30 μl of the mixture in the well that is positioned above the strip. No protein standard is needed for this procedure.

(d) Load the sample mixture in the well on top of the strip and run the first dimension at 180 V for about an hour, or until the dye front reaches the bottom of the gel.

(e) At the end of the run, separate the strip and rinse it in 8 M urea-containing gel buffer for 45 min at room temperature.

2. Second dimension

(a) Prepare the 8 M urea gel by mixing 1.5 mL of urea containing resolving buffer, 1.5 mL of 8 M urea containing acrylamide mixture, and 3 mL water in a 50-mL conical flask. Add 45 μL of ammonium persulfate, and 6 μL of TEMED, and cast a rectangular gel within a 7.25 cm × 10 cm × 1 mm (see Note 10) gel cassette as shown on Fig. 3 (see Note 11). No stacking gel is needed for this dimension (see Note 12).

(b) Insert the gel strip from the first dimension on top of the rectangular gel. Care should be taken to avoid any remaining air bubbles between the two gels (see Note 13).

(c) Run the second dimension at the same voltage as the first dimension, but 2.5 times longer (see Note 14).

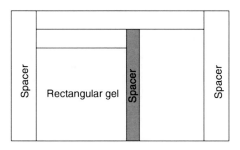

Fig. 3. Casting of the second-dimension, rectangular gel.

3. Staining

    (a) Pour some deionized water on the gel and microwave at maximum capacity for 60 s. Repeat this step after changing of the water (see Note 15).

    (b) Add 20 mL coomassie stain and heat the gel in microwave oven for 30 s. Shake in the stain for 2–5 min.

    (c) Destain the gel with deionized water (see Note 16).

4. Notes

Here are a few practical notes, which help avoid most of the problems when using this method.

1. Wear a mask when weighing acrylamide. Transfer the weighed acrylamide to the cylinder inside the fume hood and mix on a stirrer placed inside the hood. Unpolymerized acrylamide is neurotoxic and care should be exercised to avoid skin contact.

2. The acrylamide solution can be stored at 4°C for 1 month. The acrylamide mixture and buffer can be prepared in large batches, frozen in aliquots, and used indefinitely.

3. Dissolving 8 M urea can be a difficult task, gentle heating of the solution might be necessary. Also, solvation of urea involves a large increase of volume, so it is advisable to take that into account when preparing the buffers. Add small amounts of solvent until urea is completely dissolved then adjust the volume to the desired level.

4. All solutions containing urea must be prepared freshly and stored at room temperature until use (8 M urea precipitates in the cold room, which can be very annoying). Solutions older than 3 days should be discarded. On gels prepared with old urea solutions heavy smearing can be experienced.

5. Running buffer is the same for the first and second dimensions, urea is only polymerized into the gel.

6. Since in the second dimension the strip from the first dimension is used for sample loading, no sample buffer for the second dimension is needed.

7. This coomassie stain cannot be reused and should be discarded after staining each gel.

8. This method is also suitable for separating whole cell extracts. For this purpose lyse the cells in appropriate lysis buffer (50 mM Tris/HCl, 150 mM NaCl, 2 mM dithioerythritol, pH 7.5 should be a good starting buffer). After disruption of the cell membrane by sonication, separate cell debris from proteins by centrifugating the sample at $100,000 \times g$ for 30 min at 4°C. Place the supernatant in a heat-stable tube and heat the sample for 10 min at 100°C. After boiling, separate aggregated proteins by centrifugation at $100,000 \times g$ for 30 min at 4°C. The sample can be stored in small aliquots at −80°C. Take an aliquot of 100 μg protein content and add sample buffer at appropriate volume before loading on the first dimension. No heating of the sample with the buffer is needed.

9. Make sure that one of the wells is positioned directly over the strip so that all the loaded sample enters the stip.

10. The gel strip swells significantly during this time; this is why it is necessary that the second dimension is cast with thicker separators. In our experience, a first gel of 0.7 mm swells in 8 M urea so that it fits tightly within the slabs of the second 1.0 mm apart.

11. It is a good idea to prepare a rectangular paper or plastic template for the correct positioning of the spacer.

12. Take care to leave enough space to fit the strip from the first dimension.

13. Place a few drops of running buffer between the two dimensions to ensure ideal attachment of the two gels since remaining air bubbles can impede the entering of proteins into the second dimension.

14. Since the first dimension is not stained, no dye front appears here, so the measurement of time is the only guideline for the running length of the second dimension. Adjustment of running voltage and time may be necessary for the proteins to run precisely to the diagonal of the rectangular gel.

15. Gels should be warmed but not boiled.

16. For greater sensitivity gels can be stained also by silver staining. Use any silver stain kit that is commercially available and follow the instructions provided by the manufacturer, but if proteins will be identified by mass spectrometry, MS compatible stains should be chosen.

References

1. Uversky VN (2002) Natively unfolded proteins: a point where biology waits for physics. Protein Sci 11:739–756
2. Wright PE, Dyson HJ (1999) Intrinsically unstructured proteins: re-assessing the protein structure-function paradigm. J Mol Biol 293:321–331
3. Dunker AK, Brown CJ, Lawson JD, Iakoucheva LM, Obradovic Z (2002) Intrinsic disorder and protein function. Biochemistry 41:6573–6582
4. Tompa P (2002) Intrinsically unstructured proteins. Trends Biochem Sci 27:527–533
5. Dunker AK, Obradovic Z, Romero P, Garner EC, Brown CJ (2000) Intrinsic protein disorder in complete genomes. Genome Inform Ser Workshop Genome Inform 11:161–171
6. Kim TD, Ryu HJ, Cho HI, Yang CH, Kim J (2000) Thermal behavior of proteins: heat-resistant proteins and their heat-induced secondary structural changes. Biochemistry 39:14839–14846
7. Tompa P, Csermely P (2004) The role of structural disorder in the function of RNA and protein chaperones. FASEB J 18:1169–1175
8. Hackel M, Konno T, Hinz H (2000) A new alternative method to quantify residual structure in 'unfolded' proteins. Biochim Biophys Acta 1479:155–165
9. Weinreb PH, Zhen W, Poon AW, Conway KA, Lansbury PT Jr (1996) NACP, a protein implicated in Alzheimer's disease and learning, is natively unfolded. Biochemistry 35:13709–13715
10. Belmont LD, Mitchison TJ (1996) Identification of a protein that interacts with tubulin dimers and increases the catastrophe rate of microtubules. Cell 84:623–631
11. Kriwacki RW, Hengst L, Tennant L, Reed SI, Wright PE (1996) Structural studies of p21Wafl/Cipl/Sdi1 in the free and Cdk2-bound state: conformational disorder mediates binding diversity. Proc Natl Acad Sci U S A 93:11504–11509
12. Szollosi E, Bokor M, Bodor A, Perczel A, Klement E, Medzihradszky KF, Tompa K, Tompa P (2008) Intrinsic structural disorder of DF31, a Drosophila protein of chromatin decondensation and remodeling activities. J Proteome Res 7:2291–2299
13. Csizmok V, Szollosi E, Friedrich P, Tompa P (2006) A novel two-dimensional electrophoresis technique for the identification of intrinsically unstructured proteins. Mol Cell Proteomics 5:265–273
14. Uversky VN, Gillespie JR, Fink AL (2000) Why are "natively unfolded" proteins unstructured under physiologic conditions? Proteins 41:415–427

Chapter 14

pH-Induced Changes in Intrinsically Disordered Proteins

Matthew D. Smith and Masoud Jelokhani-Niaraki

Abstract

Intrinsically disordered proteins are typically enriched in amino acids that confer a relatively high net charge to the protein, which is an important factor leading to the lack of a compact structure. There are many different approaches that can be used to experimentally confirm whether a protein is intrinsically disordered. One such approach takes advantage of the distinctive amino acid composition to test whether a protein is a genuine IDP. In particular, the conformation of the protein can be monitored at different pHs; as opposed to globular or ordered proteins, IDPs will typically gain structure under highly acidic or basic conditions. Here, we describe circular dichroism and fluorescence spectroscopic experimental approaches in which the conformation of proteins is monitored as pH is altered as a way of testing whether the protein behaves as an intrinsically disordered protein.

Key words: Intrinsically disordered proteins, pH, Conformational analysis, Circular dichroism spectroscopy, Fluorescence spectroscopy

1. Introduction

Intrinsically disordered proteins differ from ordered proteins in numerous ways. One of the most fundamental ways in which these two types of proteins differ is in their amino acid composition. Typically, IDPs contain relatively few hydrophobic residues and are enriched in amino acids carrying a net charge at physiological pH (1, 2). Indeed, it has been clearly shown that IDPs and ordered proteins contain distinct amino acid compositions (1, 3, 4), which supports the hypothesis that the amino acid sequence of a protein determines its propensity to fold or remain disordered. The electrostatic repulsion among similarly charged amino acids and weak hydrophobic interaction due to the lack of hydrophobic residues in IDPs combine to prevent a high degree of folding at physiological pH. There are numerous examples of IDPs that contain a preponderance of either basic (e.g. Mylein Basic Protein and the N-terminal domains

Vladimir N. Uversky and A. Keith Dunker (eds.), *Intrinsically Disordered Protein Analysis:*
Volume 2, Methods and Experimental Tools, Methods in Molecular Biology, vol. 896,
DOI 10.1007/978-1-4614-3704-8_14, © Springer Science+Business Media New York 2012

of the core histones) (5–9) or acidic (e.g. Toc159 A-domain and prothymosin α) (10, 11) residues and therefore have a notably high or low pI, respectively.

If intramolecular electrostatic repulsion among similarly charged residues is at least partially responsible for the lack of defined structure of IDPs at physiological pH, it follows that removal of these charges should allow for an increase in secondary structure formation, presumably by decreasing the destabilizing electrostatic interactions and allowing for increasing hydrophobic or other non-covalent interactions among the residues. Indeed, there are many examples of IDPs that have a low pI whose folding increases at correspondingly low pH (e.g. see ref. 10). Likewise, folding of IDPs with a high pI increases at high pH. For example, Mylein Basic Protein is an IDP that carries a strong positive charge (pI ~10) (6), as do the core Histone proteins (pI ~11) (8, 9). On the other extreme, the so-called Acidic domain of Toc159, a receptor for preproteins destined for the chloroplasts of plant cells, has a low theoretical pI (pI = 4.0), as does human prothymosin α (pI ~3.5), and both have been shown to be intrinsically unstructured (10, 11). Monitoring the structure of these proteins at a range of pHs served as a diagnostic to demonstrate that they are genuine IDPs.

The unique amino acid composition of IDPs and their characteristic features and behavior provide a convenient way to test experimentally, whether a protein is a genuine IDP (3). A globular, or ordered, protein will typically lose structure at extreme pH. An IDP on the other hand, will typically gain structure as the pH of its environment approaches its pI. For example, most IDPs are unstructured at neutral pH; an IDP with a high pI will remain unstructured in low pH conditions, but will gain structure as the pH increases.

A suitable and straightforward spectroscopic method to monitor the effect of pH on the conformation of IDPs is CD (circular dichroism) spectroscopy. A CD spectrum of optically active biomolecules, such as proteins, gives the difference in absorption of the left- and right-handed circularly polarized lights. CD spectroscopy is a widely used technique in estimating the secondary structure and overall conformation of proteins (12, 13). The far-UV range of the CD spectrum (below 250 nm) is particularly useful in obtaining information about the backbone structure of proteins and is most sensitive to overall conformational changes of proteins. To a lesser extent conformational changes of proteins can be also detected in the near-UV (250–300 nm) and near-UV-visible (300–700 nm) range of CD spectra. These CD spectral ranges are particularly sensitive to local environments of protein side-chain chromophores (aromatics and disulfides) and chromophoric prosthetic groups, respectively. The far-UV range of the CD spectrum can be specifically utilized to detect the pH-induced conformational changes in IDPs (for a detailed description see chapters in section VII of this volume).

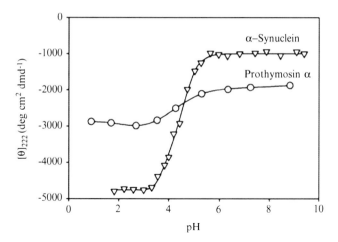

Fig. 1. The $[\theta]_{222}$ measured using far-UV CD showing the effect of pH on the structural properties of α-synuclein and prothomysin α (from Uversky, 2009) (3).

Folding or partial folding of IDPs typically occur at pH values close to their pI values. For acidic and basic IDPs, pI values are several pH units below and above the neutral pH 7, respectively. As a general trend, at pH close to pI, the less-structured IDPs gain more ordered structures. Interestingly, pH-induced folding of IDPs can be modeled as a two-state cooperative process (S-shaped graphs of ellipticity vs. pH), which is comparable to the two-state model for folding/unfolding of globular proteins (3, 14) (Fig. 1). The difference between the two processes is that the globular proteins may lose their ordered structures under highly acidic or basic conditions, whereas IDPs, depending on their acidic or basic nature, can gain more ordered structures under either acidic or basic conditions (Fig. 2). The pH-induced conformational changes in IDPs can be monitored by CD at 220 nm, the approximate maximum negative ellipticity in helical structures assigned to the electronic $n - \pi^*$ transition in the backbone amide bond of proteins (13). Acquiring dominantly ordered helical and/or β structures in IDPs under acidic or basic conditions can be attributed to overall change in the balance of non-covalent (such as electrostatic, hydrogen bonding, and hydrophobic) interactions in these proteins.

Another sensitive and simple spectroscopic technique for studying pH-dependent conformational change in IDPs is fluorescence spectroscopy (15). Generally, the conformational change can be detected by changes in the intensity and maximum emission wavelength shifts of intrinsic fluorophores in proteins. Of the three intrinsic fluorophores in proteins (Phe, Tyr, and Trp), Trp is the most sensitive and has the highest fluorescence intensity. Change in the local environment around Trp in proteins results in conspicuous changes in the emission spectrum of this amino acid;

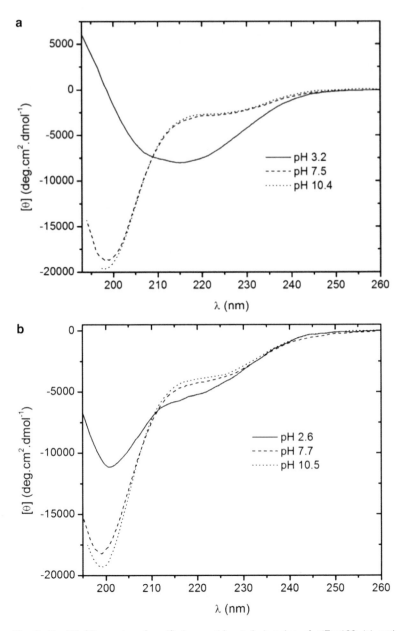

Fig. 2. Far-UV CD spectra of purified recombinant A-domains of atToc132 (**a**) and atToc159 (**b**) measured at different pHs (10).

less polar environments typically cause a blue shift in the maximum wavelength and increase in the intensity of Trp emission. Comparatively, Tyr has less fluorescence intensity than Trp and does not show considerable maximum emission wavelength shifts in different environments; Tyr emission drastically changes at high pH due to deprotonation of its phenolic group (16). Phe fluorescence is much weaker and less sensitive than Trp or Tyr. Changes in the local environment of Trp or Tyr in IDPs can be used to monitor their

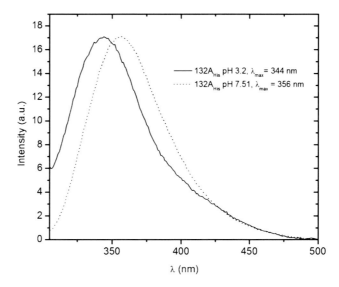

Fig. 3. Intrinsic fluorescence of the purified recombinant A-domain of atToc132 excited at 295 nm measured at pH 3.2 and 7.5 (10).

pH-induced conformational changes (Fig. 3). In the presence of both Tyr and Trp, two excitation wavelengths at 280 nm and 295 nm are typically used. The latter excitation wavelength eliminates the effect of Tyr emission and is specific to Trp. At 280 nm both Trp and Tyr are excited and the emission contains the contribution of both fluorophores, which can be resolved to its components (15, 17). Fluorescence spectroscopy can be therefore used as a complementary method to CD spectroscopy for studying the effect of pH on structural properties of IDPs. It is worth mentioning that the change in the microenvironment of the aromatic amino acids as a result of pH variation can be also monitored in the near-UV range of CD spectra (13, 18). In comparison to fluorescence, near-UV CD measurements are less sensitive and require higher sample quantities.

Overall, using small amounts of sample (1 mg/mL or less), CD and fluorescence spectroscopy are sensitive and efficient techniques for quick analysis of the effect of pH on the conformation of IDPs.

2. Materials

The method presented here was used to confirm that domains of two homologous proteins, atToc159 and atToc132, behaved as IDPs (10). The cDNAs corresponding to the domains, called acidic domains because of their preponderance of acidic amino acids, were cloned such that they could be expressed as recombinant His$_6$-tagged proteins in *Escherichia coli* (see Note 1) in the absence of the other domains of the proteins. The A-domains of atToc159 and atToc132 have theoretical pIs of 4.0 and 4.25, respectively. The overall conformation of the A-domains was then monitored using far-UV CD spectroscopy. Because atToc132A contains a single Trp residue, its intrinsic fluorescence was also monitored at a range of pHs. Fluorescence was not measured for atToc159, because this domain does not contain any Trps.

1. Purified protein(s) or peptide(s) of study (see Note 2).

2. Test solutions at a range of pHs. The exact composition of the test solution and the exact pHs will vary depending on the protein(s) under investigation. At a minimum, there should be a test solution at low pH, one at neutral pH, and one at high pH. The buffer that is used may also differ for each protein; however, if CD will be used to monitor protein conformation, the buffer salt should be CD-compatible possessing minimal or no absorbance in the range of measured CD spectrum. In addition, it is important for the test solution to have some ionic strength. At higher ionic strengths sometimes fluoride containing salts (such as NaF) can be used, since compared to other halides fluoride ions have minimal absorbance in the far-UV range of CD spectrum (19). In the case of studying the A-domains of atToc159 (pI = 4.0) and atToc132 (pI = 4.25), the "CD Buffer" contained 10 mM Tris–HCl and 50 mM NaCl and was adjusted to pHs of ~3, ~7.5, and ~10 (10).

3. Instrument for monitoring the conformation or overall structure of the protein. A CD spectrometer can be used in the far-UV range to monitor secondary structure change and to determine the phase-transition pH (see Subheading 3.2); and a fluorescence spectrometer can be used to detect conformational changes based on changes in the microenvironment of aromatic amino acids (see Note 3).

3. Methods

3.1. Protein Preparation (see Note 2)

1. Recombinant proteins expressed in *E. coli* can be purified using standard techniques if they have been produced as fusion proteins with a fusion tag that offers a convenient method of purification. For example, proteins produced with a hexahistidine tag fused to one end of the protein, can be rapidly purified using immobilized metal affinity chromatography. His-tags have the benefit of being small, and not dramatically affecting the overall charge of the protein (see Note 1). Other small fusion tags (e.g. myc-tags, FLAG-tags) can also be used to facilitate purification. Many excellent references are available

on how to purify affinity-tagged recombinant proteins, and a chapter of this volume is devoted to the production of recombinant IDPs.

2. If necessary, the purified recombinant protein can be purified further using a polishing step. For example, protein purity can be improved using gel filtration or an ion-exchange step. For detailed protocols on the production of recombinant IDPs and their purification, refer to section VII of this volume, which contains chapters devoted to the expression and purification of IDPs.

3. Purity of the protein should be confirmed using SDS-PAGE, and its concentration determined using a standard protein quantification technique (20).

4. For a protein purified in near-neutral pH solution, at least two other aliquots of the protein should be prepared and adjusted to low and high pH. This can be accomplished by adding concentrated acid (HCl) or base (KOH) directly to the protein, and monitoring the pH.

3.2. Conformational Analysis

All measurements should be taken at the same temperature, which in most cases should be room temperature.

1. The far-UV CD spectra of the test proteins are measured at each pH (broad range: extreme acidic to neutral to extreme basic). Alternatively, the absorbance at a single wavelength (e.g. 222 nm) can be monitored at a series of pH values.

2. The broad range pH-dependent overall CD spectral change can be specifically used to detect the global conformational changes of the IDPs. Depending on the nature of IDP (acidic or basic), the conformational change towards a more structured protein occurs at pH values close to the protein's pI.

3. Monitoring the pH-dependent conformational change of IDPs at specific wavelengths sensitive to ordered structures (such as 222 for helical and 215 for β-sheet structures) can be used to study the conformational transition based on a two-state (unstructured → structured) model. In this way a phase-transition pH can be defined where 50 % of the protein is transformed to its structured or partially structured form. This phase-transition pH and its changes are particularly useful for comparing the conformational change propensity of IDPs and detection of their interaction with other biomolecules or ions.

4. On the other hand, pH-dependent fluorescence spectra of IDPs can also exhibit their conformational change. The condition for detecting this conformational change is both the presence of environmentally sensitive aromatic amino acids such as Trp or Tyr in IDPs, as well as their position in the

sequence and change in their microenvironment upon the conformational change in the protein. Detection of pH-dependent change in the microenvironment of Trp and Tyr from a polar to nonpolar or less-polar environment is of particular interest, as it shows that a folded domain is forming in the IDP. Tryptophan is the most sensitive intrinsic fluorophore to environmental changes and can be potentially used to complement the pH-dependent far-UV CD monitoring of conformational change in IDPs.

4. Notes

1. A hexahistidine chain is typically located either at the N- or C-terminus of proteins and is not long enough to form a particular secondary structure on its own. Histidine side-chain is neutrally charged at high or neutral pH, but carries a positive charge at low pH. Generally it is plausible that the hexahistidine chain does not influence the overall charge of a recombinant protein at neutral and basic pH and its overall influence on the conformation of the protein in the pH range used for spectroscopic measurements is negligible.

2. Test proteins can be recombinantly expressed and purified proteins, endogenous proteins purified from tissue extracts, or synthetic peptides. What is most critical is that the protein is of high purity, so that structural changes can be definitively attributed only to the protein of interest. Furthermore, if the protein interacts with any other biomolecule thought to be important for its biological function (e.g. cofactors, lipids), it will be important to include those with the protein as well.

3. In this chapter we have focused on the use of far-UV CD spectroscopy and intrinsic fluorescence spectroscopy for monitoring the overall conformation and structure of the protein in question. However, one can theoretically use any spectroscopic method that can detect changes in protein conformation or structure if neither of these instruments is available. For example, small-angle X-ray scattering (SAXS) can be used to monitor protein dimension and shape, as can high-resolution techniques such as NMR. In principle, lower-resolution techniques such as size-exclusion chromatography and dynamic light scattering can also be used to detect global changes in protein folding at different pH levels. Refer to other chapters in this volume for detailed protocols on how these approaches that can be used to assess IDPs.

References

1. Uversky VN, Gillespie JR, Fink AL (2000) Why are natively unfolded proteins unstructured under physiologic conditions? Proteins 41:415–427

2. Tompa P (2002) Intrinsically unstructured proteins. Trends Biochem Sci 27:527–533

3. Uversky VN (2009) Intrinsically disordered proteins and their environment: effects of strong denaturants, temperature, pH, counter ions, membranes, binding partners, osmolytes and macromolecular crowding. Protein J 28:305–325

4. Lise S, Jones DT (2005) Sequence patterns associated with disordered regions in proteins. Proteins 58:144–150

5. Harauz G, Ishiyama N, Hill CMD, Bates IR, Libich DS, Fares C (2004) Myelin basic protein – diverse conformational states of an intrinsically unstructured protein. Micron 35:503–542

6. Boggs JM (2006) Myelin basic protein: a multifunctional protein. Cell Mol Life Sci 63:1945–1961

7. Hansen JC, Lu X, Ross ED, Woody RW (2006) Intrinsic protein disorder, amino acid composition, and histone terminal domains. J Biol Chem 281:1853–1856

8. Hansen JC, Tse C, Wolffe AP (1998) Structure and function of the core histone N-termini: more than meets the eye. Biochemistry 37:17637–17641

9. Munishkina LA, Fink AL, Uversky VN (2004) Conformational prerequisites for formation of amyloid fibrils from histones. J Mol Biol 342:1305–1324

10. Richardson LGL, Jelokhani-Niaraki M, Smith MD (2009) The acidic domains of the Toc159 chloroplast preprotein receptor family are intrinsically disordered protein domains. BMC Biochem 10:35

11. Uversky VN, Gillespie JR, Millet IS, Khodyakova AV, Vasiliev AM, Chernovskaya TV, Vasilenko RN, Kozlovskaya GD, Dolgikh DA, Fink AL, Doniach S, Abramov VM (1999) Natively unfolded human prothymosin a adopts partially folded collapsed conformation at acidic pH. Biochemistry 38:15009–15016

12. Berova N, Nakanishi K, Woody RW (eds) (2000) Circular dichroism: principles and applications, 2nd edn. Wiley-VCH, New York

13. Fasman GD (ed) (1996) Circular dichroism and the conformational analysis of biomolecules. Plenum, New York

14. Finkelstein A, Ptitsyn OB (2002) Protein physics. Academic, Amsterdam, pp 227–263

15. Lakowicz JR (2006) Principles of fluorescence spectroscopy, 3rd edn. Springer, New York

16. Ross JBA, Laws WR, Wyssbrod HR (1992) Tyrosine fluorescence and phosphorescence from proteins and polypeptides. In: Lakowicz JR (ed) Topics in fluorescence spectroscopy, vol. 3: biochemical applications. Plenum, New York, pp 1–63

17. Boteva R, Zlateva T, Dorovska-Taran V, Visser AJWG, Tsanev R, Salvato B (1996) Dissociation equilibrium of human recombinant interferon γ. Biochemistry 35:14825–14830

18. Seerama N, Manning MC, Powers ME, Zhang J-X, Goldenberg DP, Woody RW (1999) Tyrosine, phenylalanine, and disulfide contributions to the circular dichroism of proteins: circular dichroism spectra of wild-type and mutant bovine pancreatic trypsin Inhibitor. Biochemistry 38:10814–10822

19. Ivanova MV, Hoang T, McSorley FR, Krnac G, Smith MD, Jelokhani-Niaraki M (2010) A comparative study on conformation and ligand binding of the neuronal uncoupling proteins. Biochemistry 49:512–521

20. Pingoud A, Urbank C, Hoggett J, Jeltsch A (2002) Biochemical methods. Wiley-VCH, New York

Chapter 15

Temperature-Induced Transitions in Disordered Proteins Probed by NMR Spectroscopy

Magnus Kjaergaard, Flemming M. Poulsen, and Birthe B. Kragelund

Abstract

Intrinsically disordered proteins are abundant in nature and perform many important physiological functions. Multidimensional NMR spectroscopy has been crucial for the understanding of the conformational properties of disordered proteins and is increasingly used to probe their conformational ensembles. Compared to folded proteins, disordered proteins are more malleable and more easily perturbed by environmental factors. Accordingly, the experimental conditions and especially the temperature modify the structural and functional properties of disordered proteins. NMR spectroscopy allows analysis of temperature-induced structural changes at residue resolution using secondary chemical shift analysis, paramagnetic relaxation enhancement, and residual dipolar couplings. This chapter discusses practical aspects of NMR studies of temperature-induced structural changes in disordered proteins.

Key words: Residual structure, Polyproline II, Chemical shift, Random coil, Transient helicity

1. Introduction

Since the discovery of intrinsically disordered proteins (IDPs), considerable interest has been invested in characterizing the functional ensembles of these proteins. The goal of these investigations is to understand the connection between transient structure in IDPs and their functions. Many techniques have been used in this characterization as amply demonstrated by the chapters in this volume. Recent biophysical experiments of IDPs have demonstrated that their ensemble distributions are sensitive to the environmental parameters including temperature (1). More importantly, the structural changes induced by changes in temperature may have a significant influence on the function of the IDP. Such an effect has been seen in the fibrillation of α-synuclein, which depends strongly on temperature (2). The temperature

Vladimir N. Uversky and A. Keith Dunker (eds.), *Intrinsically Disordered Protein Analysis: Volume 2, Methods and Experimental Tools*, Methods in Molecular Biology, vol. 896, DOI 10.1007/978-1-4614-3704-8_15, © Springer Science+Business Media New York 2012

dependence of residual structure in IDPs thus allows examination of the "non-structure/function relationship" of these proteins.

In nature, proteins operate at temperatures ranging from below the freezing point of water to boiling hot deep-sea thermal vents. In thermophilic organisms, proteins have evolved to withstand the denaturing effect of high temperatures (1, 2). Generally, the thermophilic globular proteins are more rigid than their mesophilic homologues, when compared at similar temperature. At the temperatures at which the proteins natively operate, however, meso- and thermophilic proteins have comparable flexibility as assessed from amide hydrogen exchange measurements (3, 4). For intrinsically disordered proteins (IDPs), such studies of variations in flexibility as a function of temperature are not available. Recent bioinformatics analyses of genomes of archaea have shown that the species that live at very high temperatures have lower predicted disorder in their proteome than mesophilic archaea (5). This may be intuitively understandable as increased temperatures in itself leads to an increase in intramolecular flexibility. In a similar bioinformatics analysis of 300 prokaryotes, the predicted structural disorder within the proteomes was correlated to growth temperatures (6). Here both cold and hot-adapted organisms had a similar or lower predicted disorder in their proteome compared to the mesophilic organisms. Although the origins for the lower disorder within proteomes of organism living at extreme temperatures are unknown, the above analyses suggest that the degree of disorder in a given protein may be optimized for its operating temperatures.

Due to the strong dependence of structural ensembles on temperature, it is important to describe the structure ensemble at the physiological temperature of the organism. Due to practical considerations, however, this may be an intricate and not always feasible task. Therefore, biophysical measurements are often performed near room temperature. These results thus need to be extrapolated to the natural operating temperature of the protein, which requires a fundamental knowledge of how temperature changes affect the structural ensembles of disordered proteins.

Temperature-induced changes in disordered proteins can be analyzed with many different biophysical techniques including the most widely applied methods of circular dichroism (CD) spectroscopy, dynamic light scattering, small angle X-ray scattering, FRET, and molecular dynamics simulations. From far-UV CD spectroscopy data, it is apparent that increasing temperature induces a structural transition in both IDPs and chemically denatured proteins (Fig. 1) (7–15). The change in the CD spectrum is, however, ambiguous and it can be interpreted as if higher temperature induces either an increase of the α-helix content, a loss of polyproline II or a combination of the two. Dynamic light scattering (16), FRET (16) and small angle X-ray scattering (14) at various temperatures estimate the overall compactness of the

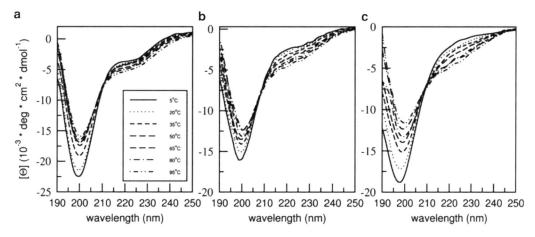

Fig. 1. Far UV CD spectra of the activation domain of ACTR (**a**), the intracellular domain of the human Na+/H+ exchanger 1 (**b**) and Spd1 (**c**) at a range of temperatures from 5 to 95°C. At higher temperatures, the negative signal at 198 nm lose intensity and the broad signal at 222 nm gain intensity, which is similar to what have been observed for many disordered proteins. Adapted from ref. 14 with permission of John Wiley & Sons, Inc.

molecules. In concert, these techniques show that disordered proteins become more compact at higher temperatures (14). Single molecule FRET experiments have demonstrated that the temperature-induced compaction does not depend on the hydrophobicity of the protein (16), and it is thus likely that compaction reports on a general property of the peptide backbone rather than on an increased hydrophobic collapse. Consistent with the spectroscopic data, molecular dynamics simulations suggested that polyproline II conformations are lost at higher temperature (16), which mirrors the behavior of small disordered peptides (17). The techniques discussed above probe the global properties of disordered proteins. The averaging that is intrinsic to global properties means that these methods possibly conceal opposite changes occurring in different parts of the protein. To understand the temperature effects in detail, it is necessary to break down the global behavior into a residue specific description.

2. Practical Aspects of Studying Temperature Effects in IDPs by NMR

2.1. Sample Preparations

NMR spectroscopy has emerged as the main tool for analysis of the conformational properties of disordered proteins (18–20). The NMR methods for characterizing disordered proteins in general are covered elsewhere in this volume, so we restrict our discussion to temperature-induced structural changes. In general, high resolution NMR characterization of a disordered protein requires stable isotope labeling with ^{15}N and ^{13}C. For proteins produced recombinantly in bacteria or yeast, this is a routine task and is discussed elsewhere

Table 1
Suggested buffer conditions for NMR
studies of temperature-dependent changes
in IDPs

0.5–2 mM protein labeled with ^{15}N and ^{13}C
20 mM NaH$_2$PO$_4$/Na$_2$HPO$_4$ buffer pH 6.5 or lower
5–10% (v/v) D$_2$O
25 μM DSS
0.02% (v/v) NaN$_3$

(21, 22). When the isotope labeled protein has been produced to high purity (>95%), a number of decisions have to be made regarding the sample conditions. Temperature changes alter the pH of a buffered solution due to the temperature dependence of pK_a. When the specific effect of temperature on an IDP is being analyzed, it is important to minimize the temperature-induced change in pH. The pK_a of sodium phosphate buffer has the lowest temperature coefficient amongst the buffers in common use in biochemistry (23). Furthermore, phosphate buffer is desirable for NMR as all its protons exchange rapidly with the solvent, and thus do not contribute to the NMR spectrum. Additionally, due to several pK_a values, this buffer is also applicable to the study of acid unfolded proteins. In general, buffer and salt concentration should be kept low as the ionic strength decreases the sensitivity of especially cryogenic NMR probes (24). This is particularly important for polyionic buffers such as phosphate (25). Amide protons in IDPs exchange rapidly with the solvent due to the lack of protection from structure formation. As the most common proton NMR experiments are detected on the exchangeable amide protons, it is necessary to minimize proton exchange. The exchange rate depends on both pH and temperature (26). To retain signals from all residues at physiological pH, it is often necessary to record NMR spectra of IDPs below room temperature. In temperature studies, where a range of temperatures are needed, it is thus essential to operate at a pH lower than 7 to observe signals at the higher temperatures. We find that a pH of approximately 6.5 is a good compromise that is close to physiological conditions and distinct from the pI of most IDPs, but slows down amide proton exchange rates enough to allow NMR detection at moderately elevated temperatures.

The NMR sample needs to contain a few more components (Table 1). The spectrometer uses the ^{2}H$_2$O signal to lock the magnetic field so it is required to include 5–10% (v/v) ^2H_2O. It is also recommended to add 0.02% (v/v) NaN_3 to the NMR sample to avoid microbial growth. This may have an effect on pH, so it is important to add all additives before the final adjustment of pH. To correctly reference the chemical shifts, small quantities of DSS should be included. Due to the absence of a packed hydrophobic core (27), IDPs do not have upfield shifted methyl groups. Accordingly the DSS signal will always have the lowest proton chemical shift in the sample and is per definition at 0 ppm (28). Studies of the BRMB database suggest that incorrect referencing of chemical shifts is common (29, 30). As the backbone chemical shifts are important structural probes of disordered proteins, it is crucial to ensure that the chemical shifts are referenced correctly. The ^{13}C and ^{15}N chemical shifts are referenced indirectly to the methyl proton signal of DSS using the ratios between the gyromagnetic ratios as described by Wishart et al. (28).

2.2. Temperature Control and Measurements

In a study of temperature effects on sample properties, it is important to carefully control the temperature of the experiment. Commercially available NMR spectrometers have built-in temperature controls, but the reported values are rarely accurate. The temperature of the sample relative to the displayed temperature should thus be calibrated prior to the NMR experiment. The temperature offset depends on the temperature and of the flow rate of the cooling air used in the spectrometer, so it is necessary to calibrate the temperature using exactly the same settings of the air cooling system as used for the recording of the protein spectra. The deviation between the actual temperature in the sample and the reading on the spectrometer is rarely constant over the temperature range; hence it is important to measure the temperature throughout the temperature range needed for the study. The temperature of the NMR sample can be determined by recording a 1D ^1H experiments of a sample of pure methanol (31). The temperature is then calculated based on the difference between the chemical shifts of the methyl protons and the hydroxyl proton, $\Delta\delta$, according to the following formula:

$$T = -16.7467 \times (\Delta\delta)^2 - 52.5130 \times \Delta\delta + 419.1381$$

Temperature calibration using methanol samples is limited by the boiling point of methanol, and is thus restricted to temperatures below approximately 45°C. Alternatively, the temperature can be measured directly by mounting a thermocouple into an NMR tube containing a small amount of water and inserting the tube into the spectrometer. In this procedure, it is of the utmost importance to ensure that the thermocouple and thermometer is compatible with the high magnetic field of the spectrometer. We favor the thermocouple method, as it is fast and directly measures the temperature within the NMR tube.

2.3. Assignment of NMR Signals

When the sample is ready, the first task in an NMR investigation is the assignment of the signals. In spite of the poor proton dispersion in disordered proteins, sequential assignment can routinely be performed using a set of backbone experiments as discussed previously (21). A convenient set of experiments for assignments of a disordered protein is HNCO (32, 33), HNCACO, HNCACB (34, 35), CBCACONH (36), and HNN (37). These experiments are available in the pulse sequence libraries that come with the spectrometer, and can routinely be acquired with the assistance of the staff of an NMR facility. The HNN experiment is not strictly necessary, but it is useful for assignments of disordered proteins due to the higher dispersion in the ^{15}N dimension. For the assignment of folded proteins, the HNCA (38) and HNCOCA (39) experiments are usually also recorded, but due to the favorable relaxation properties of disordered proteins these experiments can generally be omitted. For small peptides that cannot easily be isotopically labeled, it is possible to assign the chemical shifts using long range carbonyl HSQC spectra at natural isotope abundance (40). Recently alternative pulse sequences have been proposed that do not rely on detection on the exchangeable amides. Instead, the signals can be detected on the α-protons (41) or directly on carbons (42). These methods allow NMR experiments to be performed at physiological pH and temperature, which is not generally possible for IDPs using H^N-detected experiments due to fast hydrogen exchange rates (2). Despite the obvious advantages of these new methods, they are not yet widely available in the standard experiment libraries of spectrometer producers, and may be more laborious to set up.

2.4. Temperature-Induced Changes in NMR Parameters

Most NMR parameters change with a change in temperature either due to changes in the structural ensemble or due to an intrinsic temperature dependence of the parameter. The suitability of an NMR parameter for probing temperature-induced structural changes will thus depend on whether it is possible to separate contributions from the structural ensemble from its intrinsic temperature dependence. ^{15}N relaxation measurements, for example, are commonly used to identify regions with restricted local motion in disordered proteins (43, 44). It will, however, be difficult to compare relaxation measurements obtained at different temperatures, as the peptide chain intrinsically will be more dynamic at higher temperatures even if the structural ensemble remains unchanged. As the NMR methods used to characterize IDPs in general are described elsewhere in this volume, we will restrict this discussion to the methods especially suitable for investigation of temperature-induced structural changes.

In order to probe temperature-dependent changes in an NMR parameter as for example in the chemical shifts, it is necessary to record the experiments in question at multiple temperatures.

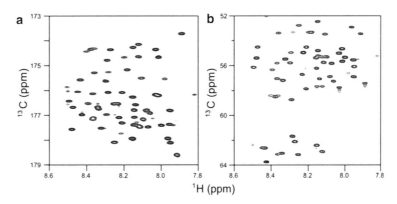

Fig. 2. 2D versions of HNCO (**a**) and HNCOCA (**b**) recorded on the 71-residue activation domain of ACTR. In small or medium sized IDPs, the triple resonance correlation experiments used for sequential assignment have sufficient dispersion in a 2D version to allow determination of almost all the chemical shifts needed for secondary structure analysis. This allows for rapid analysis of temperature-induced changes in the chemical shifts.

As most of the NMR experiments discussed are 3D experiments, this potentially requires a large amount of expensive instrument time. For small and medium sized proteins, however, 2D versions of for example the HNCO and HNCOCA experiments usually have sufficient dispersion to allow determination of the most conformationally sensitive ^{13}C backbone chemical shifts (Fig. 2). The 2D experiment omits the indirect nitrogen dimension, which cuts down the experiment time drastically.

2.5. Structural Information from Chemical Shifts

The backbone chemical shifts of disordered proteins depend on the populations of transient secondary structures in the protein (45, 46). As the backbone chemical shifts are available immediately after sequential assignment, they have become one of the main NMR parameters for characterizing disordered proteins by NMR spectroscopy (21). Accordingly, chemical shift analysis has been the principal method for investigation of temperature-induced secondary structural changes (2, 14, 47, 48). Chemical shifts are the NMR parameter that can be measured most precisely, which makes it ideally suited for detection of the subtle changes induced by temperature changes. The structural information from the chemical shifts is extracted by subtracting random coil chemical shifts from the experimentally obtained chemical shifts as follows:

$$\Delta\delta = \delta - \delta_{rc}$$

This difference is called the secondary chemical shift. Secondary chemical shift analysis can be routinely performed for all backbone nuclei, however, the C^{α}, C^{β}, and C' chemical shifts contain most secondary structure information (49). In α-helices, C^{α} and C' have positive secondary chemical shifts, while C^{β} has negative secondary chemical shifts. In β-sheets, the secondary chemical shifts are of the opposite sign. Polyproline II helices have been suggested to have C^α and C' chemical shifts similar to the random coil, but can be identified by secondary chemical shifts that are positive for N (50) and negative for C^β (51). A fully formed α-helix has C^α and C' secondary chemical shifts between 2 and 4 ppm depending on the residue type (29), and for transiently formed secondary structures in disordered proteins the secondary chemical shift is assumed to be scaled by the population of the structured element. By recording the chemical shifts as a function of temperature, it is possible to probe the effects of temperature on the populations of transiently formed secondary structures over the entire protein sequence.

2.6. Choice of Random Coil Chemical Shifts and Their Temperature Dependence

The quality of a secondary chemical shift analysis is determined by the suitability of the random coil chemical shifts. The chemical shifts of a protein can be measured very accurately provided that the spectra can be recorded with a narrow line width and a good signal to noise ratio. In practice, however, the reproducibility of the chemical shift measurements does not determine the error of the estimated secondary structure populations. Instead, the main source of errors in the secondary chemical shifts is the assumptions made in the determination of the random coil chemical shifts. It is unlikely that any protein is a true random coil under any conditions, so the random coil chemical shifts can only be approximated. There are two main methods for determining the random coil chemical shifts, one based on small unstructured peptides (52, 53) and one based on chemical shift databases (29, 54, 55). It is important to choose a random coil dataset that matches the solution conditions of the protein under investigation. A particular important parameter is solution pH, where a mismatch between the random coil dataset and the experimental conditions will lead to spikes in the secondary chemical shifts of aspartate, glutamate, and histidine residues. It should thus be noted that one of the most popular random coil dataset published by Schwarzinger et al., is intended for acid denatured proteins (52, 56), and is not appropriate for studying IDPs at neutral pH due to the effects of side chain protonation. Histidine has a side chain pK_a value in the pH range where most NMR studies are normally carried out, and it is thus important to determine its random coil chemical shift at the exact pH of the study. Recently, a pH titration of a histidine random coil peptide was reported (40), which allows the random coil chemical shifts to be matched to those of the protein sample. This approach requires that the pK_a values of the histidine side chains in the protein are comparable to the pK_a values of the histidine in the random coil peptides. This is likely to be the case for IDPs due to the absence of well-defined tertiary interactions, even though the pK_a values of histidine residues in highly charged regions may be perturbed by their sequence environment. In contrast, database derived chemical shifts do not correspond to a specific set of solution conditions, but represent the average conditions of protein NMR studies in the database. It is important to note that so far only few chemical shift dataset of IDPs are available (55). If a database derived random coil chemical shift set is used, it is important to consider if the solution conditions deviate from the database average, and if possible to correct for this deviation as described for temperature corrections below. Besides the solution conditions, random coil chemical shifts differ in whether they correct for the effects of sequential neighbors or not. Sequence correction is particularly important for residues preceding prolines, as the neighbor effect of a proline is an order of magnitude larger than for the other residue types (56).

The random coil chemical shifts also depend on the sample temperature. This means that the observed chemical shift changes as a function of temperature will be the sum of the contribution from changes in the random coil chemical shifts and from changes in the structural ensemble. To isolate the chemical shift changes due to structural transitions, the contributions from the random coil chemical shifts must be subtracted. The temperature coefficients of random coil chemical shifts derived from peptides have been reported for ^{1}H (57), ^{15}N (50), and most importantly ^{13}C (40). Using these coefficients, the temperature dependence of the random coil chemical shift can be estimated using the following equation, where a is the temperature coefficient:

$$\delta_{rc}(T) = \delta_{rc}(25°C) + a \times (T - 25).$$

Alternatively, the temperature dependence of the random coil chemical shifts can be estimated by analyzing the same protein in the presence of a high concentration of denaturant but under otherwise matching conditions. This approach is called *intrinsic random coil referencing* and effectively removes the sequence dependence of the random coil chemical shifts, which results in cleaner secondary chemical shifts (14, 58). The ^{13}C random coil chemical shifts are largely independent of the urea concentration, even though it has been suggested that urea might induce structure in disordered proteins (40). By recording the chemical shifts of the denatured states at the same temperatures and in the same buffer system as the native state, it is possible to remove the temperature dependence of the random coil chemical shifts. This method was used to evaluate temperature-dependent changes in the populations of transient α-helices in the activation domain of the intrinsically disordered ACTR (Fig. 3), and demonstrated that the α-helix populations decreased with increasing temperature (14). This observation is the opposite of what is usually concluded from far-UV CD data, but reflects the usual temperature dependence of transient structures in peptides (59). Several studies have suggested that the temperature-induced changes in disordered proteins may

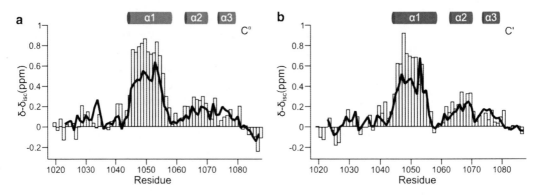

Fig. 3. Secondary chemical shifts of C$^\alpha$ (**a**) and C′ (**b**) for the activation domain of ACTR recorded at 5°C (*bars*) and 45°C (*line*). Intrinsic statistical coil chemical shifts are determined using a urea denatured state of the same protein at otherwise identical conditions in order to remove the intrinsic temperature dependence of the statistical coil chemical shifts. The cylinders show the positions of α-helices of ACTR in complex with CREB binding protein (69, 70). The secondary chemical shifts show that the population of transient α-helices decrease with increasing temperature, which is the opposite of what was concluded from far UV CD data. Adapted from ref. 14 with permission of John Wiley & Sons, Inc.

reflect a redistribution of the statistical coil ensemble at higher temperature that is general to the polypeptide chain (14, 15). This process presumably also takes place in the random coil peptides used to determine the temperature dependence of the random coil chemical shifts, and thus the temperature corrected chemical shifts cannot be used to understand this phenomenon. Temperature correction of the chemical shifts, however, allows quantification of other structural changes on the background of the redistribution of the random coil ensemble.

Cooperative unfolding transitions have sigmoidal unfolding curves due to their all-or-none nature. In contrast, the intrinsic temperature dependence of the random coil chemical shifts is linear (40). When the chemical shifts of an IDP do not depend linearly on temperature, it thus signifies a cooperative structural change in the ensemble. In α-synuclein, the temperature dependence of the backbone chemical shifts is linear for some residues, but deviates significantly from linearity for other residues (2). Most likely, this reflects the difference between the temperature dependence of the random coil chemical shifts and a local cooperative unfolding transition.

2.7. Other NMR Techniques Applied to Temperature Transitions in IDPs

In addition to chemical shift analysis, several other NMR methods seem well suited to probe temperature-induced changes in IDPs, although few studies have been conducted so far. Paramagnetic relaxation enhancement (PRE) measurements have been widely used to probe transient tertiary interactions in disordered proteins (60–63) and are discussed in detail in Chapter 7 of this volume. The PRE is based on the addition of a paramagnetic group to the protein, usually by introduction of a cysteine by site-directed mutagenesis and a subsequent reaction with a cysteine-specific paramagnetic tag. The PRE is measured by comparing the NMR peak intensities (or the relaxation rates) of nuclei under both oxidizing (diamagnetic) and reducing (paramagnetic) conditions. The ratio of the intensities (I_{para}/I_{dia}) then reports on the closeness of the tag to a particular residue of the protein sequence. In a fully random coil peptide, these effects will only be local and resemble a bell-shaped curve that can be approximated by an ensemble of peptide chains generated randomly (62). PREs are well suited for studies of temperature-induced structural changes as the relaxation enhancement is unlikely to depend on the temperature in the absence of structural changes. Thus any change in the PRE profile across the sequence will in principle report on a structural change. Increasing temperatures may however in part increase the internal dynamics of the system, thereby increasing the rotation of the paramagnetic label. This may subsequently influence the residence time of the label and affect the intensity ratios and thus require more control experiments. So far, PREs from covalently attached paramagnetic groups have not been used to probe temperature-dependent transitions neither in IDPs nor in globular proteins, however, the PREs from noncovalently bound copper ions show differences in the structural ensembles of α-synuclein at different temperatures (2). How covalently attached PREs will evolve as suitable probes for temperature-induced structural changes in IDPs remains therefore to be seen.

Another NMR parameter that has been used with great success for disordered proteins is residual dipolar couplings (RDCs) (64). The practical considerations regarding determination of RDCs of disordered proteins have been discussed in detail elsewhere (65). RDCs are measured under conditions where the protein is partially aligned, and carries information about the orientation of nuclear vectors relative to the external magnetic field. Before recording RDCs at multiple temperatures, it should be ensured that the alignment medium is temperature stable and that interactions between the medium and the protein are constant over the applied temperature range. Alcohol-polyglycol and bicelle alignment media, for instance, are only stable in a narrow temperature range (66, 67), whereas polyacrylamide gels (68) are likely to be stable at a range of temperatures.

In conclusion, a detailed structural description of the temperature dependence of IDPs will contribute to our understanding of the functions of this important class of proteins. NMR spectroscopy provides a range of methods to probe these transitions at atomic resolution. A key challenge when addressing temperature-induced structural changes is to separate the intrinsic effects of the change in temperature from changes in the structural ensemble. This can either be achieved by studying the intrinsic effect of temperature in structure-less reference systems, or by choosing observables that do not depend on temperature. So far, few studies have focused on temperature-induced structural changes in IDPs, so our knowledge of temperature-induced structural transitions is based on few examples. Future studies will reveal how general the conclusions from the currently known studies are.

Acknowledgments

Gitte Wolfsberg Haxholm and Simon Erlendsson are thanked for critical comments on this manuscript.

References

1. Uversky VN (2009) Intrinsically disordered proteins and their environment: effects of strong denaturants, temperature, pH, counter ions, membranes, binding partners, osmolytes, and macromolecular crowding. Protein J 28 (7–8):305–325

2. Hsu ST, Bertoncini CW, Dobson CM (2009) Use of protonless NMR spectroscopy to alleviate the loss of information resulting from exchange-broadening. J Am Chem Soc 131 (21):7222–7223

3. Jaenicke R, Bohm G (1998) The stability of proteins in extreme environments. Curr Opin Struct Biol 8(6):738–748

4. Jaenicke R (2000) Do ultrastable proteins from hyperthermophiles have high or low conformational rigidity? Proc Natl Acad Sci USA 97 (7):2962–2964

5. Xue B, Williams RW, Oldfield CJ, Dunker AK, Uversky VN (2010) Archaic chaos: intrinsically disordered proteins in Archaea. BMC Syst Biol 4 Suppl 1:S1

6. Burra PV, Kalmar L, Tompa P (2010) Reduction in structural disorder and functional complexity in the thermal adaptation of prokaryotes. PLoS One 5(8):e12069

7. Jarvet J, Damberg P, Danielsson J, Johansson I, Eriksson LE, Graslund A (2003) A left-handed 3(1) helical conformation in the Alzheimer Abeta(12-28) peptide. FEBS Lett 555 (2):371–374

8. Uversky VN, Li J, Fink AL (2001) Evidence for a partially folded intermediate in alpha-synuclein fibril formation. J Biol Chem 276 (14):10737–10744

9. Dawson R, Muller L, Dehner A, Klein C, Kessler H, Buchner J (2003) The N-terminal domain of p53 is natively unfolded. J Mol Biol 332(5):1131–1141

10. Gast K, Zirwer D, Damaschun G (2003) Are there temperature-dependent structural transitions in the "intrinsically unstructured" protein prothymosin alpha? Eur Biophys J 31 (8):586–594

11. Jeganathan S, von Bergen M, Mandelkow EM, Mandelkow E (2008) The natively unfolded character of tau and its aggregation to Alzheimer-like paired helical filaments. Biochemistry 47(40):10526–10539

12. Sanchez-Puig N, Veprintsev DB, Fersht AR (2005) Human full-length Securin is a natively unfolded protein. Protein Sci 14(6):1410–1418

13. Malm J, Jonsson M, Frohm B, Linse S (2007) Structural properties of semenogelin I. FEBS J 274(17):4503–4510

14. Kjaergaard M, Norholm AB, Hendus-Altenburger R, Pedersen SF, Poulsen FM, Kragelund BB (2010) Temperature-dependent structural changes in intrinsically disordered proteins: formation of alpha-helices or loss of polyproline II? Protein Sci 19(8):1555–1564

15. Yang WY, Larios E, Gruebele M (2003) On the extended beta-conformation propensity of polypeptides at high temperature. J Am Chem Soc 125(52):16220–16227

16. Nettels D, Muller-Spath S, Kuster F, Hofmann H, Haenni D, Ruegger S, Reymond L, Hoffmann A, Kubelka J, Heinz B, Gast K, Best RB, Schuler B (2009) Single-molecule spectroscopy of the temperature-induced collapse of unfolded proteins. Proc Natl Acad Sci USA 106(49):20740–20745

17. Shi Z, Olson CA, Rose GD, Baldwin RL, Kallenbach NR (2002) Polyproline II structure in a sequence of seven alanine residues. Proc Natl Acad Sci USA 99(14):9190–9195

18. Dyson HJ, Wright PE (2002) Insights into the structure and dynamics of unfolded proteins

from nuclear magnetic resonance. Adv Protein Chem 62:311–340

19. Mittag T, Forman-Kay JD (2007) Atomic-level characterization of disordered protein ensembles. Curr Opin Struct Biol 17(1):3–14

20. Eliezer D (2009) Biophysical characterization of intrinsically disordered proteins. Curr Opin Struct Biol 19(1):23–30

21. Eliezer D (2007) Characterizing residual structure in disordered protein States using nuclear magnetic resonance. Methods Mol Biol 350:49–67

22. Pickford AR, O'Leary JM (2004) Isotopic labeling of recombinant proteins from the methylotrophic yeast *Pichia pastoris*. Methods Mol Biol 278:17–33

23. Dawson RMC, Elliot DC, Elliot WH, Jones KM (1986) Data for biochemical research, 3rd edn. Oxford University Press, Oxford, UK.

24. Voehler MW, Collier G, Young JK, Stone MP, Germann MW (2006) Performance of cryogenic probes as a function of ionic strength and sample tube geometry. J Magn Reson 183(1):102–109

25. Kelly AE, Ou HD, Withers R, Dotsch V (2002) Low-conductivity buffers for high-sensitivity NMR measurements. J Am Chem Soc 124 (40):12013–12019

26. Teilum K, Kragelund BB, Poulsen FM (2005) Application of hydrogen exchange kinetics to studies of protein folding. Protein folding handbook. Wiley-VCH Verlag GmbH. doi:10.1002/9783527619498.ch18

27. Kjaergaard M, Iešmantavičius V, Poulsen FM (in press) The interplay between transient α-helix formation and side chain rotamer distributions in disordered proteins probed by methyl chemical shifts. Protein Sci 20(12):2023–34.

28. Wishart DS, Bigam CG, Yao J, Abildgaard F, Dyson HJ, Oldfield E, Markley JL, Sykes BD (1995) 1H, 13C and 15N chemical shift referencing in biomolecular NMR. J Biomol NMR 6(2):135–140

29. Zhang H, Neal S, Wishart DS (2003) RefDB: a database of uniformly referenced protein chemical shifts. J Biomol NMR 25(3):173–195

30. Wang Y, Wishart DS (2005) A simple method to adjust inconsistently referenced 13C and 15N chemical shift assignments of proteins. J Biomol NMR 31(2):143–148

31. Findeisen M, Brand T, Berger S (2007) A 1H-NMR thermometer suitable for cryoprobes. Magn Reson Chem 45(2):175–178

32. Grzesiek S, Bax A (1992) Improved 3D triple-resonance NMR techniques applied to a 31-kDa protein. J Magn Reson 96(2):432–440

33. Ikura M, Kay LE, Bax A (1990) A novel approach for sequential assignment of 1H, 13C, and 15N spectra of proteins: heteronuclear triple-resonance three-dimensional NMR spectroscopy. Application to calmodulin. Biochemistry 29(19):4659–4667

34. Wittekind M, Mueller L (1993) Hncacb, a high-sensitivity 3D NMR experiment to correlate amide-proton and nitrogen resonances with the alpha-carbon and beta-carbon resonances in proteins. J Magn Reson B 101 (2):201–205

35. Kay LE, Xu GY, Yamazaki T (1994) Enhanced-sensitivity triple-resonance spectroscopy with minimal H_2O saturation. J Magn Reson A 109(1):129–133

36. Grzesiek S, Bax A (1992) Correlating backbone amide and side-chain resonances in larger proteins by multiple relayed triple resonance NMR. J Am Chem Soc 114(16):6291–6293

37. Panchal SC, Bhavesh NS, Hosur RV (2001) Improved 3D triple resonance experiments, HNN and HN(C)N, for HN and 15N sequential correlations in (13C, 15N) labeled proteins: application to unfolded proteins. J Biomol NMR 20(2):135–147

38. Ikura M, Kay LE, Bax A (1990) A novel-approach for sequential assignment of H-1, C-13, and N-15 spectra of larger proteins – heteronuclear triple-resonance 3-dimensional NMR-spectroscopy – application to calmodulin. Biochemistry 29(19):4659–4667

39. Yamazaki T, Muhandiram R, Kay LE (1994) NMR experiments for the measurement of carbon relaxation properties in highly enriched, uniformly C-13, N-15-labeled proteins – application to C-13(alpha) carbons. J Am Chem Soc 116(18):8266–8278

40. Kjaergaard M, Brander S, Poulsen FM (2011) Random coil chemical shift for intrinsically disordered proteins: effects of temperature and pH. J Biomol NMR 49(2):139–149

41. Mantylahti S, Aitio O, Hellman M, Permi P (2010) HA-detected experiments for the backbone assignment of intrinsically disordered proteins. J Biomol NMR 47(3):171–181

42. Bermel W, Bertini I, Felli IC, Lee YM, Luchinat C, Pierattelli R (2006) Protonless NMR experiments for sequence-specific assignment of backbone nuclei in unfolded proteins. J Am Chem Soc 128(12):3918–3919

43. Ebert MO, Bae SH, Dyson HJ, Wright PE (2008) NMR relaxation study of the complex formed between CBP and the activation domain of the nuclear hormone receptor coactivator ACTR. Biochemistry 47 (5):1299–1308

44. Danielsson J, Liljedahl L, Barany-Wallje E, Sonderby P, Kristensen LH, Martinez-Yamout MA, Dyson HJ, Wright PE, Poulsen FM, Maler L, Graslund A, Kragelund BB (2008) The intrinsically disordered RNR inhibitor Sml1 is a dynamic dimer. Biochemistry 47 (50):13428–13437

45. Wishart DS, Sykes BD, Richards FM (1991) Relationship between nuclear magnetic resonance chemical shift and protein secondary structure. J Mol Biol 222(2):311–333

46. Spera S, Bax A (1991) Empirical correlation between protein backbone conformation and C.alpha. and C.beta. 13C nuclear magnetic resonance chemical shifts. J Am Chem Soc 113(14):5490–5492

47. Kim HY, Heise H, Fernandez CO, Baldus M, Zweckstetter M (2007) Correlation of amyloid fibril beta-structure with the unfolded state of alpha-synuclein. Chembiochem 8 (14):1671–1674

48. Wu KP, Kim S, Fela DA, Baum J (2008) Characterization of conformational and dynamic properties of natively unfolded human and mouse alpha-synuclein ensembles by NMR: implication for aggregation. J Mol Biol 378 (5):1104–1115

49. Marsh JA, Singh VK, Jia Z, Forman-Kay JD (2006) Sensitivity of secondary structure propensities to sequence differences between alpha- and gamma-synuclein: implications for fibrillation. Protein Sci 15(12):2795–2804

50. Lam SL, Hsu VL (2003) NMR identification of left-handed polyproline type II helices. Biopolymers 69(2):270–281

51. Iwadate M, Asakura T, Williamson MP (1999) C alpha and C beta carbon-13 chemical shifts in proteins from an empirical database. J Biomol NMR 13(3):199–211

52. Schwarzinger S, Kroon GJ, Foss TR, Wright PE, Dyson HJ (2000) Random coil chemical shifts in acidic 8 M urea: implementation of random coil shift data in NMR view. J Biomol NMR 18(1):43–48

53. Wishart DS, Bigam CG, Holm A, Hodges RS, Sykes BD (1995) 1H, 13C and 15N random coil NMR chemical shifts of the common amino acids. I. Investigations of nearest-neighbor effects. J Biomol NMR 5(1):67–81

54. De Simone A, Cavalli A, Hsu ST, Vranken W, Vendruscolo M (2009) Accurate random coil chemical shifts from an analysis of loop regions in native states of proteins. J Am Chem Soc 131 (45):16332–16333

55. Tamiola K, Acar B, Mulder FA (2010) Sequence-specific random coil chemical shifts of intrinsically disordered proteins. J Am Chem Soc 132(51):18000–18003

56. Schwarzinger S, Kroon GJ, Foss TR, Chung J, Wright PE, Dyson HJ (2001) Sequence-dependent correction of random coil NMR chemical shifts. J Am Chem Soc 123 (13):2970–2978

57. Merutka G, Dyson HJ, Wright PE (1995) 'Random coil' 1H chemical shifts obtained as a function of temperature and trifluoroethanol concentration for the peptide series GGXGG. J Biomol NMR 5(1):14–24

58. Modig K, Jurgensen VW, Lindorff-Larsen K, Fieber W, Bohr HG, Poulsen FM (2007) Detection of initiation sites in protein folding of the four helix bundle ACBP by chemical shift analysis. FEBS Lett 581(25):4965–4971

59. Munoz V, Serrano L (1995) Elucidating the folding problem of helical peptides using empirical parameters. III. Temperature and pH dependence. J Mol Biol 245(3):297–308

60. Gillespie JR, Shortle D (1997) Characterization of long-range structure in the denatured state of staphylococcal nuclease. 1. Paramagnetic relaxation enhancement by nitroxide spin labels. J Mol Biol 268(1):158–169

61. Gillespie JR, Shortle D (1997) Characterization of long-range structure in the denatured state of staphylococcal nuclease. 2. Distance restraints from paramagnetic relaxation and calculation of an ensemble of structures. J Mol Biol 268(1):170–184

62. Teilum K, Kragelund BB, Poulsen FM (2002) Transient structure formation in unfolded acyl-coenzyme A-binding protein observed by site-directed spin labelling. J Mol Biol 324 (2):349–357

63. Lietzow MA, Jamin M, Dyson HJ, Wright PE (2002) Mapping long-range contacts in a highly unfolded protein. J Mol Biol 322 (4):655–662

64. Jensen MR, Markwick PR, Meier S, Griesinger C, Zweckstetter M, Grzesiek S, Bernado P, Blackledge M (2009) Quantitative determination of the conformational properties of partially folded and intrinsically disordered proteins using NMR dipolar couplings. Structure 17(9):1169–1185

65. Gebel EB, Shortle D (2007) Characterization of denatured proteins using residual dipolar couplings. Methods Mol Biol 350:39–48

66. Rückert M, Otting G (2000) Alignment of biological macromolecules in novel nonionic liquid crystalline media for NMR experiments. J Am Chem Soc 122(32):7793–7797

67. Fleming K, Matthews S (2004) Media for studies of partially aligned states. Methods Mol Biol 278:79–88

68. Tycko R, Blanco FJ, Ishii Y (2000) Alignment of biopolymers in strained gels: a new way to create detectable dipole–dipole couplings in high-resolution biomolecular NMR. J Am Chem Soc 122(38):9340–9341

69. Demarest SJ, Martinez-Yamout M, Chung J, Chen H, Xu W, Dyson HJ, Evans RM, Wright PE (2002) Mutual synergistic folding in recruitment of CBP/p300 by p160 nuclear receptor coactivators. Nature 415 (6871):549–553

70. Demarest SJ, Chung J, Dyson HJ, Wright PE (2002) Assignment of a 15 kDa protein complex formed between the p160 coactivator ACTR and CREB binding protein. J Biomol NMR 22(4):377–378

Chapter 16

Analyzing Temperature-Induced Transitions in Disordered Proteins by NMR Spectroscopy and Secondary Chemical Shift Analyses

Magnus Kjaergaard, Flemming M. Poulsen, and Birthe B. Kragelund

Abstract

Intrinsically disordered proteins are abundant in nature and perform many important physiological functions. Multidimensional NMR spectroscopy has been crucial for the understanding of the conformational properties of disordered proteins and is increasingly used to probe their conformational ensembles. Compared to folded proteins, disordered proteins are more malleable and more easily perturbed by environmental factors. Accordingly, the experimental conditions and especially the temperature modify the structural and functional properties of disordered proteins. This chapter discusses practical aspects of NMR studies of temperature-induced structural changes in disordered proteins using chemical shifts.

Key words: Intrinsically disordered protein, Transient secondary structure, Temperature dependence, Random coil chemical shifts

1. Introduction

The structural ensembles of IDPs depend strongly on their environment, including the temperature (1). Therefore, it is important to describe the structural ensemble of IDPs at the physiologically relevant temperature of the protein. Temperature-induced changes in disordered proteins have been analyzed with many different biophysical techniques, most frequently far UV CD spectroscopy (2–10). A residue-specific picture is, however, often needed as concealed opposite changes occurring in different parts of the protein are averaged in the global properties. The only technique capable of this task is Nuclear Magnetic Resonance (NMR) spectroscopy. Of the many methods used in protein NMR spectroscopy, chemical shifts analysis has been most widely used to probe temperature-dependent changes. Below, a protocol is

Vladimir N. Uversky and A. Keith Dunker (eds.), *Intrinsically Disordered Protein Analysis:*
Volume 2, Methods and Experimental Tools, Methods in Molecular Biology, vol. 896,
DOI 10.1007/978-1-4614-3704-8_16, © Springer Science+Business Media New York 2012

described for analyzing temperature-induced structural changes in IDPs using secondary chemical shifts (10–13). For a more comprehensive description and discussion of temperature-dependent structural changes, we refer to the essay with the title *Temperature-induced transitions in disordered proteins probed by NMR spectroscopy* in Chapter 15 this book.

2. Materials

1. Prepare all solutions using ultrapure water (prepared by purifying deionized water to attain a sensitivity of 18 MΩ cm at 25°C) and analytical grade reagents or better. Prepare and store all reagents at room temperature (unless indicated otherwise). Proteins are stored at 4°C or colder. NMR tubes with a sample volume of either 600 μL or 300 μL are available from Wilmad Labglass or Shigemi Inc., respectively. ^2H_2O is obtained in a quality of at least 99.96% purity. A thermocouple compatible with the magnetic field can be obtained from T.M. Electronics (for example type KA01-3).

2. A 20 mM NaH_2PO_4/Na_2HPO_4 buffer pH 6.5, 10% (v/v) D_2O, 25 μM 4,4-dimethyl-4-silapentane-1-sulfonic acid (DSS), 0.02% NaN_3, is prepared in a small volume (typically 10 mL) and filtered (see Notes 1 and 2). NaN_3 is included to prevent microbial growth.

3. If you are using intrinsic random coil referencing (10, 14), a corresponding sample containing 8 M urea is prepared. The final concentration of urea is measured using a refractometer as described (15).

4. For temperature calibration of the spectrometer, prepare either a 600 μL sample of ultrapure water or a sample of 100% methanol.

3. Methods

3.1. Sample Preparation for NMR

Before preparing the NMR sample, consider the optimal pH, salt, and buffer to use. A volume of either 300 or 600 μL is required according to the type of NMR tube used.

1. Prepare the sample in an Eppendorf tube either from lyophilized protein or from a concentrated sample prepared using centrifugal spin filters with an appropriate molecular weight cutoff. The final concentration of ^{15}N, ^{13}C-labeled protein should be in the range of 0.5–2 mM.

2. The protein is dissolved in 20 mM NaH_2PO_4/Na_2HPO_4 buffer at pH 6.5 or lower. The solution must contain 5–10% (v/v) D_2O which is used to lock the magnetic field. For correct referencing a final concentration of 25 μM DSS is included. The pH is checked and if necessary adjusted using a minimum of HCl and NaOH.

3. The sample is centrifuged at $13,000 \times g$ at 4°C for 5 min to sediment debris. Transfer the sample carefully to the NMR tube to avoid bubbles especially in the Shigemi tubes.

4. If intrinsic random coil referencing is used, prepare a similar sample in the same volume, buffer, and pH but including 8 M urea. Measure pH and correct if necessary.

3.2. Calibrate Temperature Controller on NMR Spectrometer

The temperature in the NMR spectrometer can be measured either directly using a thermocouple or indirectly using the proton signals of a methanol sample.

3.2.1. Thermocouple Method

1. Connect a 0.2-mm thin, magnet-compatible thermocouple to a thermometer. Measure the temperature of a water–ice mixture and ensure that the thermometer reads 0°C. If not, calibrate the thermometer according to the instructions of the manufacturer.

2. Make a small hole in a plastic lid of a non-Shigemi tube. Add 600 μL of water to the NMR tube and insert the thermocouple into the tube through the hole in the lid. Secure the lid and the thermocouple using parafilm.

3. Position the NMR sample into a spinner and insert it into the NMR spectrometer.

4. Adjust the temperature controller on the spectrometer to the desired temperature and wait 5 min for equilibrium to be established. Adjust the setting of the temperature controller until the thermometer measures the desired temperature. It is usually convenient to make a plot of T(set) vs. T(actual).

3.2.2. Methanol Method

The temperature is calculated based on the difference between the chemical shifts of the methyl protons and the hydroxyl proton, $\Delta\delta$, according to the following formula (16):

$$T = -16.7467 \times (\Delta\delta)^2 - 52.5130 \times \Delta\delta + 419.1381$$

1. Prepare a sample of pure methanol in an NMR tube. Insert the sample into the magnet and set the temperature at the desired value. Record a 1D ^{1}H experiment (see Note 3).

2. Measure the difference between the signals of the methyl protons and of the hydroxyl proton. Calculate the temperature according to the formula above.

3. Set the temperature to the next value and repeat the above measurement throughout the desired temperature range.

3.3. Record Triple Resonance NMR Spectra to Assign Resonances

In spite of the poor proton dispersion in the NMR spectra of disordered proteins, sequential assignment can routinely be performed using a set of backbone experiments as discussed previously (17).

1. Set the temperature controller of the spectrometer to the desired temperature using the calibrated values from Subheading 3.2.

2. Insert the protein NMR sample in the magnet.

3. Record a 1D ^{1}H NMR spectrum to reference the chemical shift. Set the frequency of DSS to 0 ppm. Calculate the reference chemical shifts of ^{15}N and ^{13}C as described by Wishart et al. (18).

4. Record HNCO (19, 20), HNCACO, HNCACB (21, 22), CBCACONH (23), and HNN (24) spectra as recommended by in house routines and if needed with the help from an NMR manager. For especially the HNCACB and CBCACONH experiments, it is important to record many increments in the indirect ^{13}C dimension (see Notes 4 and 5).

5. Transform and analyze the spectra as described in reference (17).

6. Assign the chemical shift as described in reference (17).

7. If intrinsic random coil referencing is used, repeat the procedure using the sample containing urea. The peaks in the urea denatured states usually are close to those in the absence of urea, so assignments can usually be transferred easily.

8. Repeat the procedure for all the desired temperatures.

3.4. Secondary Chemical Shift Analysis

The secondary chemical shifts can be calculated either using a set of tabulated random coil values or by intrinsic random coil referencing.

1. Select an appropriate set of random coil chemical shifts. We recommend using either of the two datasets made specifically for IDPs (25, 26). Notice that it is important to use a random coil data set with a pH that matches the pH of the sample (see Note 6). Calculate the random coil chemical shifts of the protein in question from its sequence as described (25, 26).

2. Random coil shifts also depend on temperature and it is necessary to correct for this effect. The tabulated random coil chemical shifts can be extrapolated to other temperatures using a set of temperature coefficients reported (25). If the random coil chemical shifts are determined at 25°C, the temperature-corrected random coil chemical shifts are calculated using the equation:

$$\delta_{rc}(T) = \delta_{rc}(25°C) + a \times (T - 25).$$

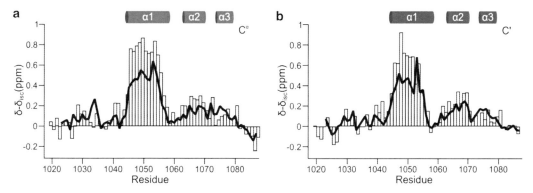

Fig. 1. Secondary chemical shifts of C^α (**a**) and C' (**b**) for the activation domain of ACTR recorded at 5°C (*bars*) and 45°C (*line*). The secondary chemical shifts are calculated using intrinsic random coil referencing, i.e., the chemical shifts of the urea denatured state are used as the random coil chemical shifts. The secondary chemical shifts show that the helices found in the complex with the CREB binding protein (34) are formed transiently in the unbound state. The populations of transient helices decrease at higher temperatures, which is the opposite of what was concluded from far UV CD data. Adapted from ref. 10 with permission of John Wiley & Sons, Inc.

3. Calculate the secondary chemical shifts for each of the ^{13}C nuclei using the following equation. In case of intrinsic random coil referencing, use the chemical shifts in urea as δ_{rc}.

$$\Delta\delta = \delta - \delta_{rc}$$

4. Plot the secondary chemical shifts as a function of residue number and evaluate the transient structure content of the IDP according to the following principles: α-helices have positive secondary chemical shifts for C^α and C' and negative chemical shifts for C^β. Extended β-strands have the secondary chemical shifts of the opposite sign to α-helices. Polyproline II helices have been suggested to have C^α and C' chemical shifts similar to the random coil, but can be identified by secondary chemical shifts that are positive for N (27) and negative for C^β (28). A fully formed α-helix has C^α and C' secondary chemical shifts between 2 and 4 ppm depending on the residue type (29). Transient secondary structures in disordered proteins are assumed to have secondary chemical shifts scaled by the population of the structured element. An example of a secondary chemical shift analysis of an IDP is shown in Fig. 1.

3.5. Mapping of Temperature-Induced Structural Changes Using Secondary Chemical Shifts

1. Record NMR spectra (e.g., 2D ^{15}N HSQC, HNCO, and HNCOCA) at multiple temperatures in the range of for example 5–50°C

2. Transfer the initial assignment to these spectra and adjust the peak positions.

3. Calculate the secondary chemical shifts at the various temperatures as described in Subheading 3.4.

4. Plot the differences in secondary shifts of C^{α}, C', and N between 5°C and 40°C (or any temperature). Secondary shift at two different temperatures are shown for an IDP in Fig. 1.

5. Evaluate the temperature-induced structural changes from analyses of the temperature-induced changes in secondary chemical shifts.

4. Notes

1. Temperature changes alter the pH of a buffered solution due to the temperature dependence of pK_a. When the specific effect of temperature on an IDP is being analyzed, it is important to minimize the temperature-induced change in pH. Phosphate buffers have low pK_a temperature dependence (30) and are thus well suited for temperature studies.

2. Amide protons in IDPs exchange rapidly with the solvent due to the lack of protection from structure formation. As the most common proton NMR experiments are detected on the exchangeable amide protons, it is necessary to minimize proton exchange. The exchange rate depends on both pH and temperature (31). To retain signals from all residues at physiological pH, it is often necessary to record NMR spectra of IDPs below room temperature. In temperature studies, where a range of temperatures is needed, it is thus essential to operate at a pH lower than 7 to observe signals at the higher temperatures. We find that a pH of approximately 6.5 is a good compromise between physiological conditions and optimal NMR conditions.

3. Temperature calibration using methanol samples is limited by the boiling point of methanol, and is thus restricted to temperatures below approximately 45°C.

4. For small and medium sized proteins, 2D versions of for example the HNCO and HNCOCA experiments usually have sufficient dispersion to allow determination of the most conformationally sensitive ^{13}C backbone chemical shifts.

5. Recently, NMR experiments that do not rely on detection of protons have been devised (32, 33). These experiments allow chemical shift measurements at physiological pH and elevated temperatures.

6. The quality of a secondary chemical shift analysis is determined by the suitability of the random coil chemical shifts. Choose your random coil shift set with great care to match pH, buffer, and temperature.

Acknowledgment

We thank our colleagues at SBiN-lab for sharing their practical experiences obtained from working with IDPs.

References

1. Uversky VN (2009) Intrinsically disordered proteins and their environment: effects of strong denaturants, temperature, pH, counter ions, membranes, binding partners, osmolytes, and macromolecular crowding. Protein J 28 (7–8):305–325

2. Jarvet J, Damberg P, Danielsson J, Johansson I, Eriksson LE, Graslund A (2003) A left-handed 3(1) helical conformation in the Alzheimer Abeta(12-28) peptide. FEBS Lett 555 (2):371–374

3. Uversky VN, Li J, Fink AL (2001) Evidence for a partially folded intermediate in alpha-synuclein fibril formation. J Biol Chem 276 (14):10737–10744

4. Dawson R, Muller L, Dehner A, Klein C, Kessler H, Buchner J (2003) The N-terminal domain of p53 is natively unfolded. J Mol Biol 332(5):1131–1141

5. Gast K, Zirwer D, Damaschun G (2003) Are there temperature-dependent structural transitions in the "intrinsically unstructured" protein prothymosin alpha? Eur Biophys J 31(8):586–594

6. Jeganathan S, von Bergen M, Mandelkow EM, Mandelkow E (2008) The natively unfolded character of tau and its aggregation to Alzheimer-like paired helical filaments. Biochemistry 47(40):10526–10539

7. Sanchez-Puig N, Veprintsev DB, Fersht AR (2005) Human full-length Securin is a natively unfolded protein. Protein Sci 14 (6):1410–1418

8. Malm J, Jonsson M, Frohm B, Linse S (2007) Structural properties of semenogelin I. FEBS J 274(17):4503–4510

9. Yang WY, Larios E, Gruebele M (2003) On the extended beta-conformation propensity of polypeptides at high temperature. J Am Chem Soc 125(52):16220–16227

10. Kjaergaard M, Norholm AB, Hendus-Altenburger R, Pedersen SF, Poulsen FM, Kragelund BB (2010) Temperature-dependent structural changes in intrinsically disordered proteins: formation of alpha-helices or loss of polyproline II? Protein Sci 19(8):1555–1564

11. Kim HY, Heise H, Fernandez CO, Baldus M, Zweckstetter M (2007) Correlation of amyloid fibril beta-structure with the unfolded state of alpha-synuclein. Chembiochem 8 (14):1671–1674

12. Wu KP, Kim S, Fela DA, Baum J (2008) Characterization of conformational and dynamic properties of natively unfolded human and mouse alpha-synuclein ensembles by NMR: implication for aggregation. J Mol Biol 378(5):1104–1115

13. Hsu ST, Bertoncini CW, Dobson CM (2009) Use of protonless NMR spectroscopy to alleviate the loss of information resulting from exchange-broadening. J Am Chem Soc 131 (21):7222–7223

14. Modig K, Jurgensen VW, Lindorff-Larsen K, Fieber W, Bohr HG, Poulsen FM (2007) Detection of initiation sites in protein folding of the four helix bundle ACBP by chemical shift analysis. FEBS Lett 581(25):4965–4971

15. Pace CN (1986) Determination and analysis of urea and guanidine hydrochloride denaturation curves. In: Hirs CHW, Timasheff SN (eds). Methods in enzymology, vol 131. Academic, Waltham, MA, pp 266–280

16. Findeisen M, Brand T, Berger S (2007) A 1H-NMR thermometer suitable for cryoprobes. Magn Reson Chem 45(2):175–178

17. Eliezer D (2007) Characterizing residual structure in disordered protein states using nuclear magnetic resonance. Methods Mol Biol 350:49–67

18. Wishart DS, Bigam CG, Yao J, Abildgaard F, Dyson HJ, Oldfield E, Markley JL, Sykes BD (1995) 1H, 13C and 15N chemical shift referencing in biomolecular NMR. J Biomol NMR 6(2):135–140

19. Grzesiek S, Bax A (1992) Improved 3D triple-resonance NMR techniques applied to a 31-kDa protein. J Magn Reson 96(2):432–440

20. Ikura M, Kay LE, Bax A (1990) A novel approach for sequential assignment of 1H, 13C, and 15N spectra of proteins: heteronuclear triple-resonance three-dimensional NMR spectroscopy. Application to calmodulin. Biochemistry 29(19):4659–4667

21. Wittekind M, Mueller L (1993) Hncacb, a high-sensitivity 3D NMR experiment to correlate amide-proton and nitrogen resonances with the alpha-carbon and beta-carbon resonances in proteins. J Magn Reson Ser B 101(2):201–205

22. Kay LE, Xu GY, Yamazaki T (1994) Enhanced-sensitivity triple-resonance spectroscopy with minimal H_2O saturation. J Magn Reson Ser A 109(1):129–133

23. Grzesiek S, Bax A (1992) Correlating backbone amide and side-chain resonances in larger proteins by multiple relayed triple resonance NMR. J Am Chem Soc 114(16):6291–6293

24. Panchal SC, Bhavesh NS, Hosur RV (2001) Improved 3D triple resonance experiments, HNN and HN(C)N, for HN and 15N sequential correlations in (13C, 15N) labeled proteins: application to unfolded proteins. J Biomol NMR 20(2):135–147

25. Kjaergaard M, Brander S, Poulsen FM (2011) Random coil chemical shift for intrinsically disordered proteins: effects of temperature and pH. J Biomol NMR 49(2):139–49.

26. Tamiola K, Bi A, Mulder FAA (2010) Sequence-specific random coil chemical shifts of intrinsically disordered proteins. J Am Chem Soc 132(51):18000–18003

27. Lam SL, Hsu VL (2003) NMR identification of left-handed polyproline type II helices. Biopolymers 69(2):270–281

28. Iwadate M, Asakura T, Williamson MP (1999) C alpha and C beta carbon-13 chemical shifts in proteins from an empirical database. J Biomol NMR 13(3):199–211

29. Zhang H, Neal S, Wishart DS (2003) RefDB: a database of uniformly referenced protein chemical shifts. J Biomol NMR 25(3):173–195

30. Dawson RMC, Elliot DC, Elliot WH, Jones KM (1986) Data for biochemical research, 3rd edn. Oxford University Press, Oxford, UK

31. Teilum K, Kragelund BB, Poulsen FM (2005) Application of hydrogen exchange kinetics to studies of protein folding. Protein folding handbook. Wiley-VCH Verlag GmbH, Hoboken, New Jersey

32. Mantylahti S, Aitio O, Hellman M, Permi P (2010) HA-detected experiments for the backbone assignment of intrinsically disordered proteins. J Biomol NMR 47(3):171–181

33. Bermel W, Bertini I, Felli IC, Lee YM, Luchinat C, Pierattelli R (2006) Protonless NMR experiments for sequence-specific assignment of backbone nuclei in unfolded proteins. J Am Chem Soc 128(12):3918–3919

34. Demarest SJ, Martinez-Yamout M, Chung J, Chen H, Xu W, Dyson HJ, Evans RM, Wright PE (2002) Mutual synergistic folding in recruitment of CBP/p300 by p160 nuclear receptor coactivators. Nature 415(6871):549–553

Chapter 17

Osmolyte-, Binding-, and Temperature-Induced Transitions of Intrinsically Disordered Proteins

Allan Chris M. Ferreon and Ashok A. Deniz

Abstract

Structural studies of intrinsically disordered proteins (IDPs) entail unique experimental challenges due in part to the lack of well-defined three-dimensional structures exhibited by this class of proteins. Although IDPs can be studied in their native disordered conformations using a variety of ensemble and single-molecule biophysical techniques, one particularly informative experimental strategy is to probe protein disordered states as part of folding–unfolding transitions. In this chapter, we describe solution methods for probing conformational properties of IDPs (and unfolded proteins, in general), including the use of naturally occurring osmolytes to force protein folding, the quantification of coupled folding and ligand binding of IDPs, and the structural interrogation of solvent- and/or binding-induced folded conformations by thermal perturbations.

Key words: Intrinsically disordered proteins, Protein folding, Protein thermodynamics, Thermal transitions, Ligand binding, Osmolytes

1. Introduction

Approximately 30% of the human proteome is either partially or predominantly unstructured (1). These intrinsically disordered proteins (IDPs) play significant roles in both human normal physiology and diseases (2–4). They function in vivo either by exhibiting disordered conformations or by adopting ordered ligand-bound structures. In addition, misfolding and aggregation of some IDPs such as α-synuclein have been linked with the pathogenesis of several diseases, including Alzheimer's and Parkinson's, the two most common neurodegenerative disorders. Because IDP function and dysfunction are intricately related to protein structural features, a clear understanding of IDP conformational properties is critical to understanding IDP biology. Here, we present protocols for probing

Vladimir N. Uversky and A. Keith Dunker (eds.), *Intrinsically Disordered Protein Analysis:*
Volume 2, Methods and Experimental Tools, Methods in Molecular Biology, vol. 896,
DOI 10.1007/978-1-4614-3704-8_17, © Springer Science+Business Media New York 2012

IDP structure and conformation through thermodynamic analyses of their folding–unfolding properties, specifically by observing conformational transitions induced by protecting osmolytes, binding partners, and temperature.

2. Materials

Obtain purified protein sample (see Note 1) and, depending on the technique that will be employed for monitoring protein structural changes (see Note 2), apply additional necessary sample treatments (e.g., for fluorescence measurements that involve FRET [Förster resonance energy transfer], label the protein with donor and acceptor fluorescent dyes) before the preparation of the final protein solution (see Note 3). High-purity osmolytes like TMAO (trimethylamine N-oxide), sarcosine, and sucrose, three of the most commonly used osmolytes in forced folding experiments, can be purchased commercially. Additional purification (e.g., by recrystallization and/or treatment with activated carbon [and mixed bed ion-exchange resin]) may be necessary, depending on the application.

2.1. Osmolyte Forced Folding of IDPs

1. Buffer solution: e.g., 0.2 M NaCl, 10 mM sodium acetate, 10 mM NaH$_2$PO$_4$, 10 mM glycine, pH 7.0 (see Note 3).

2. Stock osmolyte solution in buffer: e.g., 4 M TMAO [or 8 M sarcosine or 2 M sucrose] (see Note 4), 0.2 M NaCl, 10 mM sodium acetate, 10 mM NaH$_2$PO$_4$, 10 mM glycine, pH 7.0.

3. Stock protein solution in buffer: e.g., 0.1 mM protein [or 1 mg/mL protein], 0.2 M NaCl, 10 mM sodium acetate, 10 mM NaH$_2$PO$_4$, 10 mM glycine, pH 7.0 (see Note 5).

2.2. Ligand Binding-Induced Folding of IDPs

1. Buffer solution.

2. Stock ligand solution in buffer: e.g., 10 mM ligand, 0.2 M NaCl, 10 mM sodium acetate, 10 mM NaH$_2$PO$_4$, 10 mM glycine, pH 7.0 (see Note 5).

3. Stock protein solution in buffer.

2.3. Thermal Unfolding of Structured IDP Conformations

1. Buffer solution.

2. Stock solution of folding agent (e.g., osmolyte or ligand) in buffer.

3. Stock protein solution in buffer.

3. Methods

Because individual IDPs will inevitably exhibit unique peculiarities in terms of their folding–unfolding and binding properties, the protocols we outline here are presented more as a guide, with aspects that are generally applicable and ones that are more or less protein-specific (e.g., the choice of IDP binding partner or ligand).

Before proceeding to performing the experiments described in the subsections below, preliminary structural and thermodynamic characterization of the protein of interest (in specific solution conditions) may be necessary to verify that the protein is indeed intrinsically disordered or unstructured:

1. Acquire far-UV CD spectra for the protein in buffer and confirm demonstration of a "random coil" signature (5, 6).

2. Perform a protein unfolding measurement using chemical denaturants [e.g., urea or guanidine hydrochloride] (7, 8) and/or thermal perturbation (see below and other chapters in this volume). One signature of an IDP, by definition, is the lack of thermodynamically stable structure or absence of unfolding transition in denaturation experiments.

3.1. Osmolyte Forced Folding of IDPs

1. Prepare individual protein solutions in varying concentrations of osmolyte by mixing the appropriate volumes of buffer, stock osmolyte and stock protein solutions (see Notes 6–8).

2. Equilibrate the solutions at the experimental temperature (see Note 9).

3. Measure and record the experimental observable Υ (see Notes 2 and 9).

4. Plot the observable Υ against [osmolyte]. Add additional experimental data points as appropriate to fully define the folding transition/s (see Note 8).

5. Perform thermodynamic analysis of the protein forced folding data (see Note 10).

3.2. Ligand Binding-Induced Folding of IDPs

1. Mix appropriate volumes of buffer, stock ligand, and stock protein solutions to set up individual protein–ligand solutions with different concentrations of ligand (see Note 11).

2. Equilibrate the protein–ligand solutions at the experimental temperature (see Note 9).

3. Measure the experimental signal Υ and plot against [ligand], adding additional data points as necessary.

4. Analyze the protein–ligand binding data (see Note 12).

3.3. Thermal Unfolding of Structured IDP Conformations

1. Induce the IDP of interest to fold completely or partially by using an osmolyte and/or by binding the protein to a ligand (see above).

2. Perform a thermal scan using an instrument that is preferably equipped with an automated temperature controller, although manual control of temperature (e.g., using a thermal water bath) can also be employed (see Notes 13 and 14).

3. Analyze the protein thermal denaturation data (see Notes 15 and 16).

4. Notes

1. The minimum sample requirement is that the purified protein presents a single band/peak in SDS-PAGE and chromatographic analyses (*unless the protein naturally exhibits anomalous behavior*), displays the expected molecular weight as measured using mass spectrometry, and is soluble and devoid of aggregates. The acceptable levels of protein, nucleic acid, and other impurities depend on the experimental observable (e.g., ensemble or single-molecule fluorescence (9)) that will be used for detecting protein conformational transitions.

2. The choice of experimental technique relies on the investigator's expertise, instrument availability, and whether significant signal changes are linked to the protein transitions of interest. Different techniques can probe different aspects of protein structural changes, and in some cases, the detection of one experimental observable will be preferable over another. Examples of biophysical techniques that can be used to monitor protein transitions include ensemble and single-molecule fluorescence spectroscopy, UV and CD spectroscopy, calorimetry (ITC [isothermal titration calorimetry] and DSC [differential scanning calorimetry]), analytical size exclusion chromatography, dynamic light scattering, and NMR. For complex systems, the application of multiple *nonredundant* techniques that report on different protein properties can be extremely beneficial.

3. Choose the experimental conditions on the basis of the protein's solution properties and the requirements of the biophysical technique/s to be employed. Some of the variables that can be varied include the pH and salt buffer components and concentrations, pH, temperature, and protein concentration. If appropriate and feasible, use the same solution conditions for all experiments, e.g., osmolyte forced folding, ligand binding, and thermal denaturation measurements. As a default, one can use

the following buffer conditions: 0.2 M NaCl, 10 mM sodium acetate, 10 mM NaH_2PO_4, 10 mM glycine, pH 7.0, 25.0°C; which is minimal, and is appropriate for most thermal unfolding measurements and for varying pH conditions.

4. The osmolytes TMAO, sarcosine, and sucrose are all effective in forcing proteins to adopt compact conformations, with TMAO exhibiting the highest efficacy (10). In certain cases, however, TMAO can be "too effective," and may also induce protein aggregation/oligomerization.

5. Choose the stock protein/ligand concentrations on the basis of the experimental observable (e.g., ensemble fluorescence intensity or single-molecule FRET generally utilize nM-μM and sub-nM sample concentrations, respectively), the final observation volume, and the solution properties of the protein/ligand (e.g., aggregation and oligomerization propensities).

6. Alternatively, titrate into a buffered protein solution increasing volumes of stock osmolyte solution (ideally, with the two solutions having the same concentration of protein), equilibrate, and then measure using the chosen detection methodology (see Note 2). This alternative method is easier to implement, but is less accurate. It is also impractical for systems that require longer times to equilibrate.

7. Preparation of specific osmolyte solutions by mixing stock solutions buffered at the same pH and having different osmolyte concentrations (e.g., pH 7.0; and, 0 and 4 M TMAO) can result in solutions having different pH values. The degree of the pH perturbation depends on the buffer composition and the difference between the osmolyte concentrations of the stock solutions. If the pH discrepancy is considerable, it may be necessary to prepare stock buffered osmolyte solutions at different concentrations and individually adjust them to the experimental pH. Urea, a nonprotecting osmolyte commonly used as denaturing agent for protein unfolding measurements, can also exhibit this "pH problem," although usually not at an experimentally significant degree, especially when compared to guanidine hydrochloride, the other most commonly used denaturant. This problem may be due primarily to the cosolvent concentration dependence of the proton binding properties (e.g., pKs) of the pH buffer components.

8. Prepare about 10–15 solutions with different osmolyte concentrations, spaced accordingly to sufficiently describe the pre- and posttransition baselines, and the folding transition region.

9. Solutions are equilibrated when no further change in the detected experimental signal is observed as a function of time.

10. *Analysis of osmolyte forced folding data.* Perform nonlinear least-squares (NLS) analysis of the folding data with the linear

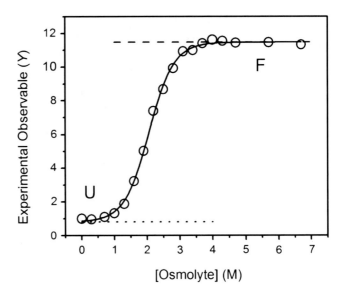

Fig. 1. Osmolyte-induced forced folding of an intrinsically disordered protein. Shown are the detected experimental signals Y (*open circle*) plotted against osmolyte concentration, the nonlinear least-squares (NLS) best fit curve to a two-state linear extrapolation model (LEM), and the pre- and posttransition (i.e., U state and F state) baselines.

extrapolation method/model (LEM) (7, 11) using data analysis software such as Origin (OriginLab), Kaleidagraph (Synergy), or Mathematica (Wolfram Research). Fit the data to the equation $\Upsilon = \Upsilon_F + (K/(1 + K))(\Upsilon_U - \Upsilon_F)$, where Υ_U and Υ_F are the pre- and postfolding transition baselines (i.e., $\Upsilon_U = \Upsilon_U^{intercept} + \Upsilon_U^{slope}$[Osmolyte] and $\Upsilon_F = \Upsilon_F^{intercept} + \Upsilon_F^{slope}$ [Osmolyte]); $\Upsilon_U^{intercept}$ and Υ_U^{slope}, and $\Upsilon_F^{intercept}$ and Υ_F^{slope} are the intercept and slope parameters of the unfolded state (U) and folded state (F) baselines, respectively; $K = \exp[-(\Delta G_{F \to U}^0 + m[Osmolyte])/RT]$; $\Delta G_{F \to U}^0$ is the Gibbs free energy of folding at 0 M osmolyte; m or the m-value represents the [osmolyte] dependence of the folding free energy; R is the ideal gas constant (1.987 cal/mol K); and, T is the absolute temperature. An example of forced folding data analyzed using the LEM is presented in Fig. 1. For a general description of the analysis of transitions involving three or more states, see ref. 6.

11. If the rate of ligand binding is not experimentally limiting, titrate into a buffered protein solution increasing volumes of stock ligand solution (with added protein) instead, maintaining constant protein concentration in both solutions. Use multiple buffered stock ligand solutions with different ligand concentrations (but same [protein]) for more accurate volume measurements.

12. *Analysis of ligand binding data.* For one-site binding of a macromolecule M to a ligand L (forming the bound form ML), the

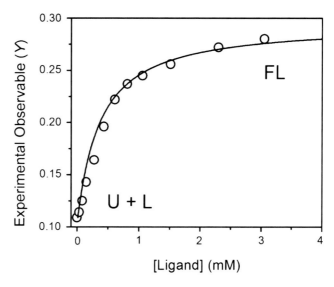

Fig. 2. Ligand binding-induced folding of an intrinsically disordered protein. Presented are the measured observable Y (*open circle*) as a function of ligand (L) concentration, and the (NLS) best fit curve to a one-site binding model. U and FL represent the disordered unbound and folded bound forms of the macromolecule.

isothermal titration data can be fit using NLS analysis to the equation $Y = ([ML]/[M_T])(Y_{ML} - Y_M) + Y_M$, where $[ML] = (-b - \sqrt{b^2 - 4c})/2$; $b = -(K_d + [M_T] + [L_T])$; $c = [M_T][L_T]$; Y_M and Y_{ML} are the experimental signals for unbound (M) and bound (ML) forms of the macromolecule; $[M_T]$ and $[L_T]$ are the total macromolecule and ligand concentrations, respectively; and, K_d is the dissociation constant, equal to the ligand concentration at which M and ML are equally populated. A sample binding data analyzed assuming one-site binding is presented in Fig. 2. For ligand binding that involves two different ligands, competitive binding can be performed (12, 13).

13. Given the ability to precisely modulate temperature (e.g., through a Peltier automated thermal control device), almost any experimental technique can be used to perform a thermal scan. The most commonly used techniques are differential scanning calorimetry (DSC) and CD, UV, and fluorescence spectroscopy. In principle, most techniques can be modified and adapted for thermal scanning.

14. An optimal scan rate balances between having high signal-to-noise ratios and equilibrated system (slow scan rate), and observing reversible thermal unfolding events free from aggregation issues (fast scan rate). As a default, use scan rates between 0.5 and 5°C/min, recording the signal every 0.5–2°C. Scan using the widest temperature range practical for the technique (e.g., 15–99°C). Check for equilibrium unfolding by using multiple scan rates,

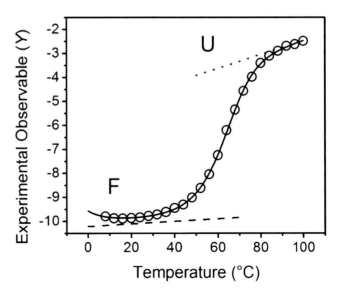

Fig. 3. Thermal unfolding of an IDP osmolyte-induced folded state. Shown as a function of temperature are the measured signal Y (*open circle*), the NLS best fit curve to a two-state denaturation model, and the pre- and posttransition (i.e., F state and U state) baselines.

which should give identical results if the measurements were both performed under equilibrium conditions. If aggregation becomes a problem at higher temperatures, use a narrower temperature range. Modulate the thermal stability of the protein by changing the concentration of the folding agent, i.e., [osmolyte] and/or [ligand].

15. *Analysis of thermal denaturation data.* For a protein system that undergoes thermal unfolding via a two-state mechanism, fit the data via NLS analysis to the equation $Y = (Y_F + Y_U K)/(1 + K)$, where $K = \exp[-\Delta G/RT]$; ΔG is the free energy of unfolding as described by the Gibbs-Helmholtz relationship, i.e., $\Delta G = \Delta H\left(1 - \frac{T}{Tm}\right) + \Delta Cp \left(T - Tm - T\ln\frac{T}{Tm}\right)$; ΔH and ΔCp are the transition enthalpy and heat capacity changes at the transition midpoint Tm; Y_F and Y_U are the pre- and postunfolding transition baselines (i.e., $Y_F = Y_F^{intercept} + Y_F^{slope} T$ and $Y_U = Y_U^{intercept} + Y_U^{slope} T$); R is the ideal gas constant; and, T is the absolute temperature. Figure 3 presents thermal scan data for the unfolding of an IDP forced folded using the osmolyte TMAO, analyzed with a two-state model. For the analysis of multistate thermal transitions, see ref. 6.

16. Global analysis of folding–unfolding transitions induced using different agents (e.g., osmolyte, ligand and temperature) will result in a more comprehensive understanding of the conformational propensities of the IDP of interest. Results can then be graphically presented as "protein phase diagrams" (14–16).

In addition, the application of the solution methods we described in this chapter in combination with single-molecule detection (17–22) is a promising experimental strategy for extracting greater information about IDP conformational distributions and dynamics.

Acknowledgment

We acknowledge support by a National Institute of General Medical Sciences grant (GM066833) from the National Institutes of Health (to A.A.D.).

References

1. Dunker AK, Lawson JD, Brown CJ, Williams RM, Romero P, Oh JS, Oldfield CJ, Campen AM, Ratliff CM, Hipps KW, Ausio J, Nissen MS, Reeves R, Kang C, Kissinger CR, Bailey RW, Griswold MD, Chiu W, Garner EC, Obradovic Z (2001) Intrinsically disordered protein. J Mol Graph Model 19:26–59

2. Wright PE, Dyson HJ (1999) Intrinsically unstructured proteins: re-assessing the protein structure-function paradigm. J Mol Biol 293:321–331

3. Uversky VN, Oldfield CJ, Dunker AK (2008) Intrinsically disordered proteins in human diseases: introducing the D2 concept. Annu Rev Biophys 37:215–246

4. Uversky VN (2011) Flexible nets of malleable guardians: intrinsically disordered chaperones in neurodegenerative diseases. Chem Rev 111:1134–1166

5. Greenfield N, Fasman GD (1969) Computed circular dichroism spectra for the evaluation of protein conformation. Biochemistry 8:4108–4116

6. Ferreon ACM, Deniz AA (2007) α-Synuclein multistate folding thermodynamics: implications for protein misfolding and aggregation. Biochemistry 46:4499–4509

7. Santoro MM, Bolen DW (1988) Unfolding free energy changes determined by the linear extrapolation method. 1. Unfolding of phenylmethanesulfonyl alpha-chymotrypsin using different denaturants. Biochemistry 27:8063–8068

8. Ferreon ACM, Bolen DW (2004) Thermodynamics of denaturant-induced unfolding of a protein that exhibits variable two-state denaturation. Biochemistry 43:13357–13369

9. Ferreon ACM, Moran CR, Gambin Y, Deniz AA (2010) Single-molecule fluorescence studies of intrinsically disordered proteins. Methods Enzymol 472:179–204

10. Auton M, Ferreon ACM, Bolen DW (2006) Metrics that differentiate the origins of osmolyte effects on protein stability: a test of the surface tension proposal. J Mol Biol 361:983–992

11. Pace CN (1986) Determination and analysis of urea and guanidine hydrochloride denaturation curves. Methods Enzymol 131:266–280

12. Wang ZX, Jiang RF (1996) A novel two-site binding equation presented in terms of the total ligand concentration. FEBS Lett 392:245–249

13. Lee CW, Ferreon JC, Ferreon ACM, Arai M, Wright PE (2010) Graded enhancement of p53 binding to CREB-binding protein (CBP) by multisite phosphorylation. Proc Natl Acad Sci U S A 107:19290–19295

14. Rosgen J, Hinz HJ (2003) Phase diagrams: a graphical representation of linkage relations. J Mol Biol 328:255–271

15. Ferreon ACM, Ferreon JC, Bolen DW, Rosgen J (2007) Protein phase diagrams II: nonideal behavior of biochemical reactions in the presence of osmolytes. Biophys J 92:245–256

16. Vandelinder V, Ferreon ACM, Gambin Y, Deniz AA, Groisman A (2009) High-resolution temperature-concentration diagram of α-synuclein conformation obtained from a single Forster resonance energy transfer image in a microfluidic device. Anal Chem 81:6929–6935

17. Ferreon ACM, Gambin Y, Lemke EA, Deniz AA (2009) Interplay of α-synuclein binding and conformational switching probed by single-molecule fluorescence. Proc Natl Acad Sci U S A 106:5645–5650

18. Ferreon ACM, Moran CR, Ferreon JC, Deniz AA (2010) Alteration of the α-synuclein folding landscape by a mutation related to Parkinson's disease. Angew Chem Int Ed 49:3469–3472

19. Mukhopadhyay S, Krishnan R, Lemke EA, Lindquist S, Deniz AA (2007) A natively unfolded yeast prion monomer adopts an ensemble of collapsed and rapidly fluctuating structures. Proc Natl Acad Sci U S A 104:2649–2654

20. Gambin Y, Vandelinder V, Ferreon ACM, Lemke EA, Groisman A, Deniz AA (2011) Visualizing a one-way protein encounter complex by ultrafast single-molecule mixing. Nat Methods 8(3):239–241

21. Trexler AJ, Rhoades E (2009) α-Synuclein binds large unilamellar vesicles as an extended helix. Biochemistry 48:2304–2306

22. Muller-Spath S, Soranno A, Hirschfeld V, Hofmann H, Ruegger S, Reymond L, Nettels D, Schuler B (2010) Charge interactions can dominate the dimensions of intrinsically disordered proteins. Proc Natl Acad Sci U S A 107:14609–14614

Chapter 18

Laser Temperature-Jump Spectroscopy of Intrinsically Disordered Proteins

Stephen J. Hagen

Abstract

Laser temperature-jump methods allow an experimenter to study the kinetics and dynamics of very rapid solution-phase processes, including conformational dynamics of biomolecules on time scales of nanoseconds and microseconds. The combination of laser temperature-jump (T-jump) excitation and appropriate optical detection techniques such as fluorescence energy transfer allows the study of intramolecular and intermolecular conformational changes and interactions that occur during protein folding and binding. This article describes the application of the laser temperature-jump method to UV–visible fluorescence studies of the coupled folding and binding of intrinsically disordered proteins. We emphasize the practical aspects of instrument alignment and optimization, sample preparation, and data collection using fluorescently labeled peptides with UV laser excitation.

Key words: Laser temperature jump, FRET, Fluorescence, Infrared, Protein folding, Intrinsically disordered, Tryptophan, Spectroscopy

1. Introduction

Relaxation methods are a powerful tool in the study of biomolecular kinetics and dynamics. By applying a physical perturbation to a biochemical system and then observing the kinetics of the return to equilibrium, an experimenter can gain information about the rates and mechanisms of the underlying chemical and physical processes. This approach only requires that the perturbation be more rapid than the relaxation response that is being investigated. Therefore the study of faster chemical and biochemical processes requires faster methods for perturbing the system and probing its response. The study of coupled folding and binding reactions of intrinsically disordered proteins (IDPs) is a case in point. Recent studies have shown that IDP binding and folding may occur on time scales of milliseconds to microseconds and in some cases nanoseconds (1–3).

Vladimir N. Uversky and A. Keith Dunker (eds.), *Intrinsically Disordered Protein Analysis: Volume 2, Methods and Experimental Tools*, Methods in Molecular Biology, vol. 896, DOI 10.1007/978-1-4614-3704-8_18, © Springer Science+Business Media New York 2012

These rates are comparable to those of the fastest elemental steps in protein folding, such as the formation of α-helix and β-hairpin structures.

Such fast processes can be triggered by chemical mixing, but the mixing of two solutions is not easily accomplished on time scales much shorter than a few milliseconds (via stopped flow) or a few microseconds (via continuous flow). However, one may trigger these reactions by applying a sudden change in solution temperature. An infrared laser can deliver a pulse of energy that will be absorbed by an aqueous solution and converted to heat within picoseconds or nanoseconds, leading to a rapid temperature jump (T-jump) that drives folding/unfolding or association/dissociation of the biomolecules. The response to this perturbation can then be monitored by a variety of optical techniques, such as IR spectroscopy, optical absorbance, fluorescence, or circular dichroism. It does not generally matter whether the T-jump drives the reaction of interest or its reverse reaction; in either case, if the kinetics are sufficiently simple (as in a two-state transition), the underlying rate constants can be extracted from the observed relaxation together with equilibrium data. Therefore the use of infrared laser pulses to induce thermal perturbations has had a large impact on the study of protein folding dynamics. It has extended the experimental window for kinetic studies down to the time scale of nanoseconds and microseconds and yielded detailed insight into the basic mechanisms of peptide folding and stability (4–7).

The technology for laser T-jump spectroscopy has developed and diversified greatly in recent years, with a number of different instrument designs currently in use. However, virtually all modern laser T-jump instruments employ a near-infrared pulsed laser to induce local heating in an aqueous sample. Typically this infrared pulse is generated from a Nd:YAG laser pulse ($\lambda = 1,064$ nm) that is shifted to longer wavelength through stimulated Raman conversion within a high pressure gas cell. The choice of the conversion gas determines the output IR pulse wavelength, which in turn determines the attenuation length of the pulse in the solvent (H_2O or D_2O). Hence the design of the rest of the instrument depends on this choice of wavelength as well as the type of experimental probe that will be applied following the T-jump. The sample may be probed by a broadband light source or a pulsed or continuous (CW) probe laser (IR, UV, or visible), and the detector employed may be a photomultiplier or semiconductor detector, a grating spectrometer and CCD camera, etc. (4). IR absorbance is very widely used to probe protein folding reactions because it allows—through selective isotopic replacements in a peptide—studies of the dynamics of individual bonds in the peptide backbone. Other probes such as optical rotatory dispersion have also been applied in T-jump studies (8). On the other hand, fluorescence measurements—especially using FRET (9)—offer the potential to time-resolve changes in the distance that separates two residues on the same molecule or on different molecules. The dynamics of these bimolecular interactions are of great interest in the study of IDPs.

Several authors have already provided detailed explanations of the design and operation principles of laser T-jump instruments currently in use (4–7, 10–13). Much of the laser T-jump instrument is assembled from commercially available equipment (lasers, detectors, optical elements, timing electronics, etc.), although a few components of the system—such as the control software and the sample cuvette holder—will likely require custom design. We do not attempt here to review or give advice on aspects of instrument design and construction. Rather we assume that the reader already has access to a well-designed, working instrument and wishes to measure the folding or binding dynamics of an intrinsically disordered protein or similar biomolecular system. Regardless of the instrument design, successful T-jump measurements require careful attention to instrument alignment and signal optimization, sample preparation, and data collection. Our goal here is to provide practical advice and guidance in regard to these matters of daily operation. For concreteness we focus on the example of laser T-jump fluorescence (FRET) spectroscopy studies of IDPs as performed in our laboratory. Below we provide a brief description of our T-jump instrument, followed by more detailed advice on the alignment and optimization procedures that are conducted prior to kinetic measurements. We then provide a description of the data collection steps. Although this protocol generally focuses on T-jump fluorescence spectroscopy using UV pulse laser excitation, many aspects of the discussion are generic to T-jump work and should be useful to researchers using other types of optical probes.

To study the folding and binding of an IDP using T-jump fluorescence spectroscopy and FRET, one must first design appropriate fluorescently labeled mutants of the protein and its interaction partner. For example, we have studied the peptide IA3, an IDP that folds into a helix as it binds to the active site of the yeast aspartic proteinase YPrA (3, 14). In order to measure the folding rate of the free IA3, we prepared a mutant IA3 that contains a fluorescence donor (tryptophan) and acceptor (dansyl cysteine). To measure the rate of interaction between IA3 and its target YPrA, we prepared a second mutant of IA3 containing only the fluorescence acceptor (tryptophan); the native tryptophans of the target act as FRET donors during the bimolecular interaction. We then used the laser T-jump method to collect the time-resolved FRET spectra under UV excitation (of the tryptophan donors) for both the free IA3 and the IA3–enzyme mixture. The T-jump shifts the equilibrium for both the folding of the IA3 and its interaction with the enzyme, leading to a kinetic shift in the fluorescence emission spectra. From the time-dependence of the FRET emission spectra, we extracted underlying rates for both folding and folding/binding.

2. Materials

We consider a laser T-jump experiment that uses a fluorescence spectroscopy configuration like that of Fig. 1: An IR laser generates a nanosecond pulse that is shifted to longer wavelength by stimulated Raman emission in a high pressure gas cell (Raman cell). This longer wavelength pulse is split into counterpropagating pulses that generate the T-jump in an aqueous sample (containing a FRET-labeled peptide). The IR pulse is followed—after a time delay t—by a UV laser pulse that excites the fluorescence of the peptide. We collect the fluorescent emission and split it between two detectors: a photomultiplier (PMT) measures the total emission intensity while a diffraction grating disperses the emission across a CCD array, allowing us to collect the fluorescence emission spectrum for each value of the delay time t.

The aqueous sample is pumped by a syringe pump (see Note 1) through a thin silica capillary of rectangular cross section (Fig. 2a). Flat surfaces are desirable for this capillary, as the IR and UV laser pulses must enter the capillary without optical distortion, and the fluorescence emission must be collected efficiently. The optimal dimensions for the capillary depend on the extinction coefficient of the IR pulses in the aqueous sample (see Note 2).

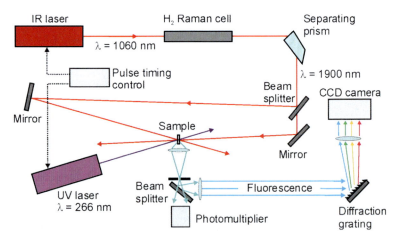

Fig. 1. Schematic of laser T-jump system for fluorescence spectroscopy. Two counter-propagating IR laser pulses at $\lambda = 1,900$ nm (separated by ~10 ns) are generated by Raman conversion of a $\lambda = 1,064$ nm pulse from a Nd:YAG laser. The IR beams are focused onto the sample, which is pumped through a fused silica capillary. A UV laser at $\lambda = 266$ nm is fired at time t and excites the fluorescent emission, which is collected by an objective lens and directed to a photomultiplier and (via a diffraction grating) dispersed onto a CCD camera array.

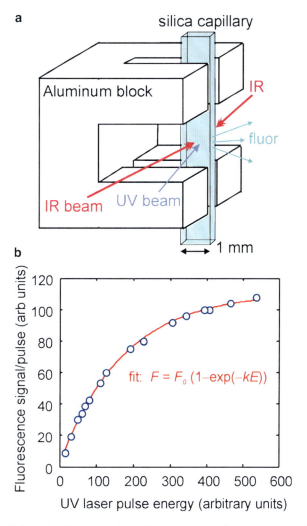

Fig. 2. (**a**) Schematic of sample holder. The sample flows through a rectangular fused silica capillary (1 mm wide × 100 μm pathlength), which is mounted on an aluminum block that is temperature-controlled by a thermoelectric stage (not shown). The counter-propagating IR laser beams and the UV excitation beam strike the capillary from the sides as shown, while fluorescent emission is collected from the front. (**b**) Adjustment of the pulse energy from a Nd:YAG laser (4th harmonic, $\lambda = 266$ nm, 5–7 ns pulse duration) to achieve saturation of the fluorescent emission from a sample of NATA (*N*-acetyl-L-tryptophanamide) in phosphate buffer. As the fluorescence lifetime of tryptophan is typically only ~1–4 ns and the quantum yield is rarely more than ~0.3 (20, 21), a UV excitation pulse (~5–7 ns duration) is unlikely to extract more than one fluorescence photon from a tryptophan residue, regardless of the pulse intensity. Therefore if the UV laser is tightly focused on the sample, the fluorescence emission F becomes insensitive to the UV pulse energy E at large values of E. This saturation improves the signal-to-noise in F in the presence of shot-to-shot fluctuations in UV laser energy. The figure shows measured data and an exponential fit, $F = F_0 (1 - \exp(-kE))$ with $k \approx 0.0059$, $F_0 \approx 110$.

| 2.1. Special Components/Equipment | The major physical components of the laser T-jump apparatus shown in Fig. 1 include the following: |

1. IR pulse laser (e.g., Nd:YAG at $\lambda = 1{,}064$ nm).
2. UV pulse laser (e.g., Nd:YAG 4th harmonic at $\lambda = 266$ nm).
3. Pressurized Raman cell (e.g., H_2 at 650 psi).
4. Electronic pulse delay timer/controller.
5. Photomultiplier with high voltage supply and amplifier.
6. Charge coupled device (CCD) detector.
7. Sample cuvette holder with thermoelectric temperature control.
8. Syringe pump.
9. Optical elements (lenses, mirrors, beam splitters, polarizer, Pellin-Broca prism, diffraction grating, translating lens mounts, etc.).
10. Digitizing oscilloscope.
11. Personal computer with instrument control software.

2.2. Other Materials and Components

1. Silicon capillary tubing with rectangular cross section,

   e.g., Vitrocom type 5010S, 100 μm path length × 1 mm width.
2. Temperature calibration reference solution,

   e.g., N-acetyl-L-tryptophanamide 30–50 μM in sample buffer.
3. Syringes and syringe filters (0.2 μm).
4. Vacuum pump and fittings for degassing samples.
5. HPLC tubing and fittings for sample flow.
6. Conductive silver paint.
7. Laser burn paper such as Kodak Linagraph type 1895.

3. Methods

The major steps to a T-jump kinetic measurement are optimization of the instrument parameters, alignment of the instrument, preparation of the sample, and data collection and analysis. These are described separately in the following sections.

3.1. Optimizing the Instrument Configuration

Even starting with a functional T-jump instrument, the experimenter still needs to choose the major operating parameters. This section gives advice on selecting these parameters in an optimal way. Most of these adjustments are made only once.

1. Select the Raman cell pressure. This cell contains a high-pressure gas in which an Nd:YAG laser pulse ($\lambda = 1{,}064$ nm)

is Raman-shifted to a longer wavelength that is more strongly absorbed by samples containing D_2O or H_2O (see Note 3). The gas pressure should be adjusted in order to optimize the efficiency of this conversion. One can slowly increase the gas pressure while measuring the energy output at the Raman-shifted wavelength. Using H_2 in a 1-m cell, we have obtained good results with a gas pressure of 650 psi. Adding up to 200 psi helium as a "buffer gas" did not improve the energy conversion efficiency in our experience.

2. Select the repetition rate of the IR laser. Reducing the time between consecutive T-jump IR pulses allows for faster signal averaging. The IR repetition rate should not however exceed the rate at which the sample temperature recovers to its initial value following each T-jump. Note 4 gives an example of using the heat diffusion geometry (15) to estimate the thermal recovery time. However, although in principle one can predict the thermal kinetics from the physics of heat diffusion, in practice a simple experimental check is still necessary. The experimenter must verify that the IR pulses come slowly enough that the sample temperature recovers completely between pulses. Methods for detecting the sample temperature are discussed below.

3.2. Alignment of the Instrument

We find that the alignment of the "back end" of the T-jump apparatus (the IR laser source, the Raman cell, separating prisms, etc.) is quite stable and rarely (if ever) needs adjustment. However the alignment of the IR and UV beams that enter the sample is affected by adjustment or replacement of the sample cell. These beams must be coaligned very precisely so that the counterpropagating IR beams coincide at the center of the sample capillary, while the UV beam passes through the center of the IR spot. This alignment should be checked and adjusted before every set of measurements. This is accomplished as follows:

1. Allow the lasers to warm up for ~1/2 h before attempting to align the instrument.

2. If a CCD detector and diffraction grating are used to collect fluorescence spectra, the wavelength scale on the CCD array should be calibrated after every major realignment of the instrument. This can be accomplished by illuminating the sample holder with a gas discharge lamp (Ne, Hg, etc.) and observing the positions of the spectral lines on the array.

3. After the lasers are warmed, coalign the counterpropagating IR beams on the sample, and align the UV laser to the center of the IR spot. Different labs use different tricks for this step. In our lab, we first adjust the UV laser focusing lens to bring the UV beam to a sharp focus on the center of the sample capillary.

We monitor the position of that beam with an inexpensive video camera that can monitor (via a flip mirror) the light reaching the photomultiplier. We then adjust the IR laser focusing lenses in order to center both IR laser beams onto the UV laser spot. We have used at least two different methods to perform this initial IR alignment:

One method is to use laser burn paper behind the capillary to align the beam roughly onto the capillary. If the capillary is filled with water, the burn paper will reveal the "shadow" of the capillary as the IR beam scans across the capillary (and is absorbed by the water in the capillary). In this way one can align the IR beam to within a few hundred microns of the center of the capillary.

Alternatively one may construct an alignment jig (e.g., a thin metal plate with a fine hole drilled at the location of the UV laser spot) that mounts in place of the sample capillary. Aligning the IR and UV lasers (all beams) through the fine hole gives a roughly correct alignment of the IR.

If the UV excitation laser is a continuous (CW) laser, then one can use thermal lensing to align the IR and UV beams. Allow the UV beam to pass through the sample and strike a white screen (such as a sheet of paper) a few meters away. If the IR beam spot is close to the UV beam focus, then the UV spot on the screen will momentarily deflect each time that the IR laser fires, owing to thermal lensing in the sample. The direction of the deflection gives a clear indication of the relative positions of the IR and UV beams. Once the beams are perfectly coaligned, the IR pulse will not cause a lateral deflection of the UV beam.

4. Once the IR beams are closely coincident on the UV spot, the photomultiplier signal should indicate a T-jump occurring after the IR laser fires, when a temperature calibration solution (see below) is present in the capillary. The objective is then to adjust the precise position of each IR beam until the T-jump signal is maximized. This adjustment is performed with the fine translating xy mounts that hold the IR focusing lenses. For our system, which uses pulsed UV laser excitation, we perform this alignment using a short time delay ~0.1 μs between the IR pulses and the UV probe pulse. Using a longer time delay (approximately μs or longer) is not advisable, as the alignment will be more subject to thermal diffusion, acoustic effects, cavitation, etc.

5. If a pulsed UV laser is used for fluorescence excitation, random fluctuations in the intensity of the pulses will lead to noise in the measured emission levels. The effect of these fluctuations on the signal-to-noise ratio can be minimized by raising the UV

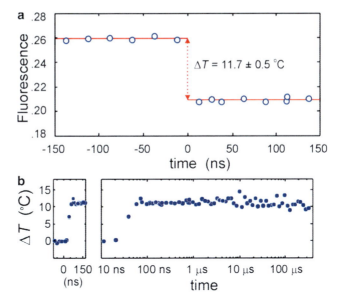

Fig. 3. (**a**) Fluorescence emission F of a sample of NATA (34 µM in 100 mM phosphate, pH 7) before and after a laser T-jump at time $t = 0$. The magnitude of the fluorescence change and the equilibrium temperature coefficient of the reference sample (d (ln$F$)/ $dT = -0.0185/°C$) indicate a T-jump amplitude $\Delta T = 11.7 \pm 0.5°C$, relative to the initial temperature $T = 22°C$. (**b**) Time-dependence of the temperature perturbation ΔT following a laser T-jump at $t = 0$, as measured from the fluorescence emission of a NATA sample. From its initial temperature $T = 50.8°C$, the sample heats to $\Delta T \approx 10°C$ within an interval ~30 ns. It remains at the elevated temperature until $t > 100$ µs.

pulse energy until the fluorescence of the sample approaches a saturating limit (Fig. 2b). Then the relative fluctuation in the emission level will be much smaller than the relative fluctuation of the excitation laser intensity (see Note 5).

6. Measure the amplitude ΔT of the T-jump. One may calibrate the T-jump using the absorbance or fluorescence of a temperature-sensitive buffer/indicator solution, or the IR absorption by the solvent itself (13). For fluorescence work many laboratories use a solution of NATA (N-acetyl-L-tryptophanamide) in experimental buffer to measure the amplitude of the laser temperature jump; the fluorescence yield of NATA declines with increasing temperature, with a temperature coefficient that is readily measureable in an equilibrium spectrofluorometer. We have found that in neutral pH buffer containing cosolutes such as guanidinium HCl (GdnHCl), the NATA fluorescence yield typically decreases by ~1–2% per °C, depending on the initial T and the cosolute concentration. Figure 3 shows a T-jump calibration in which the fluorescence intensity F of a NATA solution decreases by approximately 19% when the IR pulses arrive at $t = 0$: From this change and the equilibrium temperature dependence of the fluorescence, we calculate a T-jump amplitude $\Delta T = 11.7$

± 0.5°C. The amplitude ΔT should be checked at each experimental condition, as the IR absorption coefficient of water varies with temperature (7).

7. Adjust the power output of the IR laser until the desired ΔT is obtained. Too large a T-jump increases the uncertainty in the final (post T-jump) sample temperature and is more likely to cause cavitation (see Note 6). Too small a T-jump amplitude gives a poor signal-to-noise ratio in the relaxation kinetics. We aim for $\Delta T \approx 7$–10°C. The power from the Nd:YAG laser is best controlled by adjusting the Q-switch delay and then using neutral density filters to equalize the energy delivered by the two counterpropagating pulses.

3.3. Sample Preparation and Data Collection

1. The silica capillary should have an excellent thermal connection to the sample holder. The connection should be made as close to the observation volume as possible so that heat can diffuse quickly out of that volume after each IR pulse. We find that the thermal connection to the sample holder is noticeably improved by applying a small drop of conductive silver paint to each contact point where the sample holder (Fig. 2a) grips the silica capillary. This thermal grounding reduces temperature measurement error and allows the IR laser to be run at a higher repetition rate.

2. The window of the sample cuvette (actually a silica capillary tube) must be scrupulously clean. Dust, paper fiber, or other contaminants on the surface will burn in the IR laser beam or fluoresce under UV excitation, degrading the signal-to-noise ratio. Wiping the capillary with solvent does not necessarily clean it sufficiently. Instead we install a new, flame-cleaned silica capillary into the sample holder at least once a week (see Note 7).

3. Because the sample is irreversibly photobleached by the UV excitation laser, we use a syringe pump to drive a continuous flow of fresh sample through the observation volume. This flow rate may be adjusted to the slowest value that does not lead to photobleaching and a loss of fluorescence signal intensity. For our system we typically use a flow rate of 0.1–0.2 ml/h, which gives a flow speed of 0.03–0.06 cm/s in a cuvette of cross section 0.1×1 mm^2. The sample consumption is then 1–2 ml/day, or 50–100 nmoles of peptide at 50 μM concentration.

4. With a pulsed UV excitation source, fluorescence data should be collected in a randomized order of delay times t. If data are collected in order of increasing (or decreasing) delay time t it is much more difficult to distinguish artifactual drifts in the fluorescence emission level from real kinetic processes (see Note 8).

5. One should check the amplitude ΔT of the T-jump regularly, as the IR absorbance of water varies with temperature and solvent

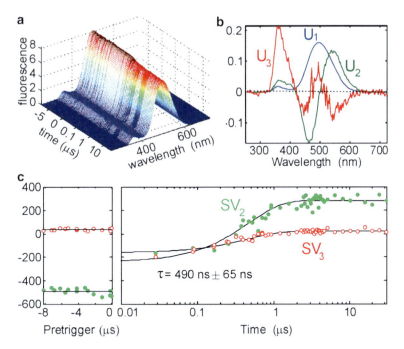

Fig. 4. Time-resolved fluorescence emission from intrinsically disordered protein IA3 interacting with proteinase YPrA (3), as studied by laser T-jump. (**a**) Fluorescence emission spectra of dansyl-labeled IA3, as excited by fluorescence energy transfer from the native tryptophan residues in the IA3 binding target (proteinase YPrA). The overall amplitude of the fluorescence emission is seen to decrease following the T-jump (which occurs at $t = 0$). (**b**) A singular value decomposition (SVD) of the time-resolved spectra yields major spectral components $U_i(\lambda)$, with most of the signal contained in the first three components $i = 1, 2, 3$. (**c**) Kinetic amplitudes $SV_i(t)$ associated with the component spectra $U_i(\lambda)$. U_1 and SV_1 primarily describe the decline in the overall amplitude of the dansyl (~500 nm) and tryptophan (~350 nm) emission following the T-jump. This is due to the intrinsic temperature dependence of the fluorescence quantum yield. U_2 describes a recovery of the tryptophan emission as well as a redshift of the dansyl emission approximately 500 ns after the T-jump. U_3 also describes a gain in the tryptophan emission as well as a broadening of the dansyl emission, at ~500 ns.

conditions. The fine alignment of the IR and UV lasers may also drift over a period of hours. Switching quickly from a protein sample to a temperature calibration sample is facilitated by installing a multiport HPLC valve into the flow line that connects the syringe pump to the sample capillary (see Note 9).

3.4. Data Analysis

1. A laser T-jump instrument that collects time-resolved fluorescence emission spectra generates a large dataset of the emission intensities $F(\lambda,t)$ vs. delay time (t) and wavelength (λ) during folding/binding of the IDP and its target. Such a dataset is best reduced and analyzed by singular value decomposition (SVD) (16). Figure 4 shows time-resolved fluorescence spectra that were collected in a laser T-jump study of the IA3 peptide. SVD decomposes the fluorescence emission

dataset into major signal components ($i = 1, 2, 3, \ldots$), each of which has a characteristic spectrum $U_i(\lambda)$ and an associated time course $SV_i(t)$. This method of signal decomposition provides many advantages in biophysical spectroscopy and is described in detail elsewhere (16).

2. Least squares fitting of the kinetic data reveals the exponential rates associated with the IDP folding/binding reaction. In general one must perform multiple kinetic and equilibrium experiments in order to determine the dominant pathway (such as folding-then-binding versus binding-then-folding) of the interaction. It has been pointed out (17) that a correct determination may require a subtle analysis, as faster rates for a given mechanism do not necessarily indicate a higher reaction flux through that mechanism. However these issues are not specific to laser T-jump method, and therefore they lie outside the scope of this article.

4. Notes

1. We use narrow gauge HPLC tubing ($0.010''$ ID) to drive the sample fluid through the observation cell. Its very small diameter minimizes the experiment dead volume, so as to allow a reasonably quick (~2–3 min) switch between the experimental sample and a calibration reference or washing buffer. Its stiffness also minimizes the mechanical elasticity ("capacitance") of the flow system, so that when the syringe pump speed is adjusted the sample flowrate responds quickly.

2. The pulse energy decreases as $\exp(-\kappa x)$ as the IR pulse penetrates a distance x into the sample, where the absorption coefficient κ for water is well-tabulated (18). If the IR beam must traverse too long a path through the sample, the heating will be greater at the entry than at the exit and so the final (i.e., post T-jump) temperature will be nonuniform throughout the sample. If the IR path is too short, then the perturbed sample volume is too small and the signal/noise ratio is degraded. One may improve the heterogeneity of the temperature perturbation by (a) pumping the sample with counterpropagating IR beams (Fig. 1), and (b) selecting a capillary with a short enough path length that the amplitude of the temperature variation is not larger than ~10% across the volume that is probed by the excitation laser. Our system uses a 1,064 nm Nd:YAG pulse laser (~7 ns pulse width) to drive a H_2 Raman cell, which shifts the laser pulse energy to a wavelength $\lambda = 1.9$ μm, for which $\kappa \sim 60$ cm^{-1} (18). If two counterpropagating

pulses at $\lambda = 1.9\mu m$ enter a capillary of width $d = 100$ μm the T-jump perturbation at the center of the capillary will be $(\exp(-\kappa d/2) + \exp(\kappa d/2))/(1 + \exp(-\kappa d)) = 96\%$ as large as at the capillary windows. Roughly ~15–20 mJ in the two pulses at $\lambda = 1.9$ μm is sufficient to generate a T-jump $\Delta T \sim 10°C$ in a ~1 mm^2 area of the sample.

3. The most commonly used gases and their associated Raman shifts are H_2 (4,155 cm^{-1}), CH_4 (2,914 cm^{-1}), D_2 (2,991 cm^{-1}), and N_2 (2,331 cm^{-1}). Although many labs use CH_4, which converts an Nd:YAG laser pulse to $\lambda = 1,540$ nm, methane can decompose at high excitation energies, depositing carbon soot on the windows of the Raman cell. This creates a very dangerous condition as the gas is contained at high pressure. In our lab D_2 (giving $\lambda = 1,560$ nm) yielded unsatisfactory conversion efficiency. By contrast, H_2 (giving $\lambda = 1,900$ nm) provided high conversion efficiency and is of course photostable. However the H_2 wavelength is very strongly absorbed in water and requires a short optical path in an aqueous sample.

4. Our instrument focuses the IR heating pulse to a spot of diameter $d \sim 1$ mm on the sample capillary, which has fused silica windows and a path length of $z = 0.1$ mm. We can approximate the thermal physics by assuming that heat exits the observation volume in the solvent only by diffusing (a) radially into the surrounding liquid or else (b) axially toward the silica windows and then radially along the windows. From the thermal diffusivity of water ($D_{water} \approx 1.4 \times 10^{-7}$ m^2/s) and the spot diameter d we can estimate the radial diffusion time as $t \sim (d/2)^2/2D_{water} = 0.9$ s, and the axial diffusion time as $t \sim (z/2)^2/2D_{water} = 0.009$ s. Heat reaching the silica window will likewise diffuse radially on a time scale $t \sim (d/2)^2/2/D_{silica} = .16$ s, where $D_{silica} \approx 8 \times 10^{-7}$ m^2/s. The slowest of these processes is the radial diffusion process (~0.9 s). This suggests that the IR laser repetition rate should not exceed roughly 1 Hz. In fact experimentation showed that a repetition rate of 2 Hz allowed sufficient thermal recovery of the sample.

5. The T-jump apparatus is relatively complex and accordingly the signal is subject to many types of noise and artifacts. These all have a physical origin (electronic, optical, thermal, etc.), and therefore, they can be eliminated, at least in principle. Some of the most common sources of nonrandom noise and other signal artifacts include instability of the IR or UV laser power, cavitation, electromagnetic noise from the circuitry in the IR laser, and unsteady flow of sample through the capillary. Small misalignment of the IR laser with respect to the UV laser can cause the sample temperature to continue rising after the initial T-jump occurs; hence, it is important to use a short value of the

IR/UV delay interval, such as $t = 0.1–0.5$ µs, while optimizing the fine coalignment of the beams. When CW excitation lasers are used, electromagnetic interference from beam shutters can generate artifacts in the PMT signal. Air bubbles in the sample flow tubing can create several types of problems and must be avoided.

6. Cavitation is a frequent source of artifacts in laser T-jump measurements. The heating IR pulse induces tensions in the sample that may intermittently exceed the tensile strength of the solution (19). This leads to nucleation of small (<100 µm) transient bubbles that form and collapse sporadically at $t \sim 1–10$ µs following the IR pulse. These bubbles generate irregular or intermittent artifacts in optical measurements on these time scales. Cavitation is understood to be facilitated by solutes or contaminants that reduce the tensile strength of the solvent. The amount of cavitation observed can vary strongly with the sample composition (e.g., cosolute concentration) and temperature. However cavitation can often be prevented or at least minimized by some combination of the following: (1) Degassing the sample briefly under vacuum in a conical vial before loading the syringe pump; (2) Inserting a submicron (e.g., 0.2 µm) syringe filter into the sample flow line; (3) Applying a back-pressure (e.g., 5–10 psi) of N_2 gas to the sample in the cuvette, (4) Using smaller T-jump amplitude.

7. One should not allow buffers or protein samples to dry inside the capillary, as any residues will be nearly impossible to remove. Once we have installed a clean capillary in the sample holder, we keep fluid moving through the capillary at all times. A steady flow of buffer at ~0.1 ml/h is easily maintained for long intervals when the system is not in immediate use.

8. Slow drifts (over minutes to hours) in the intensity of the fluorescence emission usually indicate (1) poor flushing of different solutions (e.g., NATA calibration, buffer, peptide sample), owing to excessive dead volume in the tubing system; (2) lack of temperature stability in the controller (unlikely); or (3) lack of stability in the UV laser.

9. For accurate ΔT measurement, it is important that the fluorescence signal comes from the NATA reference solution and not from contaminants on the surfaces of the sample cuvette. Hence one should periodically pump pure buffer through the capillary and check that the detected signal falls to a low background level, no more than a few percent of the level obtained from the NATA reference solution.

Acknowledgment

The author thanks Linlin Qiu and Ranjani Narayanan for their work in characterizing and optimizing the laser T-jump spectrometer at the University of Florida. The author also gratefully acknowledges funding support from the National Science Foundation, MCB 0347124.

References

1. Sugase K, Dyson HJ, Wright PE (2007) Mechanism of coupled folding and binding of an intrinsically disordered protein. Nature 447:1021–1025

2. Muralidhara BK, Rathinakumar R, Wittung-Stafshede P (2006) Folding of desulfovibrio desulfuricans flavodoxin is accelerated by cofactor fly-casting. Arch Biochem Biophys 451:51–58

3. Narayanan R et al (2008) Kinetics of folding and binding of an intrinsically disordered protein: the inhibitor of yeast aspartic proteinase YPrA. J Am Chem Soc 130:11477–11485

4. Kubelka J (2009) Time-resolved methods in biophysics. 9. Laser temperature-jump methods for investigating biomolecular dynamics. Photochem Photobiol Sci 8:499–512

5. Callender R, Dyer RB (2002) Probing protein dynamics using temperature jump relaxation spectroscopy. Curr Opin Struct Biol 12:628–633

6. Gruebele M et al (1998) Laser temperature jump induced protein refolding. Acc Chem Res 31:699–707

7. Hofrichter J (2001) Laser temperature-jump methods for studying folding dynamics. Methods Mol Biol 168:159–191

8. Chen E et al (2005) Nanosecond laser temperature-jump optical rotatory dispersion: application to early events in protein folding/unfolding. Rev Sci Instrum 76:083120

9. Van Der Meer BW, Coker G, Simon Chen SY (1994) Resonance energy transfer: theory and data. VCH Publishers, Weinheim

10. Huang CY et al (2001) Temperature-dependent helix-coil transition of an alanine based peptide. J Am Chem Soc 123:9235–9238

11. Ansari A, Kuznetsov SV, Shen YQ (2001) Configurational diffusion down a folding funnel describes the dynamics of DNA hairpins. Proc Natl Acad Sci USA 98:7771–7776

12. Ameen S (1975) Laser temperature-jump spectrophotometer using stimulated Raman effect in H_2 gas for study of nanosecond fast chemical relaxation-times. Rev Sci Instrum 46:1209–1215

13. Bremer C et al (1993) A laser temperature jump apparatus based on commercial parts equipped with highly sensitive spectrophotometric detection. Meas Sci Tech 4:1385–1393

14. Green TB et al (2004) IA(3), an aspartic proteinase inhibitor from *Saccharomyces cerevisiae*, is intrinsically unstructured in solution. Biochemistry 43:4071–4081

15. Carslaw NS, Jaeger JC (1946) Conduction of heat in solids. Oxford, New York

16. Henry ER, Hofrichter J (1992) Singular value decomposition – application to analysis of experimental-data. Methods Enzymol 210:129–192

17. Hammes GG, Chang Y, Oas TG (2009) Conformational selection or induced fit: a flux description of reaction mechanism. Proc Natl Acad Sci 106:13737–13741

18. Palmer KF, Williams D (1974) Optical properties of water in the near infrared. J Opt Soc Am 64:1107–1110

19. Wray WO, Aida T, Dyer RB (2002) Photoacoustic cavitation and heat transfer effects in the laser-induced temperature jump in water. Appl Phy B 74:57–66

20. Beechem JM, Brand L (1985) Time-resolved fluorescence of proteins. Annu Rev Biochem 54:43–71

21. Eftink MR et al (1995) Fluorescence studies with tryptophan analogs: excited state interactions involving the side chain amino group. J Phys Chem 99:5713–5723

Chapter 19

Differential Scanning Microcalorimetry of Intrinsically Disordered Proteins

Sergei E. Permyakov

Abstract

Ultrasensitive differential scanning calorimetry (DSC) is an indispensable thermophysical technique enabling to get direct information on enthalpies accompanying heating/cooling of dilute biopolymer solutions. The thermal dependence of protein heat capacity extracted from DSC data is a valuable source of information on intrinsic disorder level of a protein. Application details and limitations of DSC technique in exploration of protein intrinsic disorder are described.

Key words: Thermodynamics, Phase transitions, Thermal stability, Calorimetry, Enthalpy, Heat capacity

1. Introduction

Differential scanning calorimetry (DSC) is a unique analytical nonengaging thermophysical technique enabling to get direct information on heat effects accompanying uniform heating/cooling of dilute polymer solutions at constant pressure, which ensures that the heat effects correspond to changes in system's enthalpy, H. Being technically and experimentally sophisticated method, DSC exploits a transparent operating principle. Two nearly identical cells, called sample and reference cells, are loaded with a solvent containing dissolved sample and the same solvent without sample, respectively. Both cells are heated or cooled at a constant rate (v, see Note 1) under quasi-adiabatic conditions ensured by temperature controlled thermal jacket, which surrounds both cells (Fig. 1). The heat effects emerging in the sample cell with temperature would cause deviation in temperature of the sample cell from that of the reference cell, but a dedicated compensatory system cancels out this temperature difference, simultaneously measuring the heat power

Vladimir N. Uversky and A. Keith Dunker (eds.), *Intrinsically Disordered Protein Analysis: Volume 2, Methods and Experimental Tools*, Methods in Molecular Biology, vol. 896, DOI 10.1007/978-1-4614-3704-8_19, © Springer Science+Business Media New York 2012

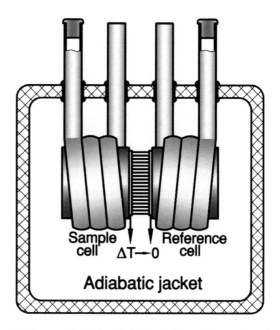

Fig. 1. Schematic diagram of calorimetric block of differential scanning microcalorimeter with capillary cells. The temperature of nonremovable calorimetric cells is uniformly changed under quasi-adiabatic conditions assured by means of temperature controlled adiabatic jacket. The heat power required for compensation of the temperature difference between the cells (ΔT) is measured.

needed for that, ΔW. The ΔW is spent for compensation of the heat capacity difference between dissolved sample ($C_{\text{total,sample}}$) and the solvent displaced by the sample molecules ($C_{\text{total,solvent}}$):

$$\Delta W = dH/dt = \left(C_{\text{total,sample}} - C_{\text{total,solvent}}\right)v \qquad (1)$$

Knowing the value of active cell volume (V, see Note 2), density (ρ_{sample}) and weight concentration of the sample at fixed temperature T_0 ($c_{w,0}$), specific heat capacity ($C_{P,\text{solvent}}$) and density of the solvent at temperature T_0 ($\rho_{\text{solvent},0}$) and arbitrary temperature (ρ_{solvent}), a temperature dependence of specific heat capacity of the dissolved sample can be derived from Eq. 1 (1):

$$C_{P,\text{sample}}(T) = \left(\rho_{\text{solvent}}/\rho_{\text{sample}}\right) \cdot C_{P,\text{solvent}}$$
$$+ \left(\rho_{\text{solvent},0}/\rho_{\text{solvent}}\right) \cdot \Delta W/\left(v \cdot c_{w,0} \cdot V\right) \qquad (2)$$

The heat capacity of a system at constant pressure as a function of temperature, $C_P(T)$, is fundamental, since it enables to get temperature dependencies of both enthalpy of the system and absolute value of its entropy, S (in the absence of chemical reactions) (2):

$$H(T) = H(0) + \int\limits_0^T C_P(T)dT, \quad S(T) = \int\limits_0^T \frac{C_P(T)}{T}dT \qquad (3)$$

Besides, the isobaric heat capacity is a measure of entropy fluctuations of the system (2):

$$C_P(T) = \left\langle [S(T) - \langle S(T) \rangle]^2 \right\rangle \qquad (4)$$

Hence, one should expect growth of C_P with temperature for any macroscopic state accessible to a biopolymer (see Note 3); thermal unfolding of a macromolecule should be accompanied with increase in its C_P value.

For a system comprising multiple interconverting polymer macrostates the total enthalpy is a superposition of their respective enthalpies, H_i:

$$H(T) = \sum_i f_i(T) \cdot H_i(T), \qquad (5)$$

where f_i is a population of the i-th state. Differentiating Eq. 5, the total isobaric heat capacity represents the sum of two terms:

$$C_P(T) = \sum_i f_i(T) \cdot C_{P,i}(T) + \sum_i f_i'(T) \cdot H_i(T)$$

$$= C_{P,\text{inner}}(T) + C_{P,\text{transitions}}(T), \qquad (6)$$

where $C_{P,i}$ is a heat capacity of the i-th state. The first term, $C_{P,\text{inner}}$, is a heat capacity inherent to all macrostates available in the system, while the second term, $C_{P,\text{transitions}}$, arises due to the enthalpy changes linked to interconversions between the macrostates (Fig. 2b). Both terms are equally important for probing of intrinsic disorder in proteins.

The globally disordered proteins lack rigid tertiary structure and therefore do not reveal cooperative endothermic thermal denaturation (positive $C_{P,\text{transitions}}$ term), characteristic for folded proteins (3) (see Note 4) (Fig. 2). At the same time, disordered proteins exhibit specific values of $C_{P,\text{inner}}$ exceeding those inherent to folded proteins, reaching in the case of extensively unfolded proteins the values characteristic for fully unfolded protein (4). The relatively high specific heat capacity values and absence of positive $C_{P,\text{transitions}}$ contribution are regarded as DSC hallmarks of globally disordered proteins. Meanwhile, hydrophobic groups of intrinsically disordered proteins (IDPs) are exposed to solvent and therefore may induce protein oligomerization or/and aggregation. Since both phenomena induce formation of protein structure, they are exothermic (see Note 4) and accompanied with a decline in the $C_P(T)$ values (negative $C_{P,\text{transitions}}$) (Fig. 2b). Some of IDPs demonstrate cooperative exothermic thermal folding (5) likely due to strengthening of hydrophobic interactions with temperature (6), eventually resulting in formation of a structure. Such behavior may be considered as an additional signature of an IDP.

In large multidomain proteins the structural cooperativity of different domains may substantially differ: some of them may be

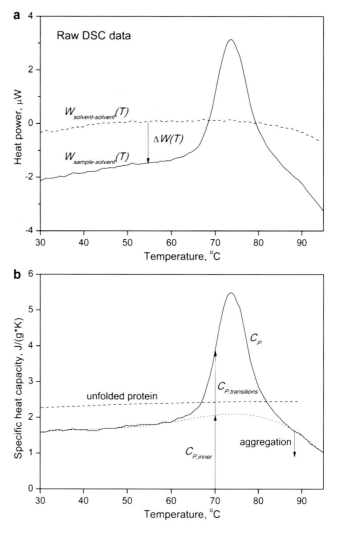

Fig. 2. Thermal denaturation of hen egg-white lysozyme (HEWL, 0.17 mg/mL) in deionized water measured at heating rate 1.5°C/min using Nano DSC scanning microcalorimeter (TA Instruments). (**a**) shows raw DSC data: two consecutive thermal scans, corresponding to baseline (*dashed line*) and melting of HEWL (*solid curve*). (**b**) shows the specific heat capacity of HEWL (C_P) calculated from data in Fig. 2a using Eq. 2. Different components of C_P are illustrated (see the text for details). The specific heat capacity of fully unfolded HEWL is estimated according to Häckel et al. (11). The high temperature region of the C_P curve is bent down due to aggregation effects.

disordered, while others can be highly stable. The DSC analysis in this case does not ensure readily extractable structural information on disordered protein domains. The method of DSC is most suited for studies of relatively small, structurally cooperative proteins. In this case DSC provides a unique opportunity for discrimination between the two opposite situations: folded protein state and globally disordered state. In a particular case of the maximal value of standard free energy of protein unfolding (ΔG^0) close to zero, the equilibrium constant of thermal unfolding ($\exp[-\Delta G^0/(\text{RT})]$) is about 1. Hence, the protein is partially unfolded even at temperature ensuring its maximal stability and exhibits a poorly defined heat effect upon thermal unfolding. Thermodynamic analysis of DSC data may give an exhaustive description of system behavior in this case.

Accurate estimations of absolute values of specific heat capacity of a biopolymer (C_P) in dilute solution using the DSC measurement principle outlined above would require the use of calorimetric cells with unrealistically close thermophysical and thermochemical characteristics. To eliminate the differences in the properties of sample and reference cells, the measurements are actually performed using just one of the cells (we will denote it as a "sample" cell), while the second cell is permanently filled with a solvent. In order to get the temperature dependence of compensatory heat power, $\Delta W(T)$ (see Eq. 1), required for estimation of C_P (Eq. 2), two consecutive thermal scans should be run (Fig. 2a). The first scan is measured upon refilling of the sample cell with the solvent yielding a heat power curve $W_{\text{solvent-solvent}}(T)$, referred to as "baseline." The second scan is performed in an identical manner but the cell is refilled with the sample dissolved in the same solvent. The resulting heat power curve $W_{\text{sample-solvent}}(T)$ is subjected to subtraction of the baseline to produce the $\Delta W(T)$ function:

$$\Delta W(T) = W_{\text{sample-solvent}}(T) - W_{\text{solvent-solvent}}(T) \qquad (7)$$

The knowledge of $\Delta W(T)$ allows the estimation of temperature dependence of C_P value according to the Eq. 2 (see Fig. 2b).

It should be emphasized that this principle of DSC measurements is applicable only provided that sufficiently high reproducibility of baseline in cyclic operating mode is achieved. To this end, the basic algorithm should be adhered to the following:

1. The working surfaces of calorimetric cells should be thoroughly cleaned. This is an absolutely necessary condition for achievement of proper performance of any DSC instrument.

2. The cells are loaded with solvent and equilibrated with it in a cyclic operating mode.

3. Upon achievement of reproducible baseline (see Note 5) the sample cell is refilled with solvent in each thermal cycle. This procedure is performed in a highly consistent manner, providing identical thermal history of the instrument from one cycle to another.

4. If two consecutive baselines measured upon refilling of the cell coincide with each other (see Note 5), the sample cell should be refilled with a sample of interest dissolved in the solvent, strictly adhering to the procedure used for refilling of the cell with the solvent.

2. Materials

Prepare all solutions using analytical grade reagents and water with specific resistance of 18 MΩ cm.

2.1. Instrumentation

1. Since IDPs are susceptible to aggregation the ultrasensitive DSC instruments designed for studies of dilute protein solutions (down to 0.05–0.1 mg/mL) should be used. Besides, capillary design of calorimetric cells (see Fig. 1) is preferred over the coin or lollipop shaped cells since the capillaries provide lower sensitivity to aggregation processes. The capillary cells are more conveniently and thoroughly washed due to their flow through design, which also lowers the chances of introduction of air bubbles during the cell loading. Two commercial instruments meet these requirements: Nano DSC (platinum 300 μL cells) from TA Instruments-Waters LLC and Manual Capillary DSC (Tantalum 61™ 130 μL cells) from GE Healthcare (MicroCal Inc.). Both instruments can be equipped with autosampler. Use an online or line interactive/sine wave uninterruptable power supply. Calorimeter should be placed in a clean, vibration-free, temperature- and humidity-controlled area. Direct air drafts should be avoided.

2. Sample degassing and cell washing station, or autosampler system (provided by DSC manufacturer).

3. Balances ensuring measurement accuracy of 0.01 mg.

4. Analytical UV-Vis spectrophotometer with quartz cells of different path lengths.

5. Benchtop centrifuge.

2.2. Sample Preparation Materials

1. At least 0.1 mg of lyophilized or dissolved protein sample of the highest available purity.

2. Buffers with low absolute value of ionization enthalpy and respectively weak temperature dependence of their pK_a values. For example, glycine ($pK_a = 2.35$ at 25°C (SC-Database)), acetic acid ($pK_a = 4.76$), cacodylic acid ($pK_a = 6.28$), phosphate ($pK_a = 7.20$).

3. Hydrochloric acid, potassium or sodium hydroxide.

4. Some substances should be avoided (see Note 5).

5. Desalting chromatographic column or dialysis membrane with appropriate molecular weight cutoff value.

2.3. Solutions for Cleaning of Platinum Cells

1. Protein removal: 1 mg/mL pepsin, 0.1 M acetic acid, 0.5 M NaCl. Prepare fresh solution each time.

2. Removal of mineral deposits: 4 M NaOH and 50% formic acid.

3. Removal of grease: HPLC grade tetrahydrofuran and 0.5% SDS solution.

2.4. Solutions for Cleaning of Tantalum™ Cells

1. Detergent cleaning: 20% v/v Contrad 70 (Decon 90) (Decon Laboratories Ltd) solution. Alternatively, 5% v/v Liqui-Nox® (Alconox, Inc.) solution.

2. Cleaning solution: weight 5 g of NaOH pellets and dissolve in 25 mL of water in a 100-mL volumetric flask. Add 50 mL of NaOCl solution (reagent grade, 10–15% chlorine). Add water to the 100 mL mark and mix the solution. Store in a dark, airtight container.

3. Concentrated nitric acid.

4. Pepsin solution (see above), 10% EDTA solution at neutral pH; methanol, hexane, toluene. See also Note 5.

3. Methods

The experimental procedures for manual filling of calorimetric cells are mostly described. In the case of autosampler, adhere to the described methods as close as possible.

3.1. Cell Cleaning: General Considerations

Reproducibility of the baseline scan critically depends upon cleanness of inner surface of calorimetric cells. The cleaning procedure depends upon the type of cell contamination and material of the cells. If the type of contamination is known, use the cleaning solution assuring solubility of the contaminant. Begin with the least aggressive agents. For protein samples resistant to precipitation cell wash with water or buffer may be sufficient. Otherwise stick to solutions and procedures recommended by instrument manufacturer. Avoid contact of off-site areas with corrosive chemicals (upon contact thoroughly rinse the area). If baseline reproducibility is not restored (can be judged from multiple water–water scans), proceed to more aggressive means, advised by manufacturer. If all recommended cleaning procedures do not restore baseline repeatability, the use of alternative reagents is possible. Refer to the manual on the corrosion resistance properties of Tantalum™ and platinum provided by MicroCal Inc. (see also Note 5). Prior to use of alternative cleaning solutions contact support center of the manufacturer for advice. If contamination is unknown combine several cleaning procedures, beginning with least aggressive. The failure to restore baseline reproducibility via cell cleaning, as confirmed by multiple water-water scans, may evidence malfunction of the calorimeter. Refer to support center for instructions.

The following chapters describe the cell cleaning procedures, recommended by TA Instruments and MicroCal Inc.

3.2. Cleaning of Platinum Cells

1. Regular cleaning: fill the cell at room temperature with 50% formic acid, set the pressure to 3 bar, followed by a scan to 75°C at 1°C/min. Upon cooling rinse the cell with 1 L of water (see Note 6).

2. Removal of protein deposits: soak the cell at 30°C in pepsin solution for 3 h. Rinse the cell with 1–2 L of water (see Note 6).

3. Removal of mineral deposits. Fill the cell at room temperature with 4 M NaOH solution; run a scan to 90°C at 2°C/min. Leave the cell at 90°C overnight. Upon cooling rinse the cell with 1–2 L of water (see Note 6). Fill the cell with 50% formic acid; run a scan to 65°C at 2°C/min. Leave the cell at 65°C for 20 min. Upon cooling flush the cell with 1–2 L of water.

4. Cleaning of grease. Fill the cell at room temperature with HPLC grade tetrahydrofuran; run a scan to 50°C at 2°C/min. Leave the cell at 50°C for 20 min. Upon cooling flush the cell with 100 mL of 0.5% SDS solution (see Note 6). Rinse the cell with 1–2 L of water.

3.3. Cleaning of Tantalum™ Cells

1. Regular cleaning: fill the cell at room temperature with 20% Contrad 70; leave the cell at 50°C for 10 min. Upon cooling rinse thoroughly the cell with water (see Note 6).

2. Removal of protein deposits. A soaking step: fill the cell at 50°C with the dedicated cleaning solution and wait for several minutes. Remove the cleaning solution and inspect it for presence of residual protein. If protein is revealed, repeat the soaking step until no signs of protein are found in the used cleaning solution. Upon cooling to room temperature rinse thoroughly the cell with water (see Note 6).

3. Nitric acid soak. The cell port, pressure transducer and its sealing should be protected from nitric acid. Fill the cell at room temperature with concentrated nitric acid and place a damp paper towel over the cell port. Leave the cell at 80°C for 1–24 h. Upon cooling the cell to room temperature remove the nitric acid and rinse thoroughly the cell with water (see Note 6).

4. Refer to Subheading 2 for other possible cleaning solutions. See also Note 5.

3.4. Establishment of Thermal Cycle of Calorimeter

1. Fill both calorimetric cells with freshly degassed buffer solution (see Note 5). In the case of manual filling of the cells use the sample degassing station (vacuum of 15–25 in Hg for 10–15 min with stirring is sufficient); empty the capillary cell without drying using micropipette (via a short silicone tubing); fill the cell with buffer slowly (5–10 s) pressing the plunger avoiding introduction of air bubbles into the cell; pull a dedicated cap or insert a plug (see Fig. 1) into one of the cell access tubes (otherwise, liquid oscillations are possible during thermal scan).

2. Choose the instrument settings most closely meeting the experiment requirements: temperature range (the wider the better; may be limited due to aggregation effects), scan rate (v) and response time (see Note 1), data collection rate, pre-run delay (10–15 min is optimal), excess pressure settings (see Note 7).

3. Run the calorimeter in cyclic operating mode using the selected presets. Wait until reproducible baseline is achieved (typically, requires five to ten scans). The inability to reach reproducible baseline may be related to continuous excess pressure drop during scanning (check sealing of the pressure handle), incomplete cleaning of the cells (see Subheadings 3.1–3.3) or incompatibility of some component(s) of the buffer solution with material of the cells (reconsider the choice of the buffer components; see Note 5).

3.5. Measurement of Thermograms for Buffer and Sample Solutions

1. Without stopping the thermal cycle repeat refilling of the sample cell with freshly degassed buffer in each cycle, following an identical procedure. Each action should be tied to a specific value of cell temperature readings. Start and finish the cell refilling procedure at cell temperatures close to room temperature. Pay special attention to the stages greatly affecting thermal equilibrium between the cells: the solution removal and filling of the cell. Minimize the time gap between these two operations. Less thermal effect causes (de)pressurization of the cells. Repeat the baseline measurements until two consecutive baselines coincide with each other (three baseline scans is generally enough). The inability to achieve reproducible baseline upon cell refilling is likely to be related to incompatibility of some component(s) of the buffer solution with material of the cells (see Note 5). The result of the last baseline scan will be referred to as "baseline."

2. Prepare 0.2–1.5 mg/mL protein solution in the buffer. For a dissolved protein sample it should be transferred into the buffer solution. Use chromatographic desalting or dialysis against the buffer (plan in advance). For exhaustively dialyzed and freeze-dried protein sample, the weighted amount of protein is dissolved in the required volume of the buffer. Remove the precipitates and insoluble residuals via centrifugation. Determine the resulting protein concentration spectrophotometrically using known extinction coefficient (preferred) or the molar extinction coefficient calculated according to Pace et al. (7) (more reliable for Trp-containing proteins):

$$\varepsilon_{280nm} = 5,500 \cdot N_{Trp} + 1,490 \cdot N_{Tyr} + 125 \cdot N_{cystine} \cdot M^{-1} \, cm^{-1},$$

$$(8)$$

where N refers to number of the respective amino acids within the protein. For extensively disordered proteins the following coefficients are suitable: 5,660, 1,292, and 120, for Trp, Tyr, and cystine, respectively (average of the values for models of these residues measured in 8 M urea or 6 M guanidine hydrochloride (7)). The weight concentration of the protein is calculated from Beer–Lambert–Bouguer law:

$$c_{w,0} = M \cdot D_{280nm}/(\varepsilon_{280nm} \cdot l) \text{ mg/mL}, \qquad (9)$$

where M is molar mass of the protein, D_{280nm} is its absorbance at optical path length of the cell l (in cm). To ensure analytical level of protein heat capacity measurements the absolute error in protein concentration estimate should be below 3%.

3. In the thermal cycle following the baseline scan refill the sample cell with the freshly degassed protein solution. Strictly follow the cell refilling procedure used in the baseline measurement. Upon completion of the measurement (and possible repeats, for control of reversibility of the process) an immediate cleaning of the sample cell is recommended (see Subheadings 3.1–3.3).

3.6. Calculation of Protein Specific Heat Capacity

1. Calculate the compensatory heat power ($\Delta W(T)$) using the last baseline scan and the sample scan (see Fig. 2a and Eq. 7).

2. Determine temperature dependence of specific heat capacity of the protein using Eq. 2 (see Note 2). For dilute aqueous buffer solutions the temperature dependencies of specific heat capacity ($C_{P,solvent}$) and density ($\rho_{solvent}$) of the solvent can be approximated by respective values for distilled water. Alternatively, $\rho_{solvent}$ is measured by densimetry, while $C_{P,solvent}$ is measured via subtraction of DSC scans ran with the sample cell filled with the solvent or distilled water (differential power $\Delta W_{solvent-water}(T)$) (1):

$$C_{P,solvent} = \left[C_{P,water} \cdot \rho_{water} + \Delta W_{solvent-water}(T)/(v \cdot V) \right] / \rho_{solvent} \qquad (10)$$

Temperature dependence of protein density (ρ_{sample}) is estimated within additive approximation based upon known partial molar volumes of protein constituents (8, 9) or can be measured by densimetry.

3.7. Analysis of Protein Specific Heat Capacity

1. Globally disordered proteins lack cooperative endothermic thermal denaturation (positive $C_{P,transitions}$ term), characteristic for folded proteins (3) (see Note 4) (Fig. 2). On the contrary, some of IDPs demonstrate cooperative exothermic thermal folding (negative $C_{P,transitions}$) (5). In the case of formation of amorphous protein aggregates a decline in protein specific heat

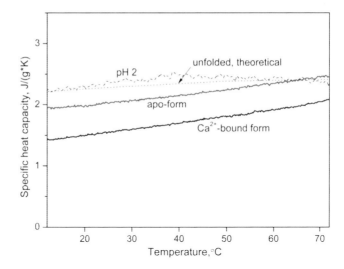

Fig. 3. Temperature dependencies of specific heat capacity of folded Ca^{2+}-bound (*black solid curve*), disordered Ca^{2+}-free (*grey solid curve*) and unfolded acidic (pH 2; *dashed curve*) forms of pike α-parvalbumin (1.5–2 mg/mL), estimated from DSC data (12). Heating rate was 0.9°C/min; DASM-4 M microcalorimeter (IBI RAS, Pushchino, Russia). The specific heat capacity of the fully unfolded protein estimated according to Häckel et al. (11) is shown (*dotted curve*) for comparison.

capacity is observed (see Fig. 2b). If DSC analysis evidences that protein is intrinsically disordered, recalculate protein concentration based upon alternative molar extinction coefficients indicated in Subheading 3.5, step 2 and recalculate the heat capacity values (Subheading 3.6).

2. Calculate temperature dependence of specific heat capacity of the ideal fully unfolded protein state using additive approximation based upon known partial molar heat capacities of protein constituents (4, 10, 11) (see Figs. 2b and 3). Alternatively, measure heat capacity of the fully unfolded protein at pH values, low enough for protein denaturation (though this state may be not fully unfolded) (see Note 8). Specific heat capacity values (C_P) of extensively unfolded protein are close (within several percent) to the values estimated for the fully unfolded protein. More compact disordered protein states possess lower C_P values (see Fig. 3). The lowest C_P values are observed for folded protein states. The latter values are relatively conservative at low temperatures (138–176 J(K·mol-residue) at 0°C), but temperature dependencies of C_P for folded proteins substantially differ (3): dC_P/dT ranges from 0.65 to 2.73 J/(K^2·mol-residue). An apparent very steep temperature dependency of C_P in some cases is a consequence of several overlapping temperature transitions, affecting different subdomains of a small globular protein (3). For proper discrimination between

folded and unfolded protein states on the basis of C_P measurements it is highly desirable to measure also the folded protein state as a reference. Binding of target molecules may induce folding of IDPs, which can be exploited for achievement of folded protein state. An example is shown in Fig. 3. The difference between C_P values for folded and thermally unfolded protein states is lower at higher temperature. Therefore the low temperature region is the most advantageous for reliable discrimination between these states.

3. The absolute values of protein specific heat capacity markedly below the values characteristic for folded proteins (see Subheading 3.7, step 2) in a wide range of temperatures are likely due to remains of contaminating low molecular weight compounds in the protein solution. Remove them via desalting or dialysis and repeat the measurement. Another possible cause of unreasonably low C_P values is protein aggregation, though this effect is generally more cooperative. Finally, an error in determination of protein concentration is possible, which may result also in overestimation of C_P values.

4. In the case of inability to get reasonable values of specific heat capacity of the protein, it is recommended to measure C_P values for some standard DSC sample (for example, hen egg-white lysozyme or other proteins listed in review by Privalov (3)) at higher sample concentration (1.5–3 mg/mL). Evident deviation of the results from published data may evidence malfunction of the instrument.

4. Notes

1. In commercial DSC instruments the scan rates lie in the range ±2–4°C/min. The larger the protein, the lower the scan rate should be chosen to overcome the kinetic limitations of the processes under study. For small globular proteins the scan rates of ±1–2°C/min are generally suited. DSC studies of fibrillar proteins may require an order of magnitude lower scan rates. Higher scan rates require lower response times. For scan rate of 1°C/min, response time of 15 s is a proper choice.

2. The active volume of calorimetric cell, V, significantly differs from the total inner volume of the cell, since presence of inlet/outlet capillar(ies) entails nonuniform heating of the liquid within the cell during thermal scanning. The V value is provided by instrument manufacturer. Temperature dependence of V can be neglected.

3. The isobaric heat capacity of denatured protein in aqueous medium may decrease with temperature at elevated

temperatures (above 75°C), since heat capacities of solvated side-chains for about half of amino acids (mostly hydrophobic) follow the unusual trend, declining with temperature (10, 11).

4. A cold denatured protein state possesses lowered enthalpy with respect to the folded state. Hence, protein folding with heating in this case is an endothermic phenomenon accompanied with positive $C_{P,\text{transitions}}$ term (see Eq. 6) (3). Meanwhile, cold denaturation of proteins typically occurs at temperatures barely accessible under aqueous conditions, so this phenomenon is not considered here.

5. Some of solvent components may be chemically incompatible with material of calorimetric cells, which may result in irreproducible DSC scans upon refilling of sample cell. For example, HEPES and related buffers are incompatible with platinum DSC cells. Some of reducing agents (for instance, DTT) may cause poor reproducibility of baseline due to irreproducible heat contribution of oxidation processes. Fluorinated solutions may cause irreparable damage to Tantalum™ cells. Similarly, use of basic solutions with Tantalum™ cells should be limited. In all these cases the choice of solvent components or their concentrations should be reconsidered.

6. Whenever you pass large volumes of solution through a cell it is recommended to turn off the calorimeter or exit from its software. Use a dedicated cell washing station.

7. At high temperature limit of 100°C an excess pressure about 1 bar is needed to avoid boiling of the aqueous solution. The further increase in the high temperature limit should be accompanied with the pressure increase of 1 bar per each 10°C. The excess pressure also suppresses the highly unfavorable formation of air bubbles with heating.

8. Use of basic pH values is undesirable due to high deprotonation enthalpies of basic protein groups (25–54 kJ/mol (3)).

Acknowledgment

This work was supported by a grant from the Program of the Russian Academy of Sciences "Molecular and Cellular Biology." The author would like to thank Ekaterina E. Klam for preparation of Fig. 1, and Prof. Eugene A. Permyakov and Dr. Alexei S. Kasakov for critically reading the manuscript.

References

1. Privalov PL, Potekhin SA (1986) Scanning microcalorimetry in studying temperature-induced changes in proteins. Methods Enzymol 131:4–51
2. Landau LD, Lifshitz LP (1980) Statistical physics, part 1, 3rd edn. Pergamon, Oxford
3. Privalov PL (2009) Microcalorimetry of proteins and their complexes. Methods Mol Biol 490:1–39
4. Privalov PL, Makhatadze GI (1990) Heat capacity of proteins. II. Partial molar heat capacity of the unfolded polypeptide chain of proteins: protein unfolding effects. J Mol Biol 213:385–391
5. Dzwolak W, Ravindra R, Lendermann J, Winter R (2003) Aggregation of bovine insulin probed by DSC/PPC calorimetry and FTIR spectroscopy. Biochemistry 42:11347–11355
6. Chandler D (2005) Interfaces and the driving force of hydrophobic assembly. Nature 437:640–647
7. Pace CN, Vajdos F, Fee L, Grimsley G, Gray T (1995) How to measure and predict the molar absorption coefficient of a protein. Protein Sci 4:2411–2423
8. Makhatadze GI, Medvedkin VN, Privalov PL (1990) Partial molar volumes of polypeptides and their constituent groups in aqueous solution over a broad temperature range. Biopolymers 30:1001–1010
9. Häckel M, Hinz HJ, Hedwig GR (1999) Partial molar volumes of proteins: amino acid side-chain contributions derived from the partial molar volumes of some tripeptides over the temperature range 10–90 degrees C. Biophys Chem 82:35–50
10. Makhatadze GI, Privalov PL (1990) Heat capacity of proteins. I. Partial molar heat capacity of individual amino acid residues in aqueous solution: hydration effect. J Mol Biol 213:375–384
11. Häckel M, Hinz HJ, Hedwig GR (1999) A new set of peptide-based group heat capacities for use in protein stability calculations. J Mol Biol 291:197–213
12. Permyakov SE, Bakunts AG, Denesyuk AI, Knyazeva EL, Uversky VN, Permyakov EA (2008) Apo-parvalbumin as an intrinsically disordered protein. Proteins 72:822–836

Chapter 20

Identifying Disordered Regions in Proteins by Limited Proteolysis

Angelo Fontana, Patrizia Polverino de Laureto, Barbara Spolaore, and Erica Frare

Abstract

Limited proteolysis experiments can be successfully used to detect sites of disorder in otherwise folded globular proteins. The approach relies on the fact that the proteolysis of a polypeptide substrate requires its binding in an extended conformation at the protease's active site and thus an enhanced backbone flexibility or local unfolding of the site of proteolytic attack. A striking correlation was found between sites of limited proteolysis and sites of enhanced chain flexibility of the polypeptide chain, this last evaluated by the crystallographically determined B-factor. In numerous cases, it has been shown that limited proteolysis occurs at chain regions characterized by missing electron density and thus being disordered. Therefore, limited proteolysis is a simple and reliable experimental technique that can detect sites of disorder in proteins, thus complementing the results that can be obtained by the use of other physicochemical and computational approaches.

Key words: Intrinsically disordered proteins, Protein structure, Limited proteolysis, B-factor, Protein dynamics, Proteases

1. Introduction

Nowadays there is a strong interest in partly folded or even fully disordered or natively unfolded proteins, since these protein states can have a role in the proper functioning of proteins (1–5). Proteins can be fully folded into a fixed, native-like three-dimensional (3D) structure and these proteins are the most well-characterized and abundant in the PDB database (http://www.pdb.org). However, even otherwise folded globular proteins can have chain regions of moderate length (10–50 residues) that are disordered, as given by the lack of backbone coordinates in their X-ray structures (6) or diminished chemical shift dispersion in

Vladimir N. Uversky and A. Keith Dunker (eds.), *Intrinsically Disordered Protein Analysis:*
Volume 2, Methods and Experimental Tools, Methods in Molecular Biology, vol. 896,
DOI 10.1007/978-1-4614-3704-8_20, © Springer Science+Business Media New York 2012

NMR measurements (7). Moreover, folded proteins normally contain rather flexible or disordered chain segments at their N- and C-terminal ends (6). In the case of multidomain proteins, flexible linker segments are joining the domains, thus allowing the domain movements that determine protein dynamics and function (8). Finally, a protein can adopt a partly folded state, nowadays named molten globule (MG), characterized by significant secondary structure, but lacking substantial tertiary interactions (9–11). Therefore, proteins appear to exhibit a continuum of structures, ranging from a tightly folded state, an otherwise folded state but containing local flexible segments or disordered regions, a MG state, or highly extended states as the intrinsically denatured (ID) proteins (12, 13).

The analysis of protein structural disorder is quite problematic and, to this aim, numerous physicochemical (14) and computational techniques (15, 16) and approaches for identifying and characterizing protein disorder are being explored. A variety of experimental techniques can be used (see their list in DisProt, http://www.disprot.org), but every technique for protein structure (or non structure) determination has both strengths and weaknesses (14). Here, we show that a classical biochemical method such as limited proteolysis (17–22) can be used to detect disordered regions in globular proteins, providing easy-to-obtain experimental results which closely complement the results that can be obtained from other, more classical, physicochemical methods and approaches. Here we show that proteolytic probes can pinpoint the sites of local unfolding or protein disorder in a protein chain (23).

1.1. Rationale of the Technique

The term "limited proteolysis" indicates the specific fission of only one peptide bond (or a few) among the many present in a protein molecule substrate (17–22, 24, 25). In numerous cases, it has been found that limited proteolysis of a globular protein results in cleaving only one peptide bond, thus leading to a nicked protein given by two fragments remaining associated in a stable and often functional complex (26–30). By using proteases of different specificities (Table 1) usually there is a narrow region of the polypeptide chain that is attacked by the various proteases (31). Of course, the fact that a protease can cut a large protein molecule at a single peptide bond encompassed by a short chain segment is intriguing and it is immediately evident that the higher order structure of the protein and not its amino acid sequence is dictating this extraordinary selectivity of proteolytic attack.

In current and past literature often it is suggested that limited proteolysis results from the fact that a specific chain segment of the compact, folded protein substrate is sufficiently exposed to bind to the active site of the protease (32–35). However, while exposure/accessibility is necessary, it is clearly not sufficient to explain the phenomenon of limited proteolysis, since it is evident that even in a

Table 1
Proteases commonly used for protein structure analysis

Protease[a]	EC number[b]	Type	Mol mass (Da)	pH optimum	Specificity[c]	Inhibitors[d]	Notes
Trypsin (bovine)	3.4.21.4	Ser	23,500	8.0–9.0	Arg, Lys	PMSF, DFP, TLCK, aprotinin, leupeptin, soybean trypsin inhibitor	Due to its narrow specificity, it can cut at disordered regions of proteins only if these contain Lys or Arg residues. It does not cleave the Lys-Pro or Arg-Pro peptide bond. If an acidic residue is on either side of the cleavage site, the rate of hydrolysis is much reduced. Trypsin preparations may contain traces of "pseudo-trypsin" that cleaves at hydrophobic amino acid residues as α-chymotrypsin
Subtilisin (*B. subtilis*)	3.4.21.62	Ser	27,300	7.0–11.0	Nonspecific	PMSF, DFP, α_2-macroglobulin	It belongs to the nonspecific endoproteinases and thus it is most useful for detecting sites or regions of protein disorder
Thermolysin (*B. thermoproteolyticus*)	3.4.24.27	Zn	34,500	7.0–9.0	Leu, Phe, Ile, Val, Met, Ala	EDTA, citrate, phosphate, 1,10-phenathroline, phosphoramidon, α_2-macroglobulin	Rather nonspecific endoproteinase that cuts at the N-terminus of mostly hydrophobic amino acid residues. For optimal stability it may be used in the presence of 1–10 mM $CaCl_2$ and, being a thermostable enzyme, at rather high temperature (up to 80°C)

(continued)

Table 1
(continued)

Protease[a]	EC number[b]	Type	Mol mass (Da)	pH optimum	Specificity[c]	Inhibitors[d]	Notes
α-Chymotrypsin (bovine)	3.4.21.1	Ser	25,000	7.0–9.0	Phe, Tyr, Trp, Leu, Ile	DFP, PMSF, TPCK, soybean trypsin inhibitor, chymostatin, heavy metals	Endoproteinase displaying moderate specificity for aromatic and hydrophobic amino acid residues; moderately useful for detecting disordered sites in proteins
Proteinase K (*Tritirachium album*)	3.4.21.64	Ser	28,900	7.5–11.0	Nonspecific	DFP, PMSF, Hg^{2+}	The most nonspecific endoproteinase that usually cuts at several amino acid residues within a disordered region of a protein; it is not inactivated by EDTA, sulfhydryl reagents and TPCK
Endoproteinase Glu-C (V8-protease) (*Staphylococcus aureus*)	3.4.21.19	Ser	27,000	7.8 (4.0)	Glu (Asp)	DFP, 3,4-dichloro-isocoumarin	Cuts specifically at Glu in ammonium bicarbonate, pH 8.0, or ammonium acetate, pH 4.0; in phosphate buffer, pH 7.8, cuts at both Glu and Asp
Endoproteinase Arg-C (Clostripain) (*Clostridium histolyticum*)	3.4.22.8	Cys	43,000 12,500 Two-chains	7.1–8.5	Arg	EDTA, Cu^{2+}, sulfhydryl reagents, citrate, borate	This is a quite specific endoproteinase and thus it will cut at disordered regions of a protein substrate only if there an Arg residue is present. It cleaves also the Arg-Pro peptide bond. The enzyme is activated by dithiothreitol or cysteine and Ca^{2+} is essential
Endoproteinase Lys-C (*Lysobacter enzymogenes*)	3.4.21.50	Ser	33,000	7.5–8.5	Lys	DFP, TLCK, aprotinin, leupeptin, benzamidine	Rather specific for Lys and thus it will cut at disordered region of a protein only if there is a Lys residue

Pepsin (porcine)	3.4.23.1	Asp	34,600	1.0–4.0	Nonspecific	Pepstatin, substrate-like epoxides, 4-bromophenacyl bromide	Cleaves preferentially at hydrophobic amino acid residues. It can be active up to pH 4.0, whereas it denatures if exposed to neutral pH
Endoproteinase Asp-N (*Pseudomonas fragi*)	3.4.24.33	Zn	27,000	6.0–8.0	Asp (Glu)	EDTA, 1,10-phenanthroline	Quite specific for Asp and thus it will cut disordered regions only if there an Asp residue is present
Papain (*Carica papaya*)	3.4.22.2	Cys	23,400	6.0–7.0	Nonspecific	PMSF, TLCK, TPCK, cystatin, leupeptin, α_2-macroglobulin, sulfhydryl alkylating agents	Nonspecific endoproteinase, classically used to produce active immunoglobulin fragments. Before use, papain preparations must be activated by incubation in 1 mM EDTA and 5 mM cysteine for 30 min
Elastase	3.4.21.36	Ser	25,900	7.5–8.8	Ala, Val, Ile, Leu, Gly, Ser, Thr	DFP, PMSF, soybean trypsin inhibitor, α_2-macroglobulin	Nonspecific endoproteinase useful for detecting sites or regions of protein disorder

[a]The source of the various proteases is shown in parenthesis

[b]The enzyme nomenclature can be accessed at the Web site http://www.expasy.ch/enzyme. The MEROPS database (http://www.merops.ac.uk) provides a structure-based classification of all known proteases. BRENDA (http://brenda-enzymes.org) is a very comprehensive and updated web site for properties of enzymes, including proteases. This database contains information on function, structure, preparation, stability and application of proteases. Moreover, BRENDA provides detailed information on substrates, inhibitors, cofactors, activating compounds and kinetic parameters, as well as tools for displaying 3D protein models of proteases listed in PDB. The majority of data are manually extracted from the primary literature and references linked to PubMed database are listed (see the recent summary of BRENDA given by Scheer et al. (2011) Nucleic Acids Res 39(Database issue):D670–D676)

[c]The proteases herewith listed are endopeptidases that cleave at the C-terminus of the indicated amino acid residue(s), with the exception of thermolysin and Asp-N protease that instead cleave at the N-terminus. Pepsin is rather nonspecific and cleaves C- or N-terminally of mostly hydrophobic or aromatic amino acid residues. As indicated in the text, "specificity" refers to peptide bonds that are preferentially attacked by the various proteases

[d]Abbreviations: DFP diisopropyl fluorophosphate (extremely toxic), PMSF phenylmethylsulphonyl fluoride, TPCK tosylamido-2-phenylethyl chloromethyl ketone, TLCK tosyllysine chloromethyl ketone

small globular protein there are many exposed sites which could be the targets of proteolysis. Instead, enhanced chain flexibility (segmental mobility) is the key feature of the site(s) of limited proteolysis, as shown by the results of systematic experiments of proteolysis and autolysis conducted on the thermophilic protease thermolysin (17). In this study, it was demonstrated that there is a clear-cut correlation between sites of limited proteolytic attack and sites of enhanced chain flexibility, the latter reflected by the values of the crystallographic temperature factors (B-factors) along the polypeptide chain of thermolysin. In crystal structures of proteins, the B-factor reflects the uncertainty in atom positions in the 3D protein model and often represents the combined effects of thermal vibrations and static disorder (36–39). When plotted against residue number, B-factor values provide a graphic image of the degree of mobility existing along the chain (17).

The fact that a flexible/unfolded polypeptide substrate is required for proteolysis is in keeping with the idea that the protein substrate must undergo considerable structural changes in order to bind effectively at the protease's active site in order to form the optimal transition state needed for hydrolysis. The key role of substrate flexibility/disorder is substantiated by the recent systematic analysis of the recognition mechanism of proteases for polypeptide substrates and inhibitors (40, 41). This analysis, made possible by the recent availability of many protease-inhibitor crystallographic structures, revealed that an almost universal recognition mechanism by all proteolytic enzymes implies that the binding of a polypeptide substrate at the active site of the protease occurs in an extended conformation, as schematically shown in Fig. 1a. Therefore, a protease can select from a conformational ensemble of polypeptide structures only an extended strand (41). Consequently, we may understand why a folded and quite rigid globular protein is usually resistant to proteolysis and why limited proteolysis does not occur at the level of hydrogen bonded segments of the polypeptide chain, such as α-helices and β-strands (17, 21, 22). Instead, limited proteolysis of a globular protein occurs only at those loops which display inherent conformational flexibility (17). Usually it is a region, rather than a specific site, which is the target of limited proteolysis, since it was observed that proteolytic cleavage takes place over a stretch of nearby peptide bonds, if several proteases are used (31).

A widely used nomenclature to describe the interaction of a polypeptide substrate at the protease's active site is that introduced by Schechter and Berger (42). As shown in Fig. 1b, it is considered that the amino acid residues of the polypeptide substrate bind at subsites of the protease's active site. By convention, these subsites on the protease are called S (for subsites) and the substrate amino acid residues are called P (for peptide). The amino acid residues of the N-terminal side of the scissile bond are numbered P3, P2, P1

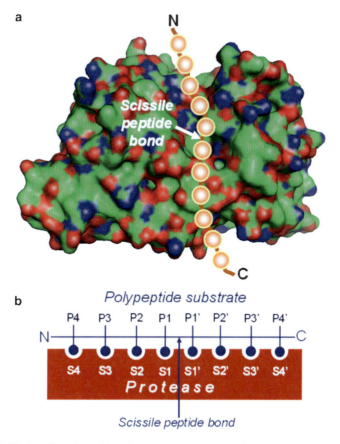

Fig. 1. Binding of a polypeptide substrate at the active site of a protease. (**a**) Schematic view of a polypeptide substrate binding at the active site of a protease in an extended conformation, as deduced from numerous crystal structures of complexes of proteases with their substrate-like peptide inhibitors (40, 41). N and C indicate the N- and C-terminus of the peptide chain, respectively. (**b**) Binding of a polypeptide at the protease's active site according to the nomenclature of Schechter and Berger (42). A polypeptide substrate interacts with its side-chain residues (P) at a series of subsites (S) of the protease. The interaction of the substrate at the active site of the protease requires a specific stereochemical adaptation of the substrate, and thus likely a significant degree of chain mobility, in order to form the idealized transition state of the hydrolytic reaction. The P1 side-chain residue interacting with the S1 binding site of the protease is usually the major determinant for the protease's specificity, but is not unique. Usually, proteases cleave at the carboxy-terminal side of the scissile bond, while thermolysin and Asp-N endopeptidase cleave at the amino-side.

and those residues of the C-terminal side are numbered P1′, P2′, P3′…The P1 or P1′ residues are those residues located near the scissile bond. The substrate residues around the cleavage site can then be numbered up to P8. The subsites on the protease that complement the substrate binding residues are numbered S3, S2, S1, S1′, S2′, S3′… In several studies it has been established that the protease–peptide/protein interaction involves a stretch of up to 12 amino acid residues (33–35).

1.2. The Sites of Limited Proteolysis Occur at Disordered Regions of Protein Chains

The fact that limited or site-directed proteolysis occurs at flexible/disordered sites or regions of a polypeptide chain is amply demonstrated by analyzing the results of the many experiments conducted over the years on proteins for which the 3D structure was solved by crystallography (21, 22). In Fig. 2 we show that limited proteolysis of DOPA decarboxylase (43, 44) and of the α-subunit of bacterial luciferase (45, 46) specifically occurs at the level of disordered regions of their polypeptide chains, while in Table 2 we provide additional case studies of limited proteolysis of proteins of known 3D structure. The results of these experimental studies strongly substantiate that site-specific proteolysis occurs at disordered protein regions for which no recognizable signal appears in the electron density maps. Here, we refer only to the results of the few experimental studies summarized in Fig. 2 and Table 2, but the generality of the fact that limited proteolysis occurs at chain regions characterized by missing electron density has been substantiated by us examining the results of ~65 cases (18, 21, 22) and unpublished results). Therefore, we may conclude that indeed limited proteolysis is a useful and reliable experimental technique for locating disordered regions in globular proteins (23).

2. Materials

The most used proteases listed in Table 1 can be purchased from numerous commercial sources, such as Sigma-Aldrich (http://www.sigmaaldrich.com), Promega (http://www.promega.com), Calbiochem (http://www.calbiochem.org), Pierce Chemical Co. (http://www.piercenet.com), Roche Molecular Biochemicals (http://www.biochem.roche.com), Biozyme Laboratories (http://www.biozyme.com), Worthington Biochemical Corp. (http://www.worthington-biochem.com), and Protea (http://www.proteabio.com). In particular, thermolysin from *Bacillus thermoproteolyticus*, subtilisin from *Bacillus subtilis*, trypsin from bovine pancreas, proteinase K from *Tritirachium album*, elastase from pig pancreas, and papain from *Carica papaya* can be obtained from Sigma-Aldrich.

Besides the information given in Table 1, some general properties of most commonly used proteases, details of experimental procedures, as well as suggestions for the optimal storage conditions of proteases can be found in web site of Sigma-Aldrich (http://www.sigmaaldrich.com), Promega (http://www.promega.com), Worthington Biochemical Corp. (http://www.worthington-biochem.com), and Protea (http://www.proteabio.com).

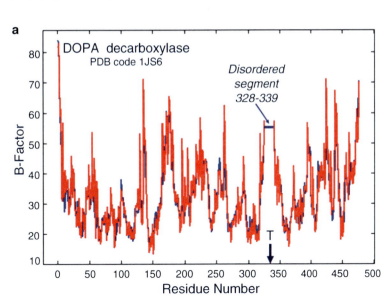

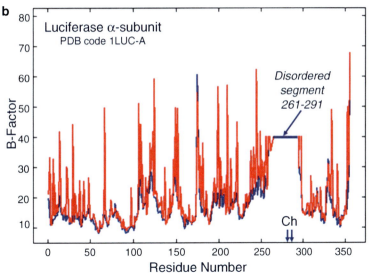

Fig. 2. Limited proteolysis occurs at disordered regions of globular proteins. (**a**) The *B*-factor profile along the 486-residue chain of DOPA decarboxylase taken from PDB (code 1JS6). The chain segment 328–339 is not visible in the electron density map of the protein and thus is disordered (43). The *Arrow* indicates the site of tryptic hydrolysis (T) at Lys334 (44), located at the disordered region of the protein. The nicked enzyme given by the two complementing fragments retains a native-like and functional structure (44). (**b**) The *B*-factor profile of the 355-residue chain of the α-subunit of bacterial luciferase taken from PDB (code 1LUC-A). The graph shows a discontinuity at chain segment 261–291 due to missing electron density and thus this segment is disordered (45). The sites of limited proteolysis of the α-subunit with chymotrypsin (Ch) at Phe280 and Leu282 (46) occur at the disordered region of the protein. Bacterial luciferase is a heterodimeric enzyme of 77 kDa, composed of α and β subunits with molecular masses of 40 and 37 kDa. Only the α subunit is susceptible to limited proteolysis, in agreement with the fact that in the X-ray structure of the luciferase only the α subunit displays a disordered chain region (see PDB code 1LUC).

Table 2
Disordered chain regions in proteins detected by X-ray methods correlate with sites of limited proteolysis

Protein	X-ray crystallography				Limited proteolysis			Notes
	Chain length	PDB code	Disordered region[a]	References	Protease[b]	Nicksites[c]	References	
Diphteria toxin	544	1DDT	188–199	(47)	T	Arg190, Arg192, Arg193	(48)	Tryptic cleavages occur internally at the chain segment containing the Cys186-Cys201 disulfide bridge and, therefore, the fragments of the nicked toxin remain covalently liked
Flavocytochrome b₂	511	1FCB	299–317	(49)	V8 Ch YP	Glu314 Met307 Lys305, Thr310	(50)	The X-ray structure 1FCB is that of the baker yeast protein
D-Amino acid oxidase from R. gracilis	368	1C0P	315–326* 361–368	(51)	T Ch Th V8	Arg305, Arg318, Arg364 Leu312 Leu312, Ala362 Glu365	(52)	The chain segment 315–326 corresponds to the highest peak (Å^2 up to 120) in the plot of B-factor values versus residue number of the protein chain and, therefore, it can be considered disordered
Calmodulin	148	1CDM	74–83	(53)	T	Lys77	(54)	The calmodulin 3D structure 1CDM is that of the protein complex with a peptide of brain protein kinase II-α

Protein		PDB		Chain regions	Protease	Cleaved residue		Comments
Staphylococcal nuclease	149	1SNC	(55)	1–7, 43–55*	T	Lys6, Lys48, Lys49	(56)	The chain segment 46–53 with B-factor values higher than 50 Å^2 is the highest peak in the plot of B-factor versus residue number of the 149-residue chain of nuclease
$E.$ $coli$ β-galactosidase	1024	1PX4	(57)	730–737[a]	E	Ala732	(58)	The chain segment from residue 730–737 corresponds to the highest peak in the plot of B-factor values versus residue number of the 1,024-residue protein chain; this segment displays B-factor values higher than 50 Å^2 and up to 98 Å^2 and thus is disordered
$E.$ $coli$ Glutathione synthetase	316	2GLT	(59)	164–167, 226–241	T	Arg233	(60)	Trypsin does not cleave at the disordered segment 164–167, since Lys or Arg residues are not occurring in this segment
Glutathione S-transferase P1-1	209	14GS	(61)	35–51	T, Ch	Lys44, Tyr49	(62, 63)	Limited proteolysis of the enzyme does not occur if proteolysis is conducted in the presence of glutathione. Indeed, segment 35–51 adopts an α-helical conformation in the holoenzyme (PDB code 5GSS)

[a] Disordered chain regions are those for which electron density is missing. The chain regions marked with an asterisk are those displaying the higher values in the plot of B-factor values along the protein chain and reaching a peak of at least 50 Å^2. These regions are considered the most flexible of the protein chain and thus disordered (64)

[b] T trypsin, Ch chymotrypsin, Th thermolysin, V8 staphylococcal V8-protease, E elastase, YP yeast protease

[c] The peptide bonds cleaved by a protease are indicated by the amino acid residue being the N-terminal one in the scissile peptide bond (e.g., the cleaved peptide bond Arg134-Gly135 is indicated as Arg134)

3. Methods

3.1. Choice of the Protease

Table 1 lists the endoproteinases with varying degrees of specificity that can be employed for limited digestion studies of proteins. Nowadays, ~20 different proteases are available from commercial sources and their purity usually is satisfactory for the specific needs of the limited proteolysis experiments. The basic premise of the limited proteolysis approach for probing protein structure implies that the proteolytic event should be dictated by the stereochemistry and flexibility of the protein substrate and not by the specificity of the attacking protease (18–22). Of course, a substrate-specific protease can cleave at a chain region only if there is a peptide bond involving a specific amino acid residue that the protease could target. Therefore, the most suitable proteases for probing protein structure disorder are those displaying broad substrate specificity, such as proteinase K, thermolysin, and subtilisin. These endopeptidases are relatively unspecific and display a moderate preference for hydrolysis at hydrophobic or neutral amino acid residues (see Table 1). The recommended approach is to perform trial experiments of proteolysis of the protein of interest utilizing several proteases of broad substrate specificity. Nevertheless, initial experiments can be conducted also utilizing proteases of more restricted specificity, such as trypsin, Glu-C protease from *Staphylococcus aureus* V8, Lys-C protease, Arg-C protease, or chymotrypsin (see Table 1). Using the latter proteases, the identification of protein fragments and thus the corresponding sites of proteolysis will be easier.

Since proteases often are unstable due to autolysis, usually fresh solutions of the proteases should be employed. However, resultant variations in the exact concentrations of protease solutions prepared at different times might make it difficult to reproduce the experimental results. For this reason, it is often a good compromise simply to freeze small aliquots of protease stock solutions for future use. Protease immobilized onto solid supports (e.g., Sepharose) can be used for proteolysis experiments, since after digestion the protease can be easily removed from the proteolysis mixture by centrifugation. It is important, however, to check if the immobilized protease behaves as the one in solution, since proteolysis can be controlled also by diffusion effects of the protein substrate to the immobilized protease.

3.2. Controlling the Rate of Proteolytic Digestion

The conditions used for limited proteolysis studies can vary widely. In studies of limited proteolysis the enzyme:substrate (E:S) ratio employed should be specified, because proteolysis is a bimolecular reaction dependent on the concentrations of both the proteolytic enzyme and the protein substrate. E:S ratios commonly used are 1:100, but 1:1,000 or even 1:10,000 are sometimes used. Possible ways to control proteolysis are the use of a low concentration of protease, short reaction times and low temperature (21, 22). It is not easy to predict in advance the most useful experimental conditions for conducting a limited proteolysis experiment, since these actually depend on the structure, dynamics, stability/rigidity properties of the protein substrate. When optimizing reaction conditions, it is most fruitful to focus on the E:S ratio and time dependencies of the digestion course.

3.3. Inhibition and/or Inactivation of Proteases

The limited proteolysis approach implies that the hydrolytic reaction is conducted under carefully controlled conditions, since if one waits long enough extensive proteolysis of a globular protein is expected to occur, thus minimizing the utility of the approach. Of course, the initial peptide fissions are the most informative on the structural features of the protein of interest. During the manipulation of the proteolytic mixture for the isolation and analysis of "nicked" proteins or fragment species there is the risk that the mixture is exposed to denaturing solvent conditions which would render the products of initial proteolysis vulnerable to further proteolysis. It is desirable, therefore, to inhibit rapidly and irreversibly the protease at the end of the proteolytic reaction by the use of specific inhibitors (see the list of inhibitors in Table 1) or to inactivate the protease by its denaturation in acid solution or in the presence of a detergent. Metalloproteases can be rapidly inhibited by adding a metal-chelating agent (see Table 1).

The easy way to stop proteolysis is to add to the proteolysis mixture trifluoroacetic acid (TFA) to a final concentration of 1.0–0.5% (by volume). An aliquot of this acid solution can be directly analyzed by reverse-phase (RP)-HPLC. It should be considered that some proteases, such as trypsin, are inhibited in acid, but recover their activity if the pH of the solution is brought to neutrality, thus causing further proteolysis. Alternatively, in order to stop proteolysis, the mixture can be added to the sample buffer used for sodium dodecyl sulphate (SDS)-polyacrylamide gel electrophoresis (PAGE) (65) and heated at 100°C. However, this procedure not necessarily is the most appropriate, since some proteolytic enzymes (trypsin, proteinase K, V8-protease) are not fully inactivated in the presence of SDS and there is the risk, therefore, that upon mixing the proteolytic mixture with the SDS-containing buffer proteolysis may proceed further. Therefore, proper controls should be conducted in order to check the irreversible inactivation of the protease before the SDS-PAGE analysis.

3.4. Isolation and Characterization of Protein Fragments

Since the initial cuts of a protein substrate are the most informative, the usual way to analyze the time-course of the proteolysis experiments is by SDS-PAGE. This is a method of choice, due to its simplicity, high sensitivity, high resolving power, and capability to analyze many samples in a single gel (65). Usually, 1–5 μg of a protein digest can be analyzed by SDS-PAGE. Since the protein fragments produced by limited proteolysis can remain associated in a stable complex at neutral pH, usually the electrophoretic analysis is conducted in the presence of SDS, which permits the dissociation of the fragments and their separation on the basis of their molecular masses. A quantitative analysis of the proteolytic fragments produced and separated in the gel can be achieved by densitometry of the bands stained with the Coomassie Blue dye. As stated above, before the electrophoretic analysis of the proteolytic mixture, it is essential to stop the proteolysis reaction by inactivating the protease.

Protein chemistry methods combining electrophoresis or chromatography with N-terminal sequencing by the Edman technique can be used to establish the identity of the polypeptide fragments and thus to identify the nicksites along the protein chain (21). These methods can be quite labor-intensive and relatively highly demanding in terms of protein sample requirements. On the other hand, nowadays the identification of protein fragments can be made much more easily by mass spectrometry (MS) techniques (66–70). In this case, the analysis can require minute amounts of protein sample (1–10 ng, femtomoles) and can be completed within minutes. Indeed, MS techniques provide a powerful method of fragment identification due to the mass accuracy of modern MS instruments. Tandem MS can be used also for the partial sequencing of peptides, thus leading to the unambiguous identification of the protein fragments. Of course, the time-consuming steps of electrophoretic and chromatographic separation of protein fragments and their subsequent N-terminal sequencing by the Edman technique can be eliminated, since it is possible to analyze directly proteolysis mixtures by MS methods and perform partial sequencing using tandem MS. In recent years, numerous studies have been conducted by the combined use of limited proteolysis and MS techniques (71–75).

3.5. Protocols for Limited Proteolysis

A general procedure for limited proteolysis experiments on native proteins can be as follows. Prepare a stock solution of the protease in a volatile buffer such as 0.1 M ammonium bicarbonate, pH 7.8–8.5. Some proteases may slowly inactivate over time, so it is desirable to make the stock solution from the solid powder of protease as close to the start time of the experiment as possible. Use a minimum volume in order to achieve high substrate concentration. Most of the commercially available proteases (see Table 1) can be used as supplied. For thermolysin, trypsin, chymotrypsin, and subtilisin the reaction mixtures should better include 1–5 mM Ca^{2+} in order to avoid autolysis. Incubate enzyme and protein substrate at a ratio of 1:100 (by weight) or, depending upon the specific protein under study, a molar ratio as low as 1:1,000–5,000 can be used. Incubate for several hours at room temperature or 37°C. However, some limited proteolysis experiments can require incubation as short as few minutes and a low temperature. The amount of protease can be adjusted so that the reaction can go faster or slower.

Proteolysis can be stopped by adding a suitable inhibitor of the protease (see Table 1) or simply by acidifying the reaction mixture with trifluoroacetic acid (TFA) or glacial acetic acid (final concentration 1%, by volume). Serine proteinases can be inhibited by a number of different compounds (see the list given in Table 1). Inhibitors that form tight but reversible complexes with serine proteinases (e.g., avian ovomucoids, soybean trypsin inhibitor, or Kunitz inhibitor) and peptide aldehyde inhibitors, some of which preferentially target certain serine proteinases, are also commonly used. It is recommended to take samples at intervals (for example at 5, 10, 30, and 60 min and then after several hours), stop the reaction by acidification of these aliquots and then analyze them by SDS/PAGE or RP-HPLC.

In order to analyze digestion samples by nano-LC MS or MS/MS, first separate peptides by RP-HPLC prior to MS analysis. In case, multidimensional chromatography approaches, e.g., a combination of ion exchange and subsequent RP liquid chromatography, may improve the number of separated protein fragments and their identifications. Load 5–10 μL of a sample onto a RP column at 10 μL per minute, then wash with solvent A (2% acetonitrile, 0.1% TFA) and elute over a 2 h linear gradient from 10 to 60% solvent B (80% acetonitrile, 0.1% TFA) directly into the mass spectrometer. The newly introduced Orbitrap LTQ MS/MS instrument (76–78) performs very well for protein fragment identification, although other MS/MS instruments will work as well.

3.6. Detecting Protein Disorder in Apomyoglobin

1. As a specific example of an experiment of probing protein structure by limited proteolysis, the details of limited proteolysis experiments conducted on sperm-whale apomyoglobin (apoMb) are given here. In this case, limited proteolysis was used to detect the local unfolding of the chain region 82–97 encompassing helix F in the holo protein (20, 31), in agreement with structural data obtained by means of NMR measurements (79, 80) (see Fig. 3).

2. Proteolysis of apoMb was conducted at 25°C with a variety of proteases (proteinase K, thermolysin, subtilisin, papain, elastase, chymotrypsin, V8-protease, and trypsin) (31). Limited proteolysis of apoMb was performed with the protein dissolved (0.4–0.6 mg/mL) in 20 mM Tris–HCl, 0.15 M NaCl, using an E:S ratio of 1:100 (by weight). The pH of the proteolysis mixture was 8.0 when trypsin, V8-protease, and elastase were used as proteolytic enzymes, 7.7 in proteolysis with proteinase K, and 7.5 in all other cases. The thermolysin proteolysis mixture contained also 1 mM $CaCl_2$, and with papain 1 mM cysteine.

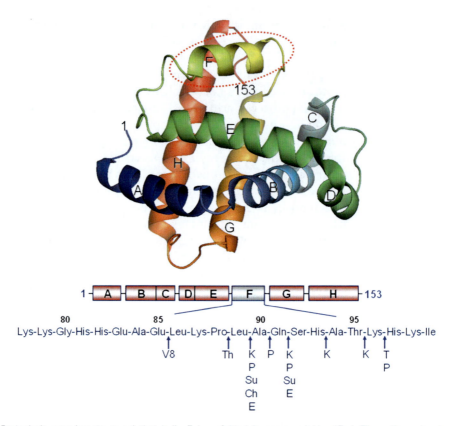

Fig. 3. Proteolysis experiments reveal that helix F is unfolded in apomyoglobin. (*Top*) Three-dimensional model of myoglobin (PDB code 1MBO) produced with the PyMol software (http:/www.pymol.org). The eight (A through H) helices of the 153-residue protein are indicated. The segment of helix F, circled in dashes, is disrupted/unfolded in apomyoglobin (apoMb) (see text). (*Bottom*) The eight helical segments of the protein are indicated by *boxes* and below them the amino acid sequence of the chain segment comprising helix F (residues 82–97) is given. The sites of proteolytic cleavage of apoMb by thermolysin (Th), subtilisin (Su), trypsin (T), papain (P), proteinase K (K), staphylococcal V8-protease (V8), chymotrypsin (Ch), and elastase (E) are indicated by arrows (adapted from ref. 31).

3. At intervals, aliquots were taken from the reaction mixture and the proteolysis was stopped by acidification of the solutions by adding TFA (final concentration 1%, by volume). The proteolysis mixtures were then separated by RP-HPLC utilizing a C18 Vydac column (4.6 × 250 mm; The Separations Group, Hesperia, CA) eluted with a linear gradient of water/acetonitrile, both containing 0.05% (v/v) TFA, from 5% to 40% in 5 min, and from 40% to 60% in 25 min, at a flow rate of 0.8 mL/min. The effluent from the column was monitored by absorption measurements at 226 nm, the fractions containing the protein fragments were pooled and then concentrated in the Speed-Vac system of Savant.

4. The identity of the fragments was established by electrospray-ionization (ESI)–MS, as well as amino-terminal sequencing. The time-course of the reaction was monitored by SDS-PAGE

analysis using the Tricine buffer system (65) in a vertical gel-slab apparatus (Miniprotean-II, Bio-Rad).

5. Initial nicking of the polypeptide chain of apoMb (153 amino acid residues, no disulfide bonds) by several proteases occurred at the level of chain segment 82–97 (see Fig. 3) (31). In contrast, holoMb was fully resistant to proteolytic digestion under identical experimental conditions. Such proteolysis experiments, therefore, indicate that helix F is highly flexible or largely disrupted in apoMb, in agreement with the results obtained analyzing the molecular features of apoMb by using both NMR (79, 80) and computational (81–83) approaches. Clearly, the local mobility/unfolding of the chain of the apoMb substrate dictates the limited proteolysis phenomenon. Indeed, when the mobile chain region encompassing helix F in apoMb is induced to adopt a quite rigid and hydrogen-bonded structure, such as that resulting from the helix-inducing Pro88 → Ala substitution, the site of limited proteolysis is fairly well protected against proteolytic attack (31).

4. Notes

1. Enhanced chain flexibility (segmental mobility) is the key feature of the site(s) of limited proteolysis of globular proteins. It is amply demonstrated the sites of limited proteolysis (nicksites) in globular proteins of known 3D structure are located at chain regions which are flexible or even fully disordered, as given by the fact that these regions do not show recognizable signals in electron density maps (see Fig. 2 and Table 2).

2. The strict requirement for chain flexibility is in keeping with the idea that the protein substrate must undergo considerable structural changes in order to bind effectively at the protease's active site in order to form the optimal transition state needed for hydrolysis. Indeed, the protease binds the polypeptide substrate in an extended conformation (see Fig. 1) (40, 41). Modeling studies of the conformational changes required for a proteolytic cleavage indicate that sites of limited proteolysis require a large conformational change (local unfolding) of a chain segment of up to 12 residues (33–35).

3. The limited proteolysis approach for detecting flexible/ unfolded sites or regions implies that the proteolytic event is determined by the stereochemistry and flexibility of the protein substrate and not by the specificity of the attacking protease. Therefore, the most suitable proteases are those displaying broad substrate specificity, such as proteinase K, subtilisin or thermolysin.

4. The most informative sites of proteolysis are those that can be classified as initial sites, since subsequent cleavages occur on a perturbed protein substrate, not necessarily retaining the overall structure and dynamics of the intact native protein. Therefore, limited proteolysis experiments should be devised in order to monitor the kinetics of proteolysis. It is clear that, if one waits long enough, many peptide bond fissions occur and the resulting proteolysis is not at all limited or selective. Under these conditions, proteolysis experiments are clearly little informative for the structural features of the protein substrate.

5. Quite frequently, in past and current literature, it is suggested that limited proteolysis results from the fact that a specific chain segment of the folded protein substrate is sufficiently exposed to bind to the active site of the protease (32). However, while exposure/accessibility is necessary, it is clearly not sufficient to explain the phenomenon of limited proteolysis, since it is evident that even in a small globular protein there are many exposed sites which could be the targets of proteolysis. Instead, chain flexibility or even local disorder can explain the site-specific proteolysis of a large protein molecule.

6. Limited proteolysis is a very useful and reliable experimental technique that can detect sites of disorder in proteins (see Fig. 2 and Table 2), thus complementing the results that can be obtained by the use of other physicochemical and computational approaches. The advantage of this simple biochemical technique resides in the fact that it is undemanding in terms of protein sample, is relatively inexpensive and generally easy. In the past, the technique was rather difficult to apply, since the analytical methods required to isolate and characterize protein fragments were labor-intensive and not sufficiently sensitive. The recent dramatic advances in MS techniques (66–70, 76) for analyzing peptides and proteins are likely to result in a much more systematic use of limited proteolysis as a simple first step for elucidating the molecular features of a novel and rare protein, especially if the protein is available in minute amounts.

Acknowledgments

This work was supported by the University of Padua (ex-60%) and by the Italian Ministry of University and Research (PRIN-2007). The former lab members Paola Picotti and Marcello Zambonin have contributed a great deal to the ideas herewith expressed.

References

1. Dunker AK, Brown CJ, Lawson JD, Iakoucheva LM, Obradovic Z (2002) Intrinsic disorder and protein function. Biochemistry 41:6573–6582

2. Tompa P (2002) Intrinsically unstructured proteins. Trends Biochem Sci 27:527–533

3. Uversky VN (2002) Natively unfolded proteins: a point where biology waits for physics. Protein Sci 11:739–756

4. Dyson HJ, Wright PE (2005) Intrinsically unstructured proteins and their functions. Nature Rev. Mol. Cell Biol. 6:197–208

5. Dunker AK, Silman I, Uversky VN, Sussman JL (2008) Function and structure of inherently disordered proteins. Curr Opin Struct Biol 18:756–764

6. Le Gall T, Romero PR, Cortese MS, Uversky VN, Dunker AK (2007) Intrinsic disorder in the Protein Data Bank. J Biomol Struct Dyn 24:325–342

7. Dyson HJ, Wright PE (2004) Unfolded proteins and protein folding studied by NMR. Chem Rev 104:3607–3622

8. Rossman MG, Liljas A (1974) Recognition of structural domains in globular proteins. J Mol Biol 85:177–181

9. Ptitsyn OB (1995) Molten globule and protein folding. Adv Protein Chem 47:83–229

10. Kuwajima K (1989) The molten globule state as a clue for understanding the folding and cooperativity of globular-protein structure. Proteins 6:87–103

11. Arai M, Kuwajima K (2000) Role of the molten globule state in protein folding. Adv Protein Chem 53:209–282

12. Uversky VN (2011) Intrinsically disordered proteins from A to Z. Int J Biochem Cell Biol 43:1090–1103

13. Dyson HJ (2011) Expanding the proteome: disordered and alternatively folded proteins. Q Rev Biophys 44:467–518

14. Uversky VN, Longhi S (eds) (2010) Instrumental analysis of intrinsically disordered proteins: assessing structure and conformation. Wiley, New York

15. Ferron F, Longhi S, Canard B, Karlin D (2006) A practical overview of protein disorder prediction methods. Proteins 65:1–14

16. Oldfield CJ, Cheng Y, Cortese MS, Brown CJ, Uversky VN, Dunker AK (2005) Comparing and combining predictors of mostly disordered proteins. Biochemistry 44:1989–2000

17. Fontana A, Fassina G, Vita C, Dalzoppo D, Zamai M, Zambonin M (1986) Correlation between sites of limited proteolysis and segmental mobility in thermolysin. Biochemistry 25:1847–1851

18. Fontana A, Polverino de Laureto P, De Filippis V (1993) Molecular aspects of proteolysis of globular proteins. In: Van den Tweel W, Harder A, Buitelaar M (eds) Protein stability and stabilization. Elsevier Science Publications, Amsterdam, pp 101–110

19. Fontana A, Polverino de Laureto P, De Filippis V, Scaramella E, Zambonin M (1997) Probing the partly folded states of proteins by limited proteolysis. Fold Des 2:R17–R26

20. Fontana A, Zambonin M, Polverino de Laureto P, De Filippis V, Clementi A, Scaramella E (1997) Probing the conformational state of apomyoglobin by limited proteolysis. J Mol Biol 266:223–230

21. Fontana A, Polverino de Laureto P, De Filippis V, Scaramella E, Zambonin M (1999) Limited proteolysis in the study of protein conformation. In: Sterchi E, Stöcker W (eds) Proteolytic enzymes: tools and targets. Springer, Heidelberg, pp 253–280

22. Fontana A, Polverino de Laureto P, Spolaore B, Frare E, Picotti P, Zambonin M (2004) Probing protein structure by limited proteolysis. Acta Biochim Pol 51:299–321

23. Fontana A, Polverino de Laureto P, Spoloare B, Frare E, Zambonin M (2010) Detecting disordered regions in proteins by limited proteolysis. In: Uversky VN, Longhi S (eds) Instrumental analysis of intrinsically disordered proteins: assessing structure and conformation. Wiley, New York, pp 569–626

24. Linderstrøm-Lang K (1950) Structure and enzymatic breakdown of proteins. Cold Spring Harb Symp Quant Biol 14:117–126

25. Neurath H (1980) Limited proteolysis, protein folding and physiological regulation. In: Jaenicke R (ed) Protein folding. Elsevier/North Holland Biomedical Press, Amsterdam, New York, pp 501–504

26. Richards FM, Vithayathil PJ (1959) The preparation of subtilisin-modified ribonuclease and the separation of the peptide and protein components. J Biol Chem 234:1459–1464

27. Vita C, Dalzoppo D, Fontana A (1985) Limited proteolysis of thermolysin by subtilisin: isolation and characterization of a partially active enzyme derivative. Biochemistry 24:1798–1806

28. Fassina G, Vita C, Dalzoppo D, Zamai M, Zambonin M, Fontana A (1986) Autolysis of thermolysin: isolation and characterization of a

folded three-fragment complex. Eur J Biochem 156:221–228

29. Musi V, Spolaore B, Picotti P, Zambonin M, De Filippis V, Fontana A (2004) Nicked apomyoglobin: a noncovalent complex of two polypeptide fragments comprising the entire polypeptide chains. Biochemistry 33:455–456

30. Spolaore B, Polverino de Laureto P, Zambonin M, Fontana A (2004) Limited proteolysis of human growth hormone at low pH: isolation, characterization and complementation of the two biologically relevant fragments 1–44 and 45–191. Biochemistry 43:6576–6586

31. Picotti P, Marabotti A, Negro A, Musi V, Spolaore B, Zambonin M, Fontana A (2004) Modulation of the structural integrity of helix F in apomyoglobin by single amino acid replacements. Protein Sci 13:1572–1585

32. Novotny J, Bruccoleri RE (1987) Correlation among sites of limited proteolysis, enzyme accessibility and segmental mobility. FEBS Lett 211:185–189

33. Hubbard SJ, Campbell SF, Thornton JM (1991) Molecular recognition. Conformational analysis of limited proteolytic sites and serine proteinase protein inhibitors. J Mol Biol 220:507–530

34. Hubbard SJ, Eisenmenger F, Thornton JM (1994) Modeling studies of the change in conformation required for cleavage of limited proteolytic sites. Protein Sci 3:757–768

35. Hubbard SJ (1998) The structural aspects of limited proteolysis of native proteins. Biochim Biophys Acta 1382:191–206

36. Ringe D, Petsko GA (1985) Mapping protein dynamics by X-ray diffraction. Prog Biophys Mol Biol 45:197–235

37. Sternberg MJE, Grace DEP, Phillips DC (1979) Dynamic information from protein crystallography: an analysis of temperature factors from refinement of the hen egg-white lysozyme. J Mol Biol 130:231–253

38. Kundu S, Melton SJ, Sorensen DC, Phillips GN Jr (2002) Dynamics of proteins in crystals: comparison of experiment with simple models. Biophys J 83:723–732

39. Radivojac P, Obradovic Z, Smith DK, Zhu G, Vucetic S, Brown CJ, Lawson JD, Dunker AK (2004) Protein flexibility and intrinsic disorder. Protein Sci 13:71–80

40. Tyndall JDA, Fairlie DP (1999) Conformational homogeneity in molecular recognition by proteolytic enzymes. J Mol Recognit 12:363–370

41. Tyndall JDA, Nall T, Fairlie DP (2005) Proteases universally recognize beta strands in their active site. Chem Rev 105:973–999

42. Schechter I, Berger A (1967) On the size of the active site in proteases. I. Papain. Biochem Biophys Res Commun 27:157–162

43. Burkhard P, Dominici P, Borri-Voltattorni C, Jansonius JN, Malashkevich VN (2001) Structural insight into Parkinson's disease treatment from drug-inhibited DOPA decarboxylase. Nat Struct Biol 8:963–967

44. Bossa F, Borri-Voltattorni C (1991) Pig kidney 3,4-dihydroxyphenylalanine (dopa) decarboxylase: primary structure and relationships to other amino acid decarboxylases. Eur J Biochem 201:385–391

45. Fisher AJ, Thompson TB, Thoden JB, Baldwin TO, Rayment I (1996) The 1.5 Å resolution crystal structure of bacterial luciferase in low salt conditions. J Biol Chem 271:21956–21968

46. Apuy JL, Park ZY, Swartz PD, Dangott LJ, Russell DH, Baldwin TO (2001) Pulsed-alkylation mass spectrometry for the study of protein folding and dynamics: development and application to the study of a folding/unfolding intermediate of bacterial luciferase. Biochemistry 40:15153–15163

47. Bennett MJ, Choe S, Eisenberg D (1994) Refined structure of dimeric diphtheria toxin at 2.0 Å resolution. Protein Sci 3:1444–1463

48. Moskaug JO, Sletten K, Sandvig K, Olsnes S (1989) Translocation of diphtheria toxin A-fragment to the cytosol: role of the site of inter-fragment cleavage. J Biol Chem 264:15709–15713

49. Xia ZX, Mathews FS (1990) Molecular structure of flavocytochrome b2 at 2.4 Å resolution. J Mol Biol 212:837–863

50. Ghrir R, Lederer F (1981) Study of a zone highly sensitive to proteases in flavocytochrome b2 from Saccharomyces cerevisiae. Eur J Biochem 120:279–287

51. Umhau S, Pollegioni L, Molla G, Diederichs K, Welte W, Pilone MS, Ghisla S (2000) The X-ray structure of D-amino acid oxidase at very high resolution identifies the chemical mechanism of flavin-dependent substrate dehydrogenation. Proc Natl Acad Sci USA 97:12463–12468

52. Campaner S, Pollegioni L, Ross BD, Pilone MS (1998) Limited proteolysis and site-directed mutagenesis reveal the origin of microheterogeneity in Rhodotorula gracilis D-amino acid oxidase. Biochem J 330:615–621

53. Meador WE, Means AR, Quiocho FA (1993) Modulation of calmodulin plasticity in molecular recognition on the basis of X-ray structures. Science 262:1718–1721

54. Draibikowski W, Brzeska H, Venyaminov SY (1982) Tryptic fragments of calmodulin. J Biol Chem 257:11584–11590

55. Loll PJ, Lattman EE (1989) The crystal structure of the ternary complex of staphylococcal nuclease, Ca^{2+}, and the inhibitor pdTp, refined at 1.65 Å. Proteins 5:183–201

56. Taniuchi H, Anfinsen CB (1968) Steps in the formation of active derivatives of staphylococcal nuclease during trypsin digestion. J Biol Chem 243:4778–4786

57. Juers DH, Hakda S, Matthews BW, Huber RE (2003) Structural basis for the altered activity of Gly794 variants of *Escherichia coli* β-galactosidase. Biochemistry 42:13505–13511

58. Edwards LA, Tian MR, Huber RE, Fowler AV (1988) The use of limited proteolysis to probe interdomain and active site regions of β-galactosidase (*Escherichia coli*). J Biol Chem 263:1848–1854

59. Matsuda K, Mizuguchi K, Nishioka T, Kato H, Go N, Oda J (1996) Crystal structure of glutathione synthetase at optimal pH: domain architecture and structural similarity with other proteins. Protein Eng 9:1083–1092

60. Tanaka T, Kato H, Nishioka T, Oda J (1992) Mutational and proteolytic studies on a flexible loop in glutathione synthetase from *Escherichia coli* B: the loop and arginine 233 are critical for the catalytic reaction. Biochemistry 31:2259–2265

61. Oakley AJ, Lo Bello M, Ricci G, Federici G, Parker MW (1998) Evidence for an induced-fit mechanism operating in pi class glutathione transferases. Biochemistry 37:9912–9917

62. Martini F, Aceto A, Sacchetta P, Bucciarelli T, Dragani B, Di Ilio C (1993) Investigation of intra-domain and inter-domain interactions of glutathione transferase P1-1 by limited chymotryptic cleavage. Eur J Biochem 218:845–851

63. Lo Bello M, Pastore A, Petruzzelli R, Parker MW, Wilce MC, Federici G, Ricci G (1993) Conformational states of human placental glutathione transferase as probed by limited proteolysis. Biochem Biophys Res Commun 194:804–810

64. Wlodawer A, Minor W, Dauter Z, Jaskolski M (2007) Protein crystallography for non-crystallographers, or how to get the best (but not more) from published macromolecular structures. FEBS J 275:1–21

65. Schägger H, von Jagow G (1987) Tricine sodium dodecyl sulfate polyacrylamide gel electrophoresis for the separation of proteins in the range from 1 kDa to 100 kDa. Anal Biochem 166:368–379

66. Aebersold R, Mann M (2003) Mass spectrometry-based proteomics. Nature 422:198–207

67. Domon B, Aebersold R (2006) Mass spectrometry and protein analysis. Science 312:212–217

68. Cravatt BF, Simon GM, Yates JR 3rd (2007) The biological impact of mass-spectrometry-based proteomics. Nature 450:991–1000

69. Yates JR, Ruse CI, Nakorchevsky A (2009) Proteomics by mass spectrometry: approaches, advances, and applications. Annu Rev Biomed Eng 11:49–79

70. Helsens K, Martens L, Vandekerckhove J, Gevaert K (2011) Mass spectrometry-driven proteomics: an introduction. Methods Mol Biol 753:1–27

71. Gheyi T, Rodgers L, Romero R, Sauder JM, Burley SK (2010) Mass spectrometry guided in situ proteolysis to obtain crystals for X-ray structure determination. J Am Soc Mass Spectrom 21:1795–1801

72. Sajnani G, Pastrana MA, Dynin I, Onisko B, Requena JR (2008) Scrapie prion protein structural constraints obtained by limited proteolysis and mass spectrometry. J Mol Biol 382:88–98

73. Gao X, Bain K, Bonanno JB, Buchanan M, Henderson D, Lorimer D, Marsh C, Reynes JA, Sauder JM, Schwinn K, Thai C, Burley SK (2005) High-throughput limited proteolysis/mass spectrometry for protein domain elucidation. J Struct Funct Genomics 6:129–134

74. Koth CM, Orlicky SM, Larson SM, Edwards AM (2003) Use of limited proteolysis to identify protein domains suitable for structural analysis. Methods Enzymol 368:77–84

75. Wernimont A, Edwards A (2009) In situ proteolysis to generate crystals for structure determination: an update. PLoS One 4:e5094

76. Mann M, Kelleher NL (2008) Precision proteomics: the case for high resolution and high mass accuracy. Proc Natl Acad Sci USA 105:18132–18138

77. Perry RH, Cooks RG, Noll RJ (2008) Orbitrap mass spectrometry: instrumentation, ion motion and applications. Mass Spectrom Rev 27:661–699

78. Makarov A, Scigelova M (2010) Coupling liquid chromatography to Orbitrap mass spectrometry. J Chromatogr A 1217:3938–3945

79. Eliezer D, Wright PE (1996) Is apomyoglobin a molten globule? Structural characterization by NMR. J Mol Biol 263:531–538

80. Eliezer D, Yao J, Dyson HJ, Wright PE (1998) Structural and dynamic characterization of partially folded states of apomyoglobin and implications for protein folding. Nat Struct Biol 5:148–155

81. Brooks CL (1992) Characterization of native apomyoglobin by molecular dynamics simulation. J Mol Biol 227:375–380

82. Hirst JD, Brooks CL (1995) Molecular dynamics simulations of isolated helices of myoglobin. Biochemistry 34:7614–7621

83. Tirado-Rives J, Jorgensen WL (1993) Molecular dynamics simulations of the unfolding of apomyoglobin in water. Biochemistry 32:4175–4184

Chapter 21

The Effect of Counter Ions on the Conformation of Intrinsically Disordered Proteins Studied by Size-Exclusion Chromatography

Magdalena Wojtas, Tomasz M. Kapłon, Piotr Dobryszycki, and Andrzej Ożyhar

Abstract

Counter ions are able to change the conformation of intrinsically disordered proteins (IDPs) to a more compact structure via the reduction of electrostatic repulsion. When the extended IDP conformation is transformed into a more ordered one, the value of the Stokes radius should decrease. Size-exclusion chromatography is a simple method for the determination of the Stokes radius, which describes the hydrodynamic properties of protein molecules. In our paper size-exclusion chromatography experiments of Starmaker (a highly acidic IDP), in the presence of various counter ions, are presented as an example of a simple experimental method, which provides valuable information about subtle counter ions-induced conformational changes in IDP.

Key words: Size-exclusion chromatography, Gel filtration, Molecular size, Stokes radius, Counter ion, Conformational change, Intrinsically disordered protein, Highly acidic protein, Electrostatic repulsion, Starmaker

1. Introduction

Intrinsically disordered proteins (IDPs) are able to adopt a more ordered and rigid structure as a consequence of different factors such as temperature, denaturants, osmolytes, binding partners, molecular crowding, or counter ions (1). IDP amino acid composition differs significantly from globular and ordered protein composition. IDPs are often enriched in polar amino acid residues, which are charged under a physiological pH. These amino acid residues give rise to the high net charge characteristic of IDPs, since they are usually exposed to a solvent. The high level of non-compensated charges combined with low hydrophobicity

Vladimir N. Uversky and A. Keith Dunker (eds.), *Intrinsically Disordered Protein Analysis: Volume 2, Methods and Experimental Tools*, Methods in Molecular Biology, vol. 896, DOI 10.1007/978-1-4614-3704-8_21, © Springer Science+Business Media New York 2012

results in the lack of a well-packed core of IDPs. However, counter ions may compensate for the high net charge of the protein and lead to a more ordered conformation (1, 2). This phenomenon was observed using several methods for IDPs (3–10). For instance, the major phosphoprotein of dentin, the aspartic acid- and phosphoserine-rich protein called phosphophoryn changes its conformation to a more compact structure upon the addition of calcium ions, as was shown by small angle X-ray scattering (SAXS) (3). The other highly acidic protein, ansocalcin, a constituent of goose eggshells, undergoes significant changes in its secondary structure in the presence of calcium ions, as determined by far UV-CD studies (4). Far- and near-UV CD measurements were also applied to make a comparison of the calcium-loaded and apo-parvalbumin, and it was shown that the protein adopts a more ordered structure in the presence of calcium ions (5). Calcium ion-induced conformational changes, leading to an ordered structure in the adenylate cyclase toxin from *Bordetella pertussis*, were studied by the fluorescence spectroscopy of tryptophan and 8-anilinonaphthalene-1-sulfonate (ANS), CD spectroscopy, and analytical ultracentrifugation (6, 7). Furthermore, the effect of other counter ions was also studied. For instance, the zinc ion-mediated structure formation of human prothymosin was analyzed by SAXS and far UV CD measurements (8). The rate of α-synuclein fibril formation in the presence of several di- and trivalent metal ions was studied by ANS fluorescence and CD spectroscopy, and it was established that aluminum, copper(II), iron(III), cobalt(II), and manganese(II) ions cause an acceleration in the rate of fibrillation (9). Another, quite simple method which can be used to monitor the conformational changes in IDPs that accompany interactions with counter ions is size-exclusion chromatography (gel filtration). This method enables one to effectively separate proteins with different hydrodynamic properties. Hence, it can be used for the determination of global conformational changes in analyzed proteins (11, 12). Global conformational changes can be analyzed by SAXS or analytical ultracentrifugation as well, but these methods require expensive and highly specialized equipment. (For more details about gel filtration, see Chapter 11.) Based on gel filtration results one can estimate the Stokes radius (R_s), which is the equivalent of the hydrated sphere and describes the hydrodynamic properties of protein molecules (13). To calculate the R_s of the target protein, a gel filtration column must be calibrated with protein standards with known R_s. The partition coefficient (K_{AV}) for each standard protein must be calculated using the formula: $K_{AV} = (V_e - V_0)/(V_t - V_0)$, where V_0 is the void volume for the column, V_e is the elution volume of the target protein, and V_t is the total volume of the column. Then, a series of gel filtrations of the target protein for several ion concentrations are carried out, and consequently, the partition coefficients are determined. The K_{AV} values of standard proteins are plotted against the logarithm of the R_s values of the protein standards. The R_s of the target protein for each counter ion concentration is determined based on the linear regression of the data points. The changes in the R_s value in an ion-dependent manner correspond to the conformational changes leading to a change in the hydrodynamic characteristic of the analyzed protein (13).

In this work, we present details of experiments where the hydrodynamic properties of the Starmaker protein (Stm) were determined by gel filtration in the presence of calcium, magnesium, and sodium ions. Stm is a highly acidic protein involved in otolith biomineralization in *Danio rerio* (14). Moreover, due to its molecular properties including being very flexible and having an extended conformation, Stm was classified as member of the family of IDPs (10, 15). Stm controls the growth of calcium carbonate crystals, so calcium ions as putative ligands of Stm were examined to see if they are able to modulate protein conformation. We observed changes in the hydrodynamic dimensions of Stm against concentrations of added ions. Increasing the calcium ion concentration caused a lowering of Stm R_s of about 9.5 Å, whereas in the case of BSA (as a control) a decrease in the R_s of only about 1 Å was observed. Magnesium ions, as another kind of divalent ion, were also examined to demonstrate that Stm interacts specifically with calcium ions. These ions also caused a decrease in the Stm; however higher concentrations of magnesium ions were required, and the change was about 7 Å. The conformational changes were not due to a simple increase in the ionic strength, since sodium ions, which were also examined, did not cause such a large effect (10). Our experiments showed that gel filtration is a simple method which can provide valuable information about subtle global conformational changes in IDPs.

2. Materials

2.1. Buffers

All buffers and solutions should be prepared at room temperature using analytical grade reagents and ultrapure water (conductivity of 18.2 MΩ cm at 25°C). It is important to select a buffer composition and pH that are compatible with the stability of the target protein. The pH of the buffer should not be close to the isoelectric point (pI) of the protein, because it may result in aggregation or even precipitation of the protein during gel filtration. If the pI is not determined experimentally, bioinformatic tools (for example: http://expasy.org/tools/protparam.html) should be used to define the theoretical pI value. The ionic strength of the buffer is also very important. At a low ionic strength, below 0.1 M, electrostatic or van der Waals interactions with the matrix may occur, which leads to an incorrect R_s (13). In the case of our Stm research,

Table 1
Preparation of buffers. To obtain the desired calcium ion concentration (the first column), mix an appropriate volume of buffer A and buffer B (or calcium chloride solution)

The final calcium ion concentration in the buffer	Calcium chloride concentration and volume	Buffer B volume	Buffer A volume
0 μM	–	–	100 mL
5 μM	1 mM, 0.5 mL	–	99.5 mL
50 μM	10 mM, 0.5 mL	–	99.5 mL
0.5 mM	–	50 μL	100 mL
5 mM	–	0.5 mL	99.5 mL
20 mM	–	2 mL	98 mL
50 mM	–	5 mL	95 mL
100 mM	–	10 mL	90 mL

calcium and magnesium ions were used to study conformational changes without the use of any phosphate-containing buffers. Calcium and magnesium phosphate are insoluble in water, thus phosphate buffers cannot be used as eluent or as a component of a sample. Here the following buffer solutions were used:

Buffer A: 10 mM Tris, 100 mM NaCl, 10% glycerol (v/v), pH 7.0.

Buffer B: 10 mM Tris, 100 mM NaCl, 1 M $CaCl_2$, 10% glycerol (v/v), pH 7.0.

Buffer C: 10 mM Tris, 100 mM NaCl, 1 M $MgCl_2$, 10% glycerol (v/v), pH 7.0.

Buffer D: 10 mM Tris, 1.1 M NaCl, 10% glycerol (v/v), pH 7.0.

Weigh required amounts of reagents and add 895 mL of water. Mix to dissolve and adjust the pH with HCl. Add 100 mL of glycerol (see Note 1) and make up to 1 L with water. All buffers should be filtrated through 0.22 μm filters (see Note 2) and degassed (see Note 3). Store at room temperature (see Note 4). Prepare buffers containing the desired concentration of ions by mixing an appropriate volume of buffer B with a buffer or solution containing calcium, magnesium, or sodium ions. Table 1 presents how buffers containing calcium ions were prepared (buffers containing other ions can be prepared in a similar way). When the calcium concentration is higher than 0.5 mM, buffer B, C, or D is used to ensure that the required ion concentration does not impair the pH. Otherwise, calcium chloride was used.

Table 2
Protein standards used for column calibration in the Stm R_s studies

Protein	Molecular mass (kDa)	Stokes radius (Å)	References
Blue dextran	2,000	–	(13)
Thyroglobulin	669	85	(13)
Apoferritin	443	67	(17)
Catalase	240	52	(13)
BSA	66.2	35.5	(13)
Ovalbumin	45	30.5	(13)
Chymotrypsinogen	25	20.9	(13)
Mioglobin	17.8	20.2	(12)
Cytochrome c	12.3	17	(12)

2.2. Proteins

1. Protein standards used for column calibration are listed in Table 2. The R_s of the target protein should be in the range of R_s values of protein standards (see Note 5). Other commonly used protein standards are described in refs. 13 and 16.

2. Recombinant, non-tagged Stm protein without a signal peptide was used. Protein preparation was performed as previously described (15). Briefly, Stm was overexpressed in BL21(DE3) pLysS *E. coli* cells (Novagen, Germany), purified to homogeneity using a three-step system: fractionation with solid $(NH_4)_2SO_4$, gel filtration, and hydroxyapatite chromatography. Stm was stored in buffer A at $-80°C$.

2.3. Equipment

2.3.1. Column

A matrix should be chosen according to the molecular mass and shape of the analyzed protein. The hydrodynamic dimensions of the protein should fall within the fractionation range of the column. It should also be taken into consideration that the molecular size of IDPs might be substantially higher than is expected for a globular protein with a similar molecular mass. For example, the expected value of the R_s for Stm was about 35 Å; however, the experimentally obtained value of R_s was about 78.6 Å (data obtained from gel filtration experiments) (15). Furthermore, it should be noted that conformational changes could cause relatively small changes in the R_s, so it is essential to choose a matrix with a sufficiently high resolution that would make it possible to observe slight changes in the V_e. Commonly used prepackaged columns and matrices for the gel filtration technique are listed in refs. (13) and (16). This information is also available on manufacturers' Web sites.

2.3.2. Instrument Type   Gel filtration requires an isocratic HPLC or FPLC system. It is critical to have a pumping system that can deliver a constant flow rate in the desired range. It is also important to ensure precise starting points for sample injection (13) (see Note 6). Protein elution is most often monitored by recording the absorbance in the ultraviolet range. Detection may be carried out at 280 nm to trace proteins with aromatic amino acid residues or at 220 nm to detect peptide bonds. In the case of Stm, absorbance at 220 nm was recorded, because this protein has no aromatic amino acid residues.

The Superdex 200 10/300 GL column (Amersham Biosciences) connected to the ÄKTAexplorer™ system (Amersham Biosciences) equilibrated with an appropriate buffer was used in this Stm study.

3. Methods

3.1. Column Equilibration

1. Connect an appropriate gel filtration column to an isocratic HPLC or FPLC system (in our study the ÄKTAexplorer™ system was used (see Note 7). Ensure that the column is in a vertical position.

2. Before use wash the column with one column volume of water followed by two column volumes of buffer A at a flow rate of 0.5 mL/min (see Notes 6 and 8).

3. Before separating each sample of the target protein, equilibrate the column with at least one column volume of a buffer containing an appropriate amount of ions. Start from the lowest ion concentration (see Note 9). Calcium, magnesium, and sodium ions were examined for the Stm protein.

3.2. Sample Preparation

1. Use the purified target protein in an appropriate buffer to prepare samples containing various ion concentrations. Calcium ions are putative ligands of the Stm protein, and this is the reason why they were used in the experiment. Magnesium ions as divalent ions were used to check if calcium ions interact specifically with Stm. Moreover, sodium ions were used to confirm that R_s changes were not caused by a simple increase in ionic strength.

2. Prepare the sample by mixing an appropriate volume of the protein solution with a buffer or solution containing ions. The final protein concentration of all samples is 0.4 mg/mL (6.2 μM) (see Note 10), and the final volume is 100 μL (see Note 11). Table 3 presents how the Stm sample containing calcium ions was prepared. When the calcium concentration is higher than 5 mM, buffer B, C, or D is used to achieve the required ion concentration without impairing the pH. Samples

Table 3
Preparation of gel filtration samples. To obtain the desired calcium ion concentration (the first column), mix an appropriate volume of the protein solution, buffer A and buffer B (or calcium chloride solution)

Final ions concentration	Volume of Stm protein	Calcium chloride concentration and volume	Buffer B volume	Buffer A volume
0	25 µL	–	–	75 µL
5 µM	25 µL	0.1 mM, 5 µL	–	70 µL
50 µM	25 µL	1 mM, 5 µL	–	70 µL
0.5 mM	25 µL	10 mM, 5 µL	–	70 µL
5 mM	25 µL	100 mM, 5 µL	–	70 µL
20 mM	25 µL	–	2 µL	73 µL
50 mM	25 µL	–	5 µL	70 µL
100 mM	25 µL	–	10 µL	65 µL

containing magnesium or sodium ions (see Note 12) were prepared in a similar way.

3. Incubate all samples about 1 h on ice (see Note 13).

4. Centrifuge the sample at $14,000 \times g$ for 10 min at 4°C (see Note 14).

5. Before injection incubate, the sample about 10 min to reach the temperature at which gel filtration will be performed (see Note 4).

3.3. Column Calibration

1. Select appropriate protein standards according to the molecular size of the target protein (Subheading 2.2).

2. Dissolve individual protein standards to a concentration of about 0.4 mg/mL in an appropriate buffer (buffer A in the Stm study).

3. Centrifuge the sample at $14,000 \times g$ for 10 min at 4°C (see Note 14).

4. Inject 100 µL of each sample (see Note 11).

5. Before injection incubate the sample about 10 min to reach the temperature at which gel filtration will be performed (see Note 4).

6. Perform gel filtration of each protein standard (see Note 15) at a flow rate 0.5 mL/min (see Note 6) and determine the elution volume (V_e).

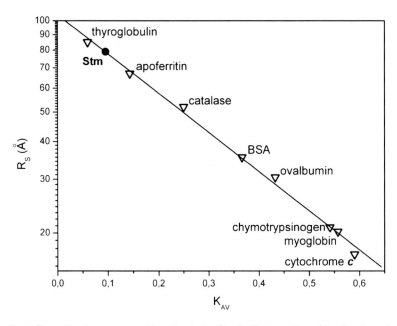

Fig. 1. The calibration curve used to estimate the Stm R_s. The logarithm of the R_s values of each standard protein were plotted against the respective K_{AV} of the proteins determined during gel filtration experiment. The equation describing the curve determined using standard protein data served to calculate the R_s of Stm. Symbols for standard proteins and Stm are open triangles and black circle, respectively.

7. Calculate the partition coefficient (K_{AV}) as: $K_{AV} = (V_e - V_0)/(V_t - V_0)$, where V_0 is the void volume of the column (V_e of blue dextran) and V_t is the total volume of the column.

8. Plot the logarithm of the R_s versus K_{AV} values of each protein standard (Fig. 1). The equation describing this curve will serve to calculate the R_s of the target protein.

3.4. Gel Filtration of the Target Protein

1. Equilibrate column (Subheading 3.1) with an appropriate buffer (buffer A in the Stm study).

2. Prepare the sample of the target protein and BSA (control) as described above (Subheading 3.2) in the same buffer.

3. Rinse the injection loop with the buffer to remove air.

4. Inject the target protein sample and perform gel filtration at a constant flow rate of 0.5 mL/min (see Notes 6 and 8). Continue gel filtration until at least one total volume of the column is reached. Determine the V_e from the chromatogram.

5. Inject the BSA sample and perform gel filtration at a constant flow rate of 0.5 mL/min. Continue gel filtration until at least one total volume of the column is reached. Determine the V_e from the chromatogram.

6. Equilibrate the column with an appropriate buffer containing the lowest ion concentration (Subheading 3.1), prepare the

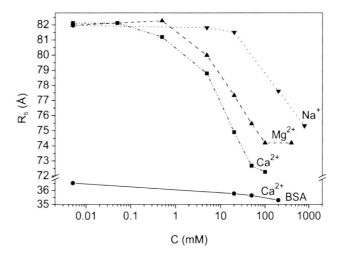

Fig. 2. The changes in Stm hydrodynamic properties induced by calcium, magnesium, and sodium ions. The R_s values of Stm at different concentrations of calcium, magnesium, and ions were estimated by a series of gel filtrations. As a reference, a globular protein with a similar mass (BSA) was used (shown on the plot as *filled circles*). Symbols for Stm are: *square*, *triangle*, and *upside-down triangle* for calcium, magnesium, and sodium ions, respectively (see Note 12), (10). Reproduced from Kapłon TM et al (2008) with permission from Elsevier.

target protein sample with the same ion concentration, and perform gel filtration as described above (steps 1–5).

7. Repeat the procedure (steps 1–6) until obtaining the V_e for Stm and BSA at all desired ion concentrations.

8. Calculate the partition coefficient (K_{AV}) as: $K_{AV} = (V_e - V_0)/(V_t - V_0)$.

9. Calculate the R_s values for each run using the standard proteins plot (Subheading 3.3)

10. Plot the R_s of the protein determined at each ion concentration versus the logarithm of the ion concentrations (Fig. 2). This plot reveals if there are any global conformational changes in the protein that were induced by the counter ions. The changes in the Stm and BSA R_s values induced by calcium, magnesium, and sodium ions are presented in Fig. 2.

4. Notes

1. Glycerol has high viscosity which hampers measuring the required volume. Avoid making bubbles when transferring the glycerol to the 100 mL cylinder. To empty the cylinder, transfer glycerol to the 1 L cylinder and use a buffer to rinse the 100 mL cylinder.

2. It is essential to filter all solutions before doing a gel filtration experiment to avoid introducing bacteria, which can degrade proteins or particulate matter which can block the column.

3. Air bubbles may affect the performance of the gel filtration. Solutions can be degassed before filtration under a vacuum. Volatile solvents, for instance ethanol, should not be degassed because of evaporation.

4. All solutions and media should reach the temperature at which the separation will be performed. A rapid change in temperature may cause air bubbles.

5. Protein standards are commercially available, for example from Sigma, Amersham Biosciences or Bio-Rad, and may be ordered individually or as a kit.

6. Successful calibration and R_s determination requires an accurate flow rate. The flow rate during calibration and the determination of the molecular size must be carefully controlled and should be the same for each sample.

7. Before connecting the column to the chromatography system, ensure there is no air in the tubing and valves. Make sure that the column inlet is filled with liquid and connect it drop-to-drop to the system to avoid introducing air bubbles.

8. The flow rate depends on the matrix and column dimensions. For example, for the Superdex 200 10/300 GL column a 0.5 mL/min. flow rate is recommended.

9. If the impact of the ions on the R_s radius is being examined, the same ion concentration must be maintained during chromatography of the sample.

10. For analytical purposes, a widely used total protein concentration is about 0.5–1 mg/mL. However, the lowest limit of the concentration depends on the sensitivity of the detector. The concentration should not be too high because of multiple factors including sample viscosity, the possibility of protein aggregation, and a reduction in the resolution between the peaks.

11. The sample volume should not exceed 0.5–1% of the V_t for high-resolution analysis. A sample volume of less than 0.5 % does not improve resolution (13). The same applied sample volume should be used for the target protein sample as for the protein standards (16).

12. In the case of sodium ions, the concentrations correspond to extra-added sodium ions and do not include those from buffer A.

13. It is possible that incubation is not necessary and incubation time may depend on the target protein. The actual incubation time may be determined by performing gel filtration after several different incubation times. In our experiment ca. 1 h incubation was sufficient.

14. The sample must be clarified before gel filtration. Centrifugation is necessary to remove particulate matter or precipitated proteins, which can contaminate and block the column.

15. Protein standards can be mixed, if there are no interactions between these proteins. Otherwise, the V_e values of the proteins may be affected.

Acknowledgment

This work was financed by the National Science Centre grant N N204 120040 (to P.D.) and by a statutory activity subsidy from the Polish Ministry of Science and Higher Education for the Faculty of Chemistry of Wrocław University of Technology.

References

1. Uversky VN (2009) Intrinsically disordered proteins and their environment: effects of strong denaturants, temperature, pH, counter ions, membranes, binding partners, osmolytes, and macromolecular crowding. Protein J 28:305–325

2. Uversky VN, Dunker AK (2010) Understanding protein non-folding. Biochim Biophys Acta 1804:1231–1264

3. He G, Ramachandran A, Dahl T, George S, Schultz D, Cookson D, Veis A, George A (2005) Phosphorylation of phosphophoryn is crucial for its function as a mediator of biomineralization. J Biol Chem 280:33109–33114

4. Lakshminarayanan R, Valiyaveettil S, Rao VS, Kini RM (2003) Purification, characterization, and in vitro mineralization studies of a novel goose eggshell matrix protein, ansocalcin. J Biol Chem 278:2928–2936

5. Permyakov SE, Bakunts AG, Denesyuk AI, Knyazeva EL, Uversky VN, Permyakov EA (2008) Apo-parvalbumin as an intrinsically disordered protein. Proteins 72:822–836

6. Chenal A, Karst JC, Pérez AC, Wozniak AK, Baron B, England P, Ladant D (2010) Calcium-induced folding and stabilization of the intrinsically disordered RTX domain of the CyaA toxin. Biophys J 99:3744–3753

7. Sotomayor Pérez AC, Karst JC, Davi M, Guijarro JI, Ladant D, Chenal A (2010) Characterization of the regions involved in the calcium-induced folding of the intrinsically disordered RTX motifs from the bordetella pertussis adenylate cyclase toxin. J Mol Biol 397:534–549

8. Uversky VN, Gillespie JR, Millett IS, Khodyakova AV, Vasilenko RN, Vasiliev AM, Rodionov IL, Kozlovskaya GD, Dolgikh DA, Fink AL, Doniach S, Permyakov EA, Abramov VM (2000) Zn(2+)-mediated structure formation and compaction of the "natively unfolded" human prothymosin alpha. Biochem Biophys Res Commun 267:663–668

9. Uversky VN, Li J, Fink AL (2001) Metal-triggered structural transformations, aggregation, and fibrillation of human alpha-synuclein. A possible molecular link between Parkinson's disease and heavy metal exposure. J Biol Chem 276:44284–44296

10. Kapłon TM, Michnik A, Drzazga Z, Richter K, Kochman M, Ozyhar A (2009) The rod-shaped conformation of Starmaker. Biochim Biophys Acta 1794:1616–1624

11. Martenson RE (1978) The use of gel filtration to follow conformational changes in proteins. Conformational flexibility of bovine myelin basic protein. J Biol Chem 253:8887–8893

12. Uversky VN (1993) Use of fast protein size-exclusion liquid chromatography to study the unfolding of proteins which denature through the molten globule. Biochemistry 32:13288–13298

13. Begg GE, Harper SL, Speicher DW (1999) Characterizing recombinant proteins using HPLC gel filtration and mass spectroscopy. In: Wingfield PT, Pain RH (eds) Current protocols in protein sciences. Wiley, USA, pp 1078–1092

14. Söllner C, Burghammer M, Busch-Nentwich E, Berger J, Schwarz H, Riekel C, Nicolson T (2003) Control of crystal size and lattice formation by

starmaker in otolith biomineralization. Science 302:282–286

15. Kapłon TM, Rymarczyk G, Nocula-Ługowska M, Jakób M, Kochman M, Lisowski M, Szewczuk Z, Ozyhar A (2008) Starmaker exhibits properties of an intrinsically disordered protein. Biomacromolecules 9:2118–2125

16. Hagel L (1998) Gel-filtration chromatography. In: Coligan JE, Dunn BM, Speicher DW, Wingfield PT (eds) Current protocols in protein science. Wiley, New York, NY, pp 8.3.1–8.3.30

17. de Haën C (1987) Molecular weight standards for calibration of gel filtration and sodium dodecyl sulfate-polyacrylamide gel electrophoresis: ferritin and apoferritin. Anal Biochem 166:235–245

Chapter 22

Mean Net Charge of Intrinsically Disordered Proteins: Experimental Determination of Protein Valence by Electrophoretic Mobility Measurements

Ana Cristina Sotomayor-Pérez, Johanna C. Karst, Daniel Ladant, and Alexandre Chenal

Abstract

Under physiological conditions, intrinsically disordered proteins (IDPs) are unfolded, mainly because of their low hydrophobicity and the strong electrostatic repulsion between charged residues of the same sign within the protein. Softwares have been designed to facilitate the computation of the mean net charge of proteins (formally protein valence) from their amino acid sequences. Nevertheless, discrepancies between experimental and computed valence values for several proteins have been reported in the literature. Hence, experimental approaches are required to obtain accurate estimation of protein valence in solution. Moreover, ligand-induced disorder-to-order transition is involved in the folding of numerous IDPs. Some of the ligands are cations or anions, which, upon protein binding, decrease the mean net charge of the protein, favoring its folding via a charge reduction effect. An accurate determination of the mean net charge of protein in both its ligand-free intrinsically disordered state and in its folded, ligand-bound state allows one to estimate the number of ligands bound to the protein in the holo-state. Here, we describe an experimental protocol to determine the mean net charge of protein, from its electrophoretic mobility, its molecular mass and its hydrodynamic radius.

Key words: Mean net charge, Protein valence, Intrinsically disordered protein, IDP, Electrophoretic mobility, Molecular mass, Hydrodynamic radius, Static light scattering, Quasi-elastic light scattering, Analytical ultracentrifugation

Abbreviations

μ_e	Electrophoretic mobility, $cm^2.V^{-1}.s^{-1}$ or $\mu m.cm.V^{-1}.s^{-1}$ (SI: $m^2.V^{-1}.s^{-1}$)
ζ	Zeta potential, V
z	Valence or "mean net charge"
e	Electronic charge, 1.602×10^{-19} coulombs, A.s
U	Applied voltage, V, $kg.m^2.s^{-3}.A^{-1}$
I	Current intensity, A
ε_0	Vacuum permittivity, $8.854 \times 10^{-12} C.V^{-1}.m^{-1}$
ε_r	Relative static permittivity (dielectric constant) of the solvent, 78.54.

Vladimir N. Uversky and A. Keith Dunker (eds.), *Intrinsically Disordered Protein Analysis:*
Volume 2, Methods and Experimental Tools, Methods in Molecular Biology, vol. 896,
DOI 10.1007/978-1-4614-3704-8_22, © Springer Science+Business Media New York 2012

ε Buffer permittivity, $\varepsilon = \varepsilon_r \varepsilon_0$, $C.V^{-1}.m^{-1}$

f/f_0 Translational frictional ratio of the protein, including shape and hydration parameters

f Frictional coefficient of the protein, $g.s^{-1}$

f_0 Frictional coefficient of an anhydrous sphere of the mass of the protein, $g.s^{-1}$

R_p Hydrodynamic radius of the protein, cm

R_b Hydrodynamic radius of the buffer, cm

R_0 Radius of an anhydrous sphere of the mass of the protein, cm

V_H Hydrodynamic volume, cm^3

D_t Translational diffusion coefficient, $cm^2.s^{-1}$

s Sedimentation coefficient obtained at the temperature of the experiment, Svedberg, 10^{-13}s

$\bar{v}$ Partial specific volume, $cm^3.g^{-1}$

η_s Viscosity of the solvent, Poise: $g.cm^{-1}.s^{-1}$

ρ Density of the solvent, $g.cm^{-3}$

M Molecular mass, $g.mol^{-1}$

T Absolute temperature, K

C Protein concentration, $mol.L^{-1}$ (M)

δ Time-averaged apparent hydration, $g_{H_2O} \times g_{protein}^{-1}$

MCR   Mean count rate

kcps   Kilo-count per second

ZQF   Zeta quality factor

FFR   Fast field reversal

SFR   Slow field reversal

DTS1060C   Malvern electrophoretic mobility cell

DTS1230   Malvern standard for electrophoretic mobility and conductivity

QELS   Quasi-elastic light scattering

k_B Boltzmann's constant, $erg.K^{-1}$; (K_B: 1.38065×10^{-16} $erg.K^{-1}$ with erg: $g.cm^2.s^{-2} = 10^{-7}$ J; 1.38065×10^{-23} $J.K^{-1}$)

N_A Avogadro's number, $molecules.mol^{-1}$

PALS   Phase shift analysis light scattering

κ Debye length, Inverse screening length, m

ZEN1010   Malvern electrophoretic mobility microcell

---

1. Introduction

Several experimental methods have been developed to measure the valence of colloids. The study of electrokinetic phenomena is the most popular approach to obtain information on the valence of macromolecules in solution. Among the various electrokinetic effects (1), electrophoresis (i.e., the motion of macromolecules under the influence of an applied electric field) allows to measure the electrophoretic mobility of a macromolecule of interest. Until recently, the investigation of the electrophoretic mobility of macromolecules was not straightforward because of the simultaneous electro-osmosis phenomenon (i.e., the motion of the buffering solution under the influence of the electric field applied to induce the macromolecule electrophoresis) affecting the electrophoretic mobility measurement of the protein. Other experimental considerations, like the high conductivity or the high viscosity of the samples, precluded accurate measurements of the electrophoretic mobility. Instrument providers have developed new generations of apparatus that allow data acquisition on a wider range of experimental conditions and different types of samples (see Note 1).

Electrophoretic mobility of a protein sample is measured by applying an electric field between two electrodes. Charged macromolecules dispersed in the solvent will migrate toward the electrode of opposite sign with a velocity that is proportional to the applied voltage, but also dependent upon the charge, mass, hydration, and shape of the macromolecule (Fig. 1).

Here we describe two procedures to determine the mean net charge of intrinsically disordered proteins (IDPs) (formally, the protein valence; see Note 2) from experimental parameters, i.e., hydrodynamic radius (R_p) and electrophoretic mobility (μ_e). The first approach, reviewed in detail by Abramson et al. (2), provides reliable results for spheres. The second approach is based on an empirical relation described by Basak and Ladisch (3), where the electrophoretic mobility is proportional to the number of charges and inversely correlated to the protein molecular mass and frictional ratio (f/f_0). This empirical approach is interesting because it takes into account the frictions between the protein and its milieu, i.e., frictions due to protein shape and time-averaged apparent hydration. Indeed, it is well known that IDPs exhibit high frictional ratio values, mostly because of their high hydration (rather than shape effects). The methodology described here has been shown to be useful for the determination of the mean net charge of IDPs, as well as for the estimation of the number of cations involved in the calcium-induced folding of intrinsically disordered RTX proteins (4).

2. Materials

1. A NanoZS instrument (Malvern company) which monitors the forward light scattering at 13° for the electrophoretic mobility and the backward light scattering at 173° for quasi-elastic light scattering (QELS data can also be acquired with forward light scattering at 13°). NanoZS instrument combines a laser Doppler velocimetry and phase analysis light scattering (see Note 1).

2. Dedicated cells such as the DTS1060C or the microcell ZEN1010 (see Note 3 and Fig. 2). These cells are designed to measure both the hydrodynamic radius (R_p) and the electrophoretic mobility (μ_e) of macromolecules in solution.

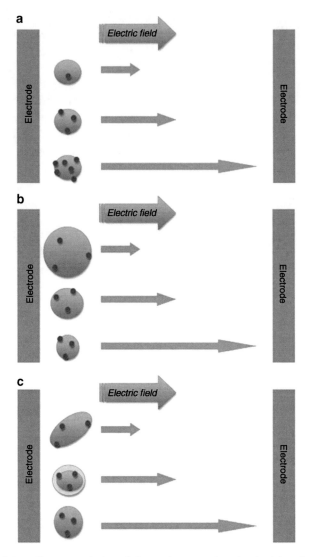

Fig. 1. Schematic representation of the main parameters involved in migration of molecules in an applied electric field. Electrophoretic mobility is a function of the number of charges (**a**), molecular mass (**b**), and frictional ratio (**c**) arising either from shape effect (see the *elliptic shape*) or hydration effect (schematized by a large concentric layer of water around the protein). The *small spots* represent the charges of the molecules and the length of the arrow is directly related to the electrophoretic mobility of the molecule in the applied electric field.

3. A buffer compatible with both the sample and the instrument (see Note 4).

4. A calibration standard, such as DTS1230, with known hydrodynamic radius (150 ± 10 nm), electrophoretic mobility (-5.4 ± 0.5 μm.cm.V^{-1}.s^{-1}), solution viscosity (0.8872 cP), and conductivity (0.3 ± 0.1 mS/cm).

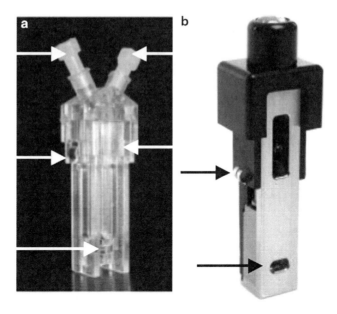

Fig. 2. Dedicated cells used for the electrophoretic mobility experiments. Both of them are specially designed to avoid the electro-osmotic effect. (**a**) DTS1060C cell is made from polycarbonate with two gold-plated electrodes. Cell must be filled with 800 μL and well closed at both extremities. (**b**) ZEN1010 is a quartz cell with an internal sample volume of 100 μL. Arrows indicate electrodes (*middle arrows*), cap (*upper arrows*), and the measurement area in the capillary (*lower arrows*).

3. Methods

3.1. Data Acquisition

1. Switch on the Nano ZS instrument at least 30 min prior to data acquisition.

2. Prior to use, wash the cell with 100% ethanol and dry it.

3. Centrifuge and filter (0.22 μm) all the solutions (buffer and samples) to eliminate aggregates and bubbles that could scatter light and interfere with the sample analysis.

4. Fill the cell with buffer (see Note 5).

5. Check first the homogeneity of the buffer. For this purpose, check that the buffer does not scatter light using the dynamic light scattering measurement type, as described below (steps 6–8).

6. Select the sample material, and define the buffer composition (the software directly compute its viscosity and its refractive index), the temperature (see Note 6), and the type of cell used.

7. Select manual measurement duration. One measurement of 10 runs of 10 s each is usually enough to check the buffer quality.

8. Run the experiment and control the quality of the buffer (see Note 7).

9. Empty the cell and fill it with the sample containing the protein of interest at an appropriate concentration (see Note 5). For correspondence between protein size and concentration, see Note 8.

10. Check the homogeneity of the sample by QELS as follows (steps 11–16 and see Note 9).

11. Verify the parameters defined in step 6.

12. For the measurement duration, select "auto" (roughly 1 measure of 18 runs of 10 s each) or select "manual" (20 measurements of 1 run of 10 s are enough to produce good data).

13. Select an analysis method (see Note 10).

14. Run the QELS experiment and check the quality of the acquisition (see Note 11).

15. Average several autocorrelation functions and analyze the results (see Note 12).

16. Once the dispersity and the hydrodynamic radius of the sample are checked, select the Zeta potential measurement type to record the electrophoretic mobility.

17. Verify that the parameters such as temperature, buffer composition, and cell are the same ones used for the QELS experiment.

18. First, the conductivity of the sample should be measured to define the maximum voltage to be applied. A typical setup to read the conductivity (preliminary test) is 1 measure of 1 run in fast field reversal (FFR) mode at 5 or 10 V (see Note 13).

19. From the sample conductivity provided by the preliminary test, adjust the voltage to be applied (see Note 14). A typical setup to measure the electrophoretic mobility of the sample is 5–10 measures of 1 run each in FFR mode at the defined voltage (see Note 14).

20. Start the acquisition of data with the lowest voltage and increase the voltage (below the voltage limit provided by the preliminary test) until satisfactory data are recorded: for this, check the Zeta Quality Factor (ZQF), the Mean Count Rate (MCR), the phase plot, and the Fourier transform of the phase plot (see Note 15).

21. At the end of the electrophoretic mobility experiment, check the status of the sample (see Note 16).

3.2. Data Processing

We describe two procedures to calculate the protein valence from the experimental parameters determined with the NanoZS instrument. The two essential parameters are the electrophoretic mobility and the hydrodynamic radius of the protein, allowing the use of the first approach. If the molecular mass of the protein is known (together with the electrophoretic mobility and the hydrodynamic radius), a second approach based on empirical observations can be employed to estimate the protein mean net charge.

3.2.1. Determination
of the Hydrodynamic Radius

The hydrodynamic radius of the protein can be determined either by QELS or by analytical ultracentrifugation (see Note 17).

QELS provides the translational diffusion coefficient (D_t) from the autocorrelation function. For a monodisperse population, the first-order autocorrelation function is processed as a single exponential decay:

$$g^1(q;\tau) = Ae^{(-\Gamma\tau)} \tag{1}$$

$g^1(q;\tau)$ is the autocorrelation function at a defined wave vector, q, and decay time, τ; A is the amplitude of the autocorrelation function; Γ is the decay rate, with $\Gamma = q^2 D_t$, where D_t is the translational diffusion coefficient. The wave vector, q, is

$$q = \frac{4\pi n_0 \sin(\theta/2)}{\lambda_{QELS}} \tag{2}$$

where n_0 is the refractive index of the sample, θ is the detector angle with respect to the incident laser beam, and λ_{QELS} is the wavelength of the incident laser beam.

Then, the hydrodynamic radius, R_p, can be deduced from the translational diffusion coefficient through the Einstein–Stokes relation:

$$R_p = \frac{k_B T}{6\pi\eta_s D_t} \tag{3}$$

where k_B is the Boltzmann's constant, T the temperature, and η_s the viscosity of the solvent.

Alternatively, analytical ultracentrifugation provides the sedimentation coefficient of the protein, s. The Svedberg equation provides the hydrodynamic radius from the sedimentation coefficient:

$$s = \frac{M(1-\bar{v}\rho)}{f \times N_A} = \frac{M(1-\bar{v}\rho)}{6\pi\eta_s R_p \times N_A} \tag{4}$$

where $\bar{v}$ is the partial specific volume, ρ is the buffer density, N_A is the Avogadro's number, and f is the frictional coefficient of the protein, $f = 6\pi\eta_s R_p$. Hence, the hydrodynamic radius is given by

$$R_p = \frac{M(1-\bar{v}\rho)}{6\pi\eta_s N_A s} \tag{5}$$

Then, the frictional ratio, f/f_0, is calculated:

$$f/f_0 = \frac{6\pi\eta_s R_p}{6\pi\eta_s R_0} = \frac{R_p}{R_0} \tag{6}$$

where R_0, the radius of a dehydrated spherical protein ($\delta = 0$ g/g) of the same molecular mass as the protein of interest, is calculated as follows:

$M(\bar{v} + \delta) = V_H N_A$ given $M\bar{v} = V_0 N_A$ and substituting V_0, the volume of the dehydrated protein, by its expression, $V_0 = 4\pi R_0^3/3$, then, R_0 becomes

$$R_0 = \left(\frac{3M\bar{v}}{4\pi N_A} \right)^{\frac{1}{3}} \qquad (7)$$

Substituting Eq. 7 into Eq. 6 gives the frictional ratio:

$$f/f_0 = R_p \left(\frac{3M\bar{v}}{4\pi N_A} \right)^{\frac{-1}{3}} \qquad (8)$$

3.2.2. Equations for the Calculation of the Valence of a Spherical Colloid

1. The equation relating the electric charge to the electrophoretic mobility for a spherical molecule is described in detail in the book of Abramson (2). Briefly, D. C. Henry described in 1931 the equation of cataphoresis (5), which integrated the diverging studies of Debye–Hückel (6) and Helmholtz–Smoluchowski (7) by the addition of the so-called Henry function (see below). A historical review on electrokinetic phenomena is presented elsewhere (8).

    The work of Henry led to the following equation relating the electrophoretic mobility to the zeta potential of a spherical colloid

$$\mu_e = \frac{\varepsilon \zeta}{6\pi \eta_s} F(\kappa R_p) \qquad (9)$$

   where μ_e is the electrophoretic mobility, ε the buffer permittivity, ζ the zeta potential, R_p the hydrodynamic radius of the protein, κ the inverse screening length, (described in Note 18), and $F(\kappa R_p)$ the Henry function (described in Note 19). These parameters and the relationship between them are described in Fig. 3. Note that Henry formulated his equation in a different way (5). Altogether, these works led to the so-called Hückel–Onsager equation:

$$\zeta = \frac{3\eta_s \mu_e}{2\varepsilon F(\kappa R_p)} \qquad (10)$$

2. In 1939, Gorin, considering the effect of the conductance of an ionic atmosphere surrounding the colloids (9), reported the relation between the zeta potential and the electric charge of the protein, Q:

$$\zeta = \frac{Q(1 + \kappa R_b)}{\varepsilon R_p (1 + \kappa(R_p + R_b))} \qquad (11)$$

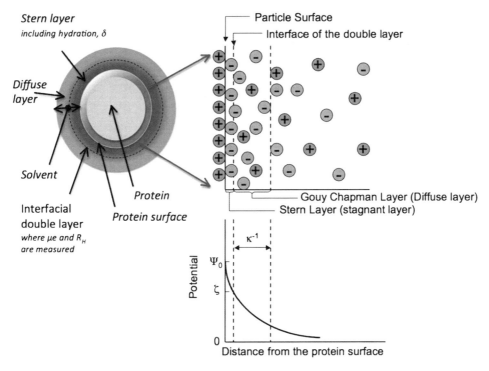

Fig. 3. Schematic representation of a protein and its potentials. On the charged surface of the protein, a first compact and stagnant layer composed of non-diffusing counter-ions constitutes the so-called Stern layer (which roughly corresponds to the inner and outer Helmholtz layers) and contains the water molecules that define the time-averaged apparent hydration (δ) of the protein. A second layer, including more diffuse ions (in exchange with the bulk buffer), constitutes the so-called Gouy–Chapman layer and corresponds to the screening length (κ^{-1}), also called the Debye length. The Stern layer is often represented by a physical thickness of 0.25–0.3 nm. The inverse screening length is dependent on the ionic strength: the Debye length decreases with the increase of the ionic strength (see Note 18). The interface between the Stern and the Gouy–Chapman layers is the slip plan, or shearing plan, which defines the boundary of the hydrodynamic volume of the molecule. The hydrodynamic radius (translational diffusion coefficient D_t), the electrophoretic mobility (and consequently the Zeta potential ζ), and the time-averaged apparent hydration (δ) are measured at this interface between the Stern and the Gouy–Chapman layers.

The substitution of the zeta potential expression (Eq. 9) into Eq. 11 provides the relation:

$$Q = \frac{6\pi\eta_s\mu_e R_p\left(1 + \kappa\left(R_p + R_b\right)\right)}{F\left(\kappa R_p\right)\left(1 + \kappa R_b\right)} \tag{12}$$

where R_b is the radius of the buffering molecule. In general, an averaged radius for the buffer is 0.25 nm. Caution should be used when considering the radius of molecules composing the buffer (see Note 20).

3. Equation 12 expresses the electrophoretic mobility and the hydrodynamic radius of the protein in terms of electric charge, which is not useful for proteins. Taking $Q = z \times e$, with Q the electric charge, z the mean net charge (valence), and e the electronic charge, Eq. 13 becomes a practical expression

for the determination of the mean net charge of a spherical protein:

$$z = \frac{6\pi\eta_s\mu_e R_p\left(1 + \kappa\left(R_p + R_b\right)\right)}{eF\left(\kappa R_p\right)\left(1 + \kappa R_b\right)} \qquad (13)$$

A similar approach can be used with Eq. 10 instead of Eq. 9. For calculations, use the SI units (m, kg, s) for the electrophoretic mobility, the viscosity, and the hydrodynamic radius rather than the CGS (cm, gram, s) system of units, because the voltage is expressed in SI units (V: $kg.m^2.s^{-3}.A^{-1}$). Despite the rigorous character of Eq. 13, its use is restricted to spherical molecules, which is not the prevalent shape of proteins (see Note 21).

4. Several authors use a correction factor, φ (10), or the frictional ratio (2, 3, 11) to correct the electrophoretic mobility for hydration and for shape deviating from the spherical envelope. With a correction factor, φ, protein valence is expressed as follows:

$$z = \frac{6\pi\eta_s\mu_e R_p\left(1 + \kappa\left(R_p + R_b\right)\right)}{\varphi eF\left(\kappa R_p\right)\left(1 + \kappa R_b\right)} \qquad (14)$$

With the frictional ratio, (f/f_0), protein valence is expressed as follows:

$$z = \frac{6\pi\eta_s(f/f_0)\mu_e R_p\left(1 + \kappa\left(R_p + R_b\right)\right)}{eF\left(\kappa R_p\right)\left(1 + \kappa R_b\right)} \qquad (15)$$

3.2.3. Empirical Relation for the Estimation of Mean Net Charge of Nonspherical and/or Hydrated Proteins

1. The approach used to convert the electrophoretic mobility into a mean net charge is based on the empirical relation described by Basak and Ladisch (3). These authors empirically determined, for a large set of proteins reviewed from the literature (2, 12–16), a relationship between the electrophoretic mobility of proteins, μ_e, the mean net charge, z, the molecular mass, M, and the frictional ratio, f/f_0. The empirical relation is the following:

$$z = \frac{\left(\left(\left(f/f_0\right) \times \mu_e\right) + B\right)}{A} \times M^{2/3} \qquad (16)$$

Equation 16 indicates that the mean net charge of the protein is proportional to its electrophoretic mobility, frictional ratio, and molecular mass. The dependency of the electrophoretic mobility on the three other parameters (molecular mass, frictional ratio, valence) is exemplified in Fig. 1. The A and B values depend on the ionic strength of the buffer. At 10 mM ionic strength, A and B values are (A: $9.5 \times 10^{-3} \pm 5 \times 10^{-4}$; B: $1.7 \times 10^{-5} \pm 1 \times 10^{-6}$); at 100 mM ionic strength, A and B values are (A: $6.5 \times 10^{-3} \pm 5 \times 10^{-4}$; B: $1.1 \times 10^{-5} \pm 2\times 10^{-6}$); at 500 mM ionic strength, A and B values are (A: $4.4 \times 10^{-3} \pm 3 \times 10^{-4}$; B: $8.3 \times 10^{-6} \pm 3 \times 10^{-7}$). A correction can be applied for intermediate ionic strengths. It is noteworthy that several authors have developed similar approaches for peptides and for partially digested proteins (see for instances refs. 11, 12, 17 and references therein).

The electrophoretic mobility units used in the calculation are in $cm^2.V^{-1}.s^{-1}$, ranging approximately from -3×10^{-4} to $+3 \times 10^{-4}\ cm^2.V^{-1}.s^{-1}$ for proteins. The NanoZS software provides the electrophoretic mobility in $\mu m.cm.V^{-1}.s^{-1}$ (ranging approximately from -3 to $+3\ \mu m.cm.V^{-1}.s^{-1}$ for proteins). The molecular mass is expressed in $g.mol^{-1}$.

2. Concerning the experimental data, the molecular mass can be determined by mass spectrometry or in native conditions by SEC-TDA, (see protocol review (18)) and (4, 19–22) the frictional ratio f/f_0 is obtained by QELS or by analytical ultracentrifugation as described above. An example of calculation of the protein valence z is provided in Note 22.

3. This approach is relevant for the determination of the mean net charge of IDPs because the frictions between the protein and its milieu are important and cannot be neglected. Indeed, IDP experience frictions with the buffer due to protein shape and mainly to time-averaged apparent hydration, both parameters contributing to the increase of the frictional ratio.

4. We recently showed that the empirical approach described here is useful to determine the mean net charge of intrinsically disordered RTX proteins (4). Moreover, the disorder-to-order transition of these proteins is induced by calcium binding. We measured the electrophoretic mobility, the hydrodynamic radius, and the molecular mass in native conditions of the RTX proteins in the absence and in the presence of calcium. We calculated the mean net charge of both states and the difference of valence between both states provided us with an estimation of the number of cations involved in the calcium-induced folding of the intrinsically disordered RTX proteins (see ref. 4 for details).

4. Notes

1. Current instruments measuring the electrophoretic mobility are derived from the laser light scattering Doppler electrophoresis; these instruments, like the NanoZS series, offer the opportunity to work in FFR mode and are embedded with the so-called phase shift analysis light scattering (PALS) technique. The field reversal is used to reduce electrode polarization induced by the conductivity of the sample solution. In

FFR mode, a high-frequency (40 Hz), alternating electric field (reversed about every 25 ms) is applied, allowing the protein to reach its velocity while electro-osmosis of the buffer solution is suppressed. Indeed, nano- to micrometer molecules reach their velocity in microseconds (electrophoresis), meanwhile the buffering solutions require approximately 100 ms to reach their actual velocity (electro-osmosis). The PALS technique measures the phase shift instead of the frequency shift, allowing the measure of electrophoretic mobility in viscous media or in buffers of relatively high conductivities at rather low voltage.

2. The expression "mean net charge" is semantically wrong. The charge, often noted Q or q, is the electric charge and corresponds to $Q = z \times e$, with the valence, z, and the electronic charge, e, which corresponds to 1.602×10^{-19} coulombs. The valence, z, is the value of interest for biologists as it corresponds to the so-called mean net charge, i.e., the sum of positive and negative charges carried by side chains of the protein at a defined pH.

3. DTS1060C is a single-use polycarbonate cell of 800 μL with two gold electrodes as shown in Fig. 2a. The main negative aspect of this cell is that under relatively high electric current, it can be damaged and a leakage of materials from the gold electrodes into the sample can be observed (see Note 4). The contaminating materials that exhibit an optical absorbance between 240 and 260 nm can be removed from the protein sample by dialysis using a membrane with a cutoff > 1 KDa.

The much more expensive microcuvette ZEN1010 (Fig. 2b) consists in a measurement block made of quartz with two palladium electrodes. The ZEN1010 cell presents two main advantages: (a) only 100 μL of sample are required to fill the measurement volume and (b) there is no leakage from the electrodes during the experiments. As a consequence of the reduced optical path and reduced measurement volume, there is less heating effect than that in DTS1060C cells making the ZEN1010 more suitable for biological materials such as proteins. Whatever the experiment duration, applied voltage, or ionic strength, we did not observe any release of contaminating materials from the ZEN1010 in the sample neither by optical absorbance nor by fluorescence.

4. It is recommended to work with buffer of rather low ionic strength (i.e., from 10 to 100 mM salts). In our laboratory, we mostly use 10–50 mM NaCl in 20 mM Hepes buffer, pH 7.5.

A low ionic strength increases the electrophoretic mobility (and increases the signal/noise ratio), stabilizes the electric field in the measurement cell, improves the repeatability of measurements, and limits the Joule effect. Indeed, the ionic strength is directly related to the conductivity of the medium σ, which is proportional to the electric current I and the applied voltage U by the following relation:

$I = \frac{U\sigma}{k}$, with $R = k\sigma^{-1}$ (R in ohms) and $k = L/S$, the dimensions (L, length and S, the surface section) between the electrodes. An upper limit for the electric current applied to protein solutions is approximately 5 mA. Even a moderated ionic strength can induce heating of the system by Joule effect, which will affect the temperature of the sample, induce electrode polarization, electrode degradation, and sample alteration, and will ultimately affect the protein mobility. From the above relation, it appears that lowering the applied voltage will reduce the electric current and the sample heating but unfortunately at the expense of the resolution. Changing the design (k factor) of the cell may also improve the system.

5. The cell must be completely filled (to avoid the electro-osmotic effect), without introducing bubbles, and carefully closed at both extremities. Leakage of solution on the external part of the electrodes should be avoided. The volume of a DTS1060C cell is 800 µL and 1 mL of solution is required to properly fill the cell. The volume of a ZEN1010 cell is 100 µL and about 150 µL of solution are required to properly fill the cell.

6. A temperature of 25°C is convenient, as some equations are described in simplified versions for this temperature in the literature.

7. To control the quality of the buffer, check the count meter set, usually between 10 and 100 kilocounts per seconds (kcps) at laser attenuation 11 (full laser power), without autocorrelation function.

8. Concentration should be around 0.5 g/L (for QELS only) and 2 g/L (for electrophoretic mobility) for proteins smaller than 10 nm; 0.2 g/L for proteins between 10 and 100 nm; and 0.02 g/L for particles ranging from 100 nm to 1 µm.

9. Electrophoretic mobility experiments are highly sensitive to contaminants that scatter light, and therefore any particle within the solution will be detected. For this reason it is important that at least 80% of the scattered intensity (as observed by static light scattering) corresponds to the protein of interest.

10. In the Malvern software, three methods can be used depending on the characteristics of the sample. For proteins of small size (below 10 nm), use either the Protein Analysis method or the Multiple Narrow Bands method. For larger proteins, use either Multiple Narrow Bands method or the General Purpose method. The General Purpose method is a good first

approximation if the characteristics of the sample are unknown, but it is less accurate in discriminating between overlapping, poorly resolved populations. The distribution analysis is based on nonnegative least squares.

11. For a protein with a hydrodynamic radius below 10 nm, at 1–2 g/L, the expected attenuator factor is 9 or 10 and the intercept of the autocorrelation function higher than 0.3.

12. To create an average of autocorrelation functions with the experimental data, it is recommended to use two of the three methods cited in Note 10 and to compare the results between them. Robust data must give similar results irrespectively of the analyzing method used. To compare them, it is not necessary to make a new experiment but just change the type of analysis method: right click on the selected data, select Edit Result, and change the analysis mode.

    Irrespectively of the method used, review the distribution plot (of hydrodynamic radius or translational diffusion coefficient) in the following order: intensity distribution, volume distribution, and, finally, number distribution. Intensity distribution is proportional to the scattered light (related to d^6) and not to the fraction—in mol—of the populations in the sample: big particles scatter more light than small particles. Data reported in volume distribution (related to $(4\pi R_p^3)/3$) give results linked to the proportion of each population. Mass distribution should be used with care, as it tends to underestimate the populations of large particles (mass distribution provides a mol: mol distribution from the intensity of the scattered static light).

13. Select Manual measurement duration. A typical setup is 1 measure of 1 run in FFR mode. FFR mode corresponds to the monomodal mode. An FFR measurement is typically a sequence of 22 repeats of fast (25 ms) field reversal (a positive followed by a negative applied voltage) for a total time of 1.2 s that should produce a phase plot with a characteristic pattern (a frequency of 25 ms) and an amplitude of at least 2 radians. The FFR mode is interesting for protein analysis because there is virtually no electro-osmosis and the applied voltage is alternatively short enough to avoid the damage of the protein sample.

    In advanced measurement, select NO for automatic voltage selection and indicate the voltage to be applied, 5 or 10 V. Then, start the acquisition.

    The software, by default, applies a voltage of 10 V. If the electric current is higher than 0.5 mA, the software performs no more test. If the electric current is lower than 0.5 mA, the software starts a new acquisition, applying a voltage of 50 V. This test provides the electric current and the conductivity of the sample, allowing to define the highest voltage value to be applied to measure the electrophoretic mobility of the sample while keeping the electric current below 5 mA. The test provides further information on the sample, such as the attenuator position (should be 11 for a protein at 1–2 g/L with a $R_p < 10$ nm, with a kcps around 10).

14. The preliminary test provides the electric current and the conductivity values of the sample. From the relation $I = U\sigma/k$, defined in Note 4, the maximum voltage to be applied is calculated. For instance, with a conductivity of 3 ms.cm^{-1} and an electric current of 2 mA for an applied voltage of 50 V, one deduces that the maximal voltage to be applied should be 125 V (to keep the electric current below 5 mA), i.e., the applied voltage should range between 10 and 125 V (high voltages alter the quality of the phase plots).

15. The quality criteria to keep a measurement are based on the following: (a) the ZQF (signal-to-noise ratio of the frequency shift) that should stay higher than 0.5–1; (b) the MCR should not change throughout the duration of the data acquisition; (c) the quality of the phase plot (at least an intensity of 1–2 radians) and (d) of the Fourier transform of the phase plot, which should provide a characteristic frequency distribution plot.

    If the value of the ZQF is above 1, no more experiments are required for the dataset. If the ZQF decreases, stop the acquisition because this could reflect sample degradation. If the ZQF is low, increase the number of runs or try to increase the voltage applied, yet keeping the intensity below 5 mA. If the MCR is fluctuating, this suggests that sample heterogeneity (aggregation; etc) is affecting the light scattering signal (see Subheading 3.1, step 3 and Note 9). Lastly, the phase plot can be noisy, but the data are acceptable if the Fourier transform of the phase plot is well resolved.

16. The status of the sample is basically checked by measuring the hydrodynamic radius of the protein as described above. The state of the sample (and its potential alteration) can be further analyzed by optical methods (absorbance, fluorescence, etc.) or functional assays.

17. At first approximation, QELS is interesting for routine experiments or for first trials because a QELS experiment is rapid and uses small quantities of protein. AUC experiments are more demanding in protein quantities and more time-consuming. Nevertheless, for final data processing (and when it is experimentally feasible) the hydrodynamic radius determined by AUC is more reliable than QELS, especially for small proteins ($\approx$2–5 nm). For proteins of higher sizes, QELS data are sufficiently reliable to be used without AUC data.

18. The inverse screening length (Debye length) is calculated as follows:

$$\kappa = \left(\frac{2 N_A e^2 I}{1,000 \varepsilon_0 \varepsilon_d k_B T} \right)^{\frac{1}{2}}$$

with ε_0, the permittivity of free space ($8,854 \ 10^{-12} C^2.J^{-1}.m^{-1}$); ε_d, the dielectric constant (78.54 at $25°C$); and I, the ionic strength ($mol.L^{-1}$). The inverse screening length at $25°C$ can be expressed as a function of the ionic strength: $\kappa = 3.27\sqrt{I}$, i.e., $\kappa R_p = 3.27 R_p \sqrt{I}$.

The ionic strength is given by $I = \frac{\Sigma(n C z^2)}{2}$, with n the number of ions, C the concentration, and z the valence.

Below is an example of ionic strength calculation with 2 mM $CaCl_2$

1. Ca^{2++}: $n = 1$; $C = 0.002$ M; $z = +2$
2. Cl^-: $n = 2$; $C = 0.002$ M; $z = -1$

$$I_{CaCl_2} = \left[\frac{(n C z^2)}{2} \right]_{Ca^{2++}} + \left[\frac{(n C z^2)}{2} \right]_{Cl^-}$$

$$= \left[\frac{(1(0.002)(2^2))}{2} \right]_{Ca^{2++}} + \left[\frac{(2(0.002)(-1^2))}{2} \right]_{Cl^-}$$

Hence, ionic strength of 2 mM $CaCl_2$ is

$$I_{CaCl_2} = \left[4.10^{-3} \right]_{Ca^{2++}} + \left[2.10^{-3} \right]_{Cl^-} = 6.10^{-3} M$$

For buffers, consider the (pH-pK_a), i.e., the fraction of charged buffer at the corresponding pH.

19. From the ionic strength and the inverse screening length, the Henry function $F(\kappa R_p)$ (5) is calculated using the simplified relation provided by D. Rodbard and A. Chrambrach, (23):

$$F(\kappa R_p) = 1 + \frac{0.5}{1 + e^{(A(1 - \log(\kappa R_p)))}}$$

for $\kappa R_p < 10$, $A = 2.8$; hence, for $\kappa R_p \to 0$, $F(\kappa R_p) \to 1$
for $\kappa R_p > 10$, $A = 2.4$; hence, for $\kappa R_p \to \infty$, $F(\kappa R_p) \to 1.5$

This expression of the Henry function is valid for the relationship $\mu_e = (\varepsilon \zeta F(\kappa R_p))/(6\pi\eta_s)$.

20. For a buffer made of charged molecules (Hepes, Tris, etc.), the dimension of the buffering molecules should be carefully considered, together with the radius of other ionic molecules or salts (NaCl, KCl, etc.). Obviously, a sodium cation is smaller than buffering molecules. A way to circumvent this problem is

to measure the electrophoretic mobility and the hydrodynamic radius of the protein in a phosphate buffer for instance, eliminating issues with buffer radius. Alternatively, one can use an averaged buffer radius (i.e., 0.3–0.4 nm).

21. A globular protein is not necessary spherical; see for instance bovine serum albumin (BSA) or α-lactalbumin used as model of globular proteins but with a nonspherical shape.

22. This note is an example describing how to determine the protein valence from electrophoretic mobility (μ_e), molecular mass (M), and frictional ratio (f/f_0).

We consider a protein with a molecular mass of 30,000 g. mol^{-1}, a sedimentation coefficient of 3 s (3×10^{-13} s), and a partial specific volume $\bar{v}$ of 0.70 $cm^3.g^{-1}$. The protein is studied at 25°C in a buffer (20 mM Hepes, 100 mM NaCl pH 7.5) of density 1 $g.cm^{-3}$ and viscosity $\eta=0.009$ $g.cm^{-1}.s^{-1}$ (equivalent to 0.9 cP). The electrophoretic mobility (μ_e) measured for this protein is -2×10^{-4} $cm^2.V^{-1}.s^{-1}$ (equivalent to -2 µm.cm. $V^{-1}.s^{-1}$).

We first calculate the anhydrous radius (R_0) of the protein, using the relationship:

$$R_0 = \left(\frac{3M\bar{v}}{4\pi N_A} \right)^{1/3}$$

with M in $g.mol^{-1}$ and $\bar{v}$ in $cm^3.g^{-1}$

$$R_0 = ((3 \times 30,000 \times 0.7)/(4 \times \pi \times N_A))^{1/3}$$

$$R_0 = 2.1 \times 10^{-7} cm$$

We then calculate the hydrodynamic radius (R_P) of the protein. R_P can be obtained either from the QELS experiments or from the sedimentation coefficient value (obtained by AUC), using the equation:

$$R_P = M(1 - \rho\bar{v})/(6\pi\eta N_A s)$$

with M in $g.mol^{-1}$, ρ in $g.cm^{-3}$, $\bar{v}$ in $cm^3.g^{-1}$, η in $g.cm^{-1}.s^{-1}$, and s in second.

$$R_P = 30,000(1 - 1 \times 0.7)/(6\pi \times 0.009 \times N_A \times 3 \times 10^{-13})$$

$$R_P = 2.95 \times 10^{-7} cm$$

Now, using the R_0 and the R_P values, we can calculate the frictional ratio f/f_0 according to the relation $f/f_0 = R_P/R_0$

$$f/f_0 = 2.95/2.1$$

$$f/f_0 = 1.4$$

Finally, using the previously measured (M and μ_e) and calculated (f/f_0) parameters, we determine the number of charges using the empirical relation:

$$z = \frac{(((f/f_0) \times \mu_e) + B)}{A} \times M^{2/3}$$

with M in g.mol^{-1}, μ_e in cm^2.V^{-1}.s^{-1}, and A and B which depend on the ionic strength of the buffer. For an ionic strength of 100 mM, A and B values are 6.5×10^{-3} and 1.1×10^{-5}, respectively.

$$z = \frac{\left((1.4 \times (-2 \times 10^{-4})) + 1.1 \times 10^{-5}\right)}{6.5 \times 10^{-3}} \times 30,000^{2/3}$$

$$z = -43$$

Taken together, this protein of 30 KDa exhibits a mean net charge of -43 at the investigated pH.

Acknowledgement

This work was supported by the Institut Pasteur (Grant PTR374), the Centre National de la Recherche Scientifique (CNRS UMR 3528), and the Agence Nationale de la Recherche, programme Jeunes Chercheurs (ANR, grant ANR-09-JCJC-0012).

References

1. Delgado AV, Gonzalez-Caballero F, Hunter RJ, Koopal LK, Lyklema J (2005) Measurement and interpretation of electrokinetic phenomena. Pure Appl Chem 77:1753–1805

2. Abramson HA, Moyer LS, Gorin MH (1942) Electrophoresis of proteins and the chemistry of cell surfaces. Reinhold Publishing Coorporation, New York

3. Basak SK, Ladisch MR (1995) Correlation of electrophoretic mobilities of proteins and peptides with their physicochemical properties. Anal Biochem 226:51–58

4. Sotomayor-Perez AC, Ladant D, Chenal A (2011) Calcium-induced folding of intrinsically disordered Repeat-in-Toxin (RTX) motifs via changes of protein charges and oligomerization states. J Biol Chem 286:16997–17004

5. Henry DC (1931) The cataphoresis of suspended particles. Part 1. The equation of cataphoresis. Proc R Soc Lond A 133:106–129

6. Debye VP, Hückel E (1924) Bemerkungen zu einem Satze über die kataphorestische Wanderungsgeschwindigkeit suspendierter teilchen. Physikalische Zeitschrift 3

7. Smoluchowski M (1903) Przyczynek do teoryi endosmozy elekrycznej i niektórych zjawisk pokrewnych. Bull Acad Sci Cracovie

8. Wall S (2010) The history of electrokinetic phenomena. Curr Opin Colloid Interface Sci 15:119–124

9. Gorin MH (1939) An equilibrium theory of ionic conductance. J Chem Phys 7:405–413

10. Winzor DJ (2004) Determination of the net charge (valence) of a protein: a fundamental but elusive parameter. Anal Biochem 325:1–20

11. Adamson NJ, Reynolds EC (1997) Rules relating electrophoretic mobility, charge and molecular size of peptides and proteins. J Chromatogr 699:133–147

12. Issaq HJ, Janini GM, Atamna IZ, Muschik GM, Lukszo J (1992) Capillary electrophoresis separation of small peptides: effect of pH, buffer additives, and temperature. J Liq Chrom Relat Tech 15:1129–1142

13. Walbroehl Y, Jorgenson JW (1989) Capillary zone electrophoresis for the determination of electrophoretic mobilities and diffusion coefficients of proteins. J Microcolumn Sep 1:41–45

14. Velick SF (1949) The interaction of enzymes with small ions. I. An electrophoretic and equilibrium analysis of ldolase in phosphate and acetate buffers. J Phys Colloid Chem 53:135–149

15. Longsworth LG (1941) The influence of pH on the mobility and diffusion of ovalbumin. Ann NY Acad Sci 41:267–285

16. Ivory CF (1990) Electrophoresis of proteins: batch and continuous methods. In: Flickinger M, Drew S (eds) The encyclopedia of bioprocess technology: fermentation, biocatalysis and bioseparations. Wiley, Chapter 9

17. Rickard EC, Strohl MM, Nielsen RG (1991) Correlation of electrophoretic mobilities from capillary electrophoresis with physicochemical properties of proteins and peptides. Anal Biochem 197:197–207

18. Karst JC, Sotomayor Perez AC, Ladant D, Chenal A (2012) Estimation of intrinsically disordered protein shape and time-averaged apparent hydration in native conditions by a combination of hydrodynamic methods. In: Uversky V, Dunker AK (eds) Intrinsically Disordered Proteins: Volume I. Experimental Techniques

19. Bourdeau RW, Malito E, Chenal A, Bishop BL, Musch MW, Villereal ML, Chang EB, Mosser EM, Rest RF, Tang WJ (2009) Cellular functions and X-ray structure of anthrolysin O, a cholesterol-dependent cytolysin secreted by *Bacillus anthracis.* J Biol Chem 284:14645–14656

20. Chenal A, Guijarro JI, Raynal B, Delepierre M, Ladant D (2009) RTX calcium binding motifs are intrinsically disordered in the absence of calcium: implication for protein secretion. J Biol Chem 284:1781–1789

21. Karst JC, Sotomayor Perez AC, Guijarro JI, Raynal B, Chenal A, Ladant D (2010) Calmodulin-induced conformational and hydrodynamic changes in the catalytic domain of *Bordetella pertussis* adenylate cyclase toxin. Biochemistry 49:318–328

22. Sotomayor Perez AC, Karst JC, Davi M, Guijarro JI, Ladant D, Chenal A (2010) Characterization of the regions involved in the calcium-induced folding of the intrinsically disordered RTX motifs from the *Bordetella pertussis* adenylate cyclase toxin. J Mol Biol 397:534–549

23. Rodbard D, Chrambach A (1971) Estimation of molecular radius, free mobility, and valence using polyacylamide gel electrophoresis. Anal Biochem 40:95–134

Chapter 23

Protein Characterization by Partitioning in Aqueous Two-Phase Systems

Larissa Mikheeva, Pedro Madeira, and Boris Zaslavsky

Abstract

Protein partitioning in aqueous two-phase systems is a technique that enables one to monitor changes in the 3D structures of proteins in solution resulting from chemical modifications, conformational changes, and interactions with other proteins and ligands. The advantage of this technique is that it may be used to monitor the aforementioned changes for purified proteins as well as for proteins in biological fluids, and that it is readily adaptable to automated high-throughput mode.

Key words: Aqueous two-phase partitioning, Protein structure in solution, Protein–solvent interactions, Aggregation, Conformation, Protein–protein interactions

1. Introduction

Partitioning in aqueous two-phase systems (ATPSs) provides a fast and simple method for characterization of protein 3D structures in solution. Protein partitioning in ATPS enables monitoring of protein–protein interactions, analysis of protein propensity for aggregation, changes in protein conformation upon ligand binding, and other structural changes (1–6).

The partitioning of a protein, or any other soluble substance in an ATPS, is characterized by the partition coefficient, K, defined as a ratio of the concentration of the protein in the top phase to the protein concentration in the bottom phase. The K-value is governed by the properties of a given protein and the ATPS composition (type and concentrations of phase-forming polymers, type and concentrations of salt additives, and pH) (1, 7, 8).

For a practitioner new to the ATPS field it is important to notice that most of the experimental procedures and protocols

Vladimir N. Uversky and A. Keith Dunker (eds.), *Intrinsically Disordered Protein Analysis:*
Volume 2, Methods and Experimental Tools, Methods in Molecular Biology, vol. 896,
DOI 10.1007/978-1-4614-3704-8_23, © Springer Science+Business Media New York 2012

reported in the literature are related to applications for separation of biological products, such as proteins, nucleic acids, biological membranes, viruses, and cells (7, 8). Analytical applications of partitioning in ATPS present rather different requirements. In the case of protein separation, the desirable conditions should provide distribution of a target protein predominantly into only one of the two phases with all the other components of the original mixture (extract, biological fluid, etc.) being distributed into the other phase. For analytical applications, the desirable conditions must meet two requirements: (a) the protein under study should partition within a reliably measurable range ($0.1 < K < 10$) (under conditions of extreme partitioning the range of concentrations of the protein in the protein-poor phase is the range of low and unreliable analytical signal) and (b) the partition coefficient K values should be sensitive to the changes of interest (change in the protein conformation, structural modification, etc.). ATPS formed by such nonionic polymers as Dextran and Ficoll, Dextran and Poly(ethylene glycol) (PEG), and Ficoll and PEG commonly meet the above requirements, provided the right type and concentration of salt additives are used (1, 9, 10).

Analytical applications of ATPS partitioning for protein analysis and characterization are listed in Table 1.

Table 1
Analytical applications of ATPS partitioning for protein analysis

Application	References	Comments
Protein aggregation	(1)	May be used in biological fluid
Protein–protein interactions	(2, 11)	May be used in biological fluid
Protein–ligand interactions	(1, 2, 7, 8)	May be used in biological fluid
Protein conformational changes	(1, 11)	May be used in biological fluid
Changes in chemical structure[a]	(5, 6, 12, 13)	May be used in biological fluid
Structure consistency	(12, 14)	For different manufacturing lots or protein preparations/isolates

[a]Mutations, glycosylation, oxidation, deamidation, and phosphorylation are possible changes in chemical structure of a protein

2. Materials

1. Stock solutions of phase-forming polymers in water. Typical polymer stock solutions: 50–60%w/w PEG-600 or 20–50%w/w. PEG-8000; 30–35%w/w Dextran-70; 30–45%w/w Ficoll-70, depending on the particular ATPS chosen. Polymer solutions are prepared in DI water. All the stock solutions are prepared by weight. The details of stock solution preparations for the above polymers can be found in refs. 15, 16 (see Note 1).

2. Stock solutions of buffers and salts. Stock solutions may be prepared by weight or volume in DI water taking into account the desirable salt and buffer concentrations in the ATPS and dilution of the stock solution when used for ATPS preparation. Sodium or potassium phosphate buffer, sodium citrate buffer, and universal buffer (composed of acetic, phosphoric, boric acids, and NaOH) are typical choices depending on the proteins examined. Typical buffer stock solution concentration is 0.5 M. Typical salt stock solution concentration is 0.5–5 M.

3. Methods

1. Into a polypropylene or siliconized glass tube of an appropriate volume (usually 2–4 mL, see Note 2), weigh out stock solutions of phase-forming polymers, salts, and buffers (see Note 3). The final volume of an ATPS and final concentrations of the phase-forming polymers and salts are achieved following addition of the protein solution in water or buffer. As a rule of thumb, the protein amount added to a given ATPS should not exceed 1–2% of the total weight of the ATPS.

   For example, to prepare a system containing 12.9%w/w - Dextran-70, 18.0%w/w Ficoll-70, 0.15 M NaCl, and 0.01 M sodium phosphate buffer (NaPB), pH 7.4, at a final weight of 1.2 g, mix 498 mg (440 µL) 31.2%w/w Dextran-70, 405 mg (335 µL) 53.5%w/w Ficoll-70, 23 mg (23 µL) of 0.5 M NaPB, pH 7.4, 60 mg (55 µL) 3.0 M NaCl, and 214 µL of protein solution in water. The final weight of the ATPS will be 1.20 g in a volume of 1,067 µL. The volumes of the separated coexisting phases will be equal (see Note 4).

   To prepare a system containing 12.4%w/w Dextran-70, 6.0%w/w PEG-8000, 0.15 M NaCl, and 0.01 M sodium phosphate buffer (NaPB), pH 7.4, at a final weight of 1.2 g, mix 477 mg (420 µL) 31.2%w/w Dextran-70, 144 mg (135 µL) 50.4%w/w PEG-8000, 60 mg (55 µL) 3 M NaCl and 23 mg (23 µL) 0.5 M NaPB, pH 7.4, and 495 µL protein solution in water. The final weight of the ATPS will be 1.20 g in a volume of 1,128 µL. The volumes of the settled coexisting phases will be equal (see Note 4).

As a final example, to prepare a system containing 15.0%w/w Ficoll-70, 7.9%w/w PEG-8000, 0.15 M NaCl, and 0.01 M sodium phosphate buffer (NaPB), pH 7.4, at a final weight of 1.2 g, mix 338 mg (280 µL) 53.5%w/w Ficoll-70, 188 mg (175 µL) 50.4%w/w PEG-8000, 60 mg (55 µL) 3 M NaCl and 23 mg (23 µL) 0.5 M NaPB, pH 7.4, and 591 µL protein solution in water. The final weight of the ATPS will be 1.20 g in a volume of 1,124 µL. The volumes of the settled coexisting phases will be equal (see Note 4).

2. Prepare a set of six tubes with identical mixtures of stock solutions (not adding the sample yet). Cap the tubes and mix if needed (see Note 5).

3. Uncap the tubes and add the appropriate volumes of the protein solution (see Note 6) to adjust sample volume so that the volumes in all the tubes are equal. For example, in a given ATPS with the overall sample volume V_{sample}, in the set of six tubes, you may add 0, 10, 20, 30, 40, and 50 µl protein solution and V_{sample}, $V_{sample} - 10$, $V_{sample} - 20$, $V_{sample} - 30$, $V_{sample} - 40$, and $V_{sample} - 50$ µL water (buffer). The particular added protein volumes depend on the available protein solution volume (and protein concentration) and the sensitivity of the analytical assay used (see below). An alternative option may be to prepare protein solutions of different concentrations and add them to the tubes in the same volume. Cap the tubes and shake them by rotating 10–20 times or using a Vortex mixer for 30–60 s until homogeneously turbid mixtures are formed in all tubes (see Note 7).

4. Place tubes into refrigerated or thermostated centrifuge (see Note 8) and spin for the time required for the ATPS to reach equilibrium. Following centrifugation, each tube should have two clear phases without any spots of turbidity. Check the volumes of the systems and bottom phases in all tubes. They should be essentially identical. Check visually for the lack of adsorption of the protein at the interface (observed as a slightly turbid thin film at the interface). If adsorption is noticed in tubes with high volumes of the protein solution added, make a note (see Note 9).

5. Place each tube in a fixed position and withdraw an aliquot from the top phase of each tube (If the volume of each phase is above 400 µl, an aliquot of up to 300 µl can be readily withdrawn. Remove an aliquot from the bottom phase, as well). To prevent contamination of the bottom phase with the top phase, first carefully remove residual volume of the top phase before

removing aliquot from the bottom phase (see Note 10). Aliquots withdrawn from both phases are placed into another set of tubes marked with the corresponding phase (top or bottom) and the particular sample volume added to the ATPS in a given tube. The aliquots should be diluted three- to tenfold or more (same dilution should be used for all aliquots) with media that is appropriate for the analytical assay that is to be used to measure the samples concentrations. Vortex each tube with the diluted aliquot with a Vortex mixer and centrifuge for a short time (e.g., 5 min at $3,000 \times g$) to ensure that the diluted aliquot is properly mixed.

6. A suitable analytical assay is used to measure the concentration of the protein in the diluted aliquots of the phases. Among available assays are the following: HPLC—sensitive but relatively time consuming (see Note 11); mass spectrometry (MS)—fast and sensitive but requiring an internal standard for concentration analysis (see Note 12); o-phthalaldehyde (OPA) assay—fast, reliable, and highly sensitive analytical assay for proteins with sufficient number of lysine residues, i.e., free NH_2 groups, requires spectrofluorometer (for details see, e.g., in ref. 18); Bradford assay—simple and fast though not as sensitive as OPA assay; measurement of UV absorbance—usually requires the large amounts of protein due to insufficient sensitivity; ELISA immunoassay—very sensitive and applicable to assaying a given protein partitioned in ATPS within biological fluid (blood, urine) or tissue or cellular extract. Calibration of the analytical assay in diluted top and bottom phases is generally recommended but not always necessary (see Note 13).

Once the protein concentrations (or analytical signal with intensity proportional to the protein concentration) in the phases are measured, the partition coefficient of the protein can be determined. For this purpose, plot the protein concentrations (analytical signals) determined in the top phases against those determined in the bottom phases. Typical plot is presented in Fig. 1a. The linear curve observed is described as

$$C_{Top}^i = a + b \times C_{Bottom}^i \qquad (1)$$

where C^i is the protein concentration (analytical signal) measured in a particular phase denoted by the subscripts "Top" and "Bottom" and a and b are constants. Coefficient a is related to the interference of the phase components with the analytical signal and may be close to 0. Coefficient b value represents the protein partition coefficient K.

The linearity of the plot in Fig. 1a indicates that the protein partition coefficient K is independent of the protein concentration. This implies that the identical species are distributed between the coexisting phases within the protein concentration range used in the experiment. Deviation of the experimentally

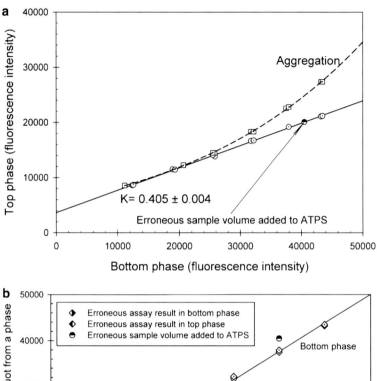

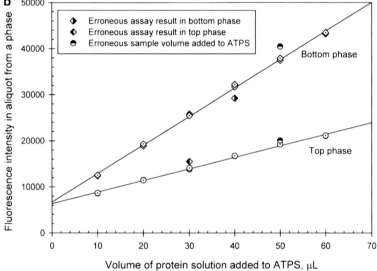

Fig. 1. (**a**) Concentration of a protein in the top phase plotted versus concentration of the protein in the bottom phase (concentration is represented by the analytical signal, in this case fluorescence in OPA assay). Deviation of the *plotted curve* from linearity indicates protein aggregation. (**b**) Concentrations of a protein in the top phase and in the bottom phase plotted versus volume of the protein solution added to ATPS (concentration is represented by the analytical signal, in this case fluorescence in OPA assay).

determined relationship between the C_{Top}^i and C_{Bottom}^i from linearity indicates that changes in the protein concentration result in appearance of species with partition behavior different from that present at lower protein concentrations generally implying formation of protein aggregates. Typical protein aggregation behavior is illustrated in Fig. 1a (nonlinear curve).

In order to check the reliability of the K-value determination, the following additional procedure may be used. All the protein concentrations (or analytical signal with intensity proportional to the protein concentration) measured in all the top and bottom phases are plotted against the protein solution volumes introduced into each ATPS. The expected relationship is illustrated in Fig. 1b. Two types of experimental error in this case are illustrated: (1) the erroneous assay result in measurements of protein concentration in one of the aliquots from the top phase and similarly in the bottom phase and (2) addition of an erroneous volume of the protein solution to ATPS in one of the six tubes. The latter case results in the "wrong" position of the corresponding point on the curve presented in Fig. 1a. The distance of the incorrect point from its neighboring points deviates from all the other points on the curve. The former case is clearly the result of measuring "wrong" analytical signal by the assay—if it happens, for example, with three points out of six on the linear curve in Fig. 1a it seriously undermines the reliability and accuracy of the determined K-value. In this case, it might be rectified by analysis of separate curves shown in Fig. 1b described as

$$C_{\text{Top}}^i = \alpha + \beta_{\text{Top}} \times V_{\text{sample}}^i \tag{2a}$$

$$C_{\text{Bottom}}^i = \alpha + \beta_{\text{Bottom}} \times V_{\text{sample}}^i \tag{2b}$$

where V_{sample}^i is the volume of the protein solution introduced to ATPS; C_{Top}^i and C_{Bottom}^i are previously defined; and α and β are constants. Partition coefficient K-value may be calculated from the ratio $K = \beta_{\text{Top}}/\beta_{\text{Bottom}}$.

As a final check of the accuracy of the experimental results obtained, it is recommended to check the material balance for the protein concentrations measured in the phases. This check tests if the total amount of protein measured in both phases equals the total amount of the protein introduced into the ATPS. The protein amount introduced into the ATPS, G_{protein}^o, is determined by the protein concentration in the original protein solution, C_{protein}^o, and the volume of this solution added to a tube with ATPS, V_{sample}, as $G_{\text{protein}}^o = C_{\text{protein}}^o \times V_{\text{sample}}$. The amount of protein in a given phase is calculated similarly taking into account the dilution of the aliquot used for the analytical assay: $G_{\text{protein}}^{\text{phase}} = C_{\text{protein}}^{\text{phase}} \times V_{\text{phase}} \times \text{dilution factor}$ (contributions of the blank phases into analytical signals must be subtracted from signals assayed in the phases with distributed protein). Sum of amounts of a given protein in the top and bottom phases should be close to the amount of the protein introduced into an ATPS (see Note 14).

It is desirable to run the analytical assay in duplicates, and if possible perform partitioning in duplicate as well. Typical accuracy of determination of K-value is better than 3%.

Finally, all the procedures described (preparation of ATPS; addition of samples; and withdrawing aliquots from the phases with consequent dilution) may be automated and performed by a regular liquid-handling workstation with improved accuracy, precision, and increased throughput (see Note 15).

4. Notes

1. Sodium azide (ca. 0.2 g azide per 1 kg of polymer stock solution) should be added to inhibit bacterial contamination. Typical shelf life of polymer stock solution is 6 months, if stored at room temperature, and more if stored at 4°C.

2. A 15-mL graduated centrifuge tube is typically recommended (15, 16) but it is commonly related to applications for separations of biological products. The ATPS volume should be decided upon depending on the amount of the protein available for analysis. Tubes or vials of smaller volumes, e.g., polypropylene cryogenic vials of 2 mL volume or polypropylene microtubes of 1.2 mL volume and suitable for centrifugation at relatively low speed (3,000 × g or less), are more convenient when working with small amounts of proteins.

3. The stock polymer solutions are commonly relatively viscous; hence, accurate dispensing of these solutions is to be performed slowly by using syringes of appropriate volume or automated devices, such as Microlab 500 (Hamilton company, Reno, NV, USA). Dispensing may be performed with the tube/vial placed on analytical balances or by volume once the density of each stock solution is established. When preparing ATPS by volume, it is recommended to aspirate 8 volumes of a given polymer stock solution needed for preparation of six tubes, dispensing (at lowest speed) one volume back into the stock solution container, dispensing one volume per tube into six tubes, and dispensing the remaining volume into the waste container. This practice ensures good reproducibility of ATPS preparation with Multipette Xstream pipette (Eppendorf, Hamburg, Germany).

4. All volumes listed are relevant for the stock polymer solutions indicated only. For polymer stock solutions prepared at different concentrations the volumes to be added must be adjusted according to their density. To ensure good reproducibility of ATPS composition, accuracy of the dispensing of each stock solution must be better than 1 mg.

5. When studying proteins prone to aggregation, it is recommended to vortex prepared mixtures of stock solutions vigorously until each mixture appears homogeneously turbid. Less vigorous mixing is required following sample addition to the ATPS. Once the homogeneity is observed, centrifuge the tubes for a short time to ensure that the mixtures are accumulated at the bottom of the tubes.

6. Protein solution may be prepared in water or buffer of the same pH as in the ATPS with the protein concentration necessary for analytical assay; see below.

7. Temperature is known to affect the composition of coexisting separated phases; therefore, it is necessary to maintain the desirable temperature (e.g., 23–25 °C) during the protein sample addition and following stages. If the desirable experimental temperature chosen is, e.g., 4 °C, the mixtures of all the system components should be equilibrated at this particular temperature, and all the other procedures, i.e., sample addition, centrifugation, and withdrawal of aliquots, must be performed at 4 °C.

8. It is important to maintain the desirable temperature (e.g., 23–25 °C) during centrifugation that may require up to 60 min (in the case of Dex-Ficoll ATPS, for example). A thermostating centrifuge is preferable, but a refrigerated centrifuge capable of maintaining room temperature is also an adequate option.

9. Adsorption of the protein in tubes with high volumes of the protein solution added may be an indication that the ATPS is overloaded with the protein. Protein concentration in the solution should be reduced or volumes of protein solution not exceeding a certain threshold should be examined.

10. Due to the viscosity of the phases, an aliquot from each phase should be withdrawn carefully and slowly. The best option is to use a syringe or a manual or an electronic micropipette placed in a holder on a Z-axe moving rail. When moving the pipette into the bottom phase do it slowly, and once the pipette is in the bottom phase check that the upper phase is not drawn, with the pipette, into the bottom phase. If this should happen, wait until the top phase and the interface return to their original position before starting to draw the bottom-phase aliquot into the pipette tip. After withdrawing an aliquot from a given phase, wipe the outside of the pipette dry, and if a syringe is used, wash it with water multiple times until no traces of viscous polymer solution remain on the tip. It is also recommended to withdraw an aliquot of a volume not exceeding 70–75% of the phase volume to ensure that the aliquot from the one phase is not contaminated by the other phase.

11. HPLC with UV detector is a convenient and readily adaptable assay for analysis of protein partitioning in ATPS. The authors used it successfully when examining aggregation of a biopharmaceutical product in formulation in more than 1,000-fold excess of albumin. However, time of analysis of 12 aliquots (from top and bottom phases of 6 ATPSs of the same composition) is directly related to the time of chromatographic run, i.e., may be relatively slow.

12. MS spectra are obtained from both phases and *K*-values are calculated as the ratios of intensities at the *m/z* values of interest (for details see in ref. 17).

13. Calibration of the assay, in this context, describes examination of analytical signal resulting from protein concentration measurement in the diluted top (or bottom) phase against the known protein concentration. It may be neglected for comparison of partition behavior of different proteins when the same analytical assay is used and the amounts of proteins are limited, since it might be reasonably assumed that the analytical assay performs similarly in both phases and identically for different proteins. Even if the assay performance is slightly different in the top and bottom phases, it would require using a certain multiplication factor to transform the determined apparent experimental *K*-value into true *K*-value. If this factor is identical for different proteins the calibration may be neglected. If, on the other hand, comparison is to be made between the *K*-values obtained for different proteins with different analytical assays calibration is a must.

14. The calculated material balance is typically within 90–100% range and depends primarily on the accuracy and reproducibility of the analytical assay used.

15. Several examples using liquid-handling workstations from different vendors are reported in the literature (18–20).

References

1. Zaslavsky B (1994) Aqueous two-phase partitioning: physical chemistry and bioanalytical applications. Marcel Dekker, New York
2. Bachman L (2000) Detection and analysis of interactions by two-phase partition. In: Hatti-Kaul R (ed) Methods in biotechnology, vol 11, Aqueous two-phase systems: methods and protocols. Humana, Totowa, pp 219–228
3. Franco TT (2000) Partitioning of chemically modified proteins. In: Hatti-Kaul R (ed) Methods in biotechnology, vol 11, Aqueous two-phase systems: methods and protocols. Humana, Totowa, pp 209–218
4. Zaslavsky B (1992) Bioanalytical applications of partitioning in aqueous polymer two-phase systems. Anal Chem 64:765A–773A
5. Zaslavsky B (1995) Characterization of proteins by aqueous two-phase partitioning. In: Rogers R, Eiteman M (eds) Aqueous biphasic separations: biomolecules to metal ions. Plenum, New York
6. Zaslavsky B, Chaiko D (1996) A new analytical methodology for quality control testing of biological and recombinant products. In: Shillenn J (ed) Validation practices for biotechnology products. American Society for Testing and Materials, Philadelphia, pp 107–122

7. Albertsson PA (1986) Partition of cell particles and macromolecules, 3rd edn. Wiley, New York

8. Walter H, Brooks DE, Fisher D (eds) (1985) Partitioning in aqueous two-phase systems: theory, methods, use, and applications to biotechnology. Academic, Orlando, FL

9. Madeira P et al (2010) Solvent properties governing solute partitioning in polymer/polymer aqueous two-phase systems: nonionic compounds. J Phys Chem B 114:457–462

10. Madeira P et al (2011) Solvent properties governing protein partitioning in polymer/polymer aqueous two-phase systems. J Chromatogr A 1218:1379–1384

11. Zaslavsky A et al (2001) A new method for analysis of components in a mixture without preseparation: evaluation of the concentration ratio and protein-protein interaction. Anal Biochem 296:262–269

12. Data unpublished due to confidentiality of studies performed in collaborations with various biopharmaceutical companies.

13. Stovsky M et al (2011) Prostate specific antigen/solvent interaction analysis (psa/sia): a preliminary evaluation of a new assay concept for detecting prostate cancer using urinary samples. Urology 78:601–605

14. Chait A, Zaslavsky B (2011) Characterization of molecules, US patent 7,968,350

15. Bamberger S et al (1985) Preparation of phase systems and measurement of their physicochemical properties. In: Walter H, Brooks DE, Fisher D (eds) Partitioning in aqueous two-phase systems: theory, methods, use, and applications to biotechnology. Academic, Orlando, FL, pp 86–95

16. Forciniti D (2000) Methods in biotechnology. In: Hatti-Kaul R (ed) Aqueous two-phase systems. Methods and protocols, vol 11. Humana, Totowa, pp 23–33

17. Chait A, Zaslavsky B (2007) Systems and methods involving data patterns such as spectral biomarkers, US patent application 20070198194

18. Gulyaeva N et al (2003) Relative hydrophobicity of di- to hexapeptides as measured by aqueous two-phase partitioning. J Peptide Res 61:129–139

19. Bensch M, Selbach B, Hubbuch J (2007) High throughput screening techniques in downstream processing: preparation, characterization and optimization of aqueous two-phase systems. Chem Eng Sci 62:2011–2021

20. Oelmeier SA, Dismer F, Hubbuch J (2011) Application of an aqueous two-phase systems high throughput screening method to evaluate mAb HCP separation. Biotechnol Bioeng 108:69–81

Part IV

Mass-Spectrometry

Chapter 24

Detection and Characterization of Large-Scale Protein Conformational Transitions in Solution Using Charge-State Distribution Analysis in ESI-MS

Rinat R. Abzalimov, Agya K. Frimpong, and Igor A. Kaltashov

Abstract

Ion charge-state distribution analysis in electro-spray ionization mass spectrometry (ESI-MS) is a robust and fast technique for direct detection and characterization of coexisting protein conformations in solution. Compact folded proteins give rise to ESI-generated ions carrying a relatively small number of charges, whereas less compact conformers accommodate upon ESI a larger number of charges depending on the extent of their unfolding. A chemometric approach [1] based upon factor analysis is applied to determine contributions from individual conformers to the overall CSD. Here we present basic guidelines for the use of this MS-based technique: from the preparation of suitable solutions for ESI-MS to the acquisition of reliable MS data and their subsequent analysis.

Key words: Charge-state distribution, Chemometric approach, Electro-spray ionization mass spectrometry, Factor analysis, Singular value decomposition, MathCAD, Gaussian function, Levenberg–Marquardt algorithm

---

1. Introduction

1.1. Protein Conformations and Extent of Charging in ESI-MS

Conformational heterogeneity of proteins often poses a challenge to the traditional biophysical methods for structural analysis. Electro-spray ionization mass spectrometry (ESI-MS) allows direct detection and characterization of distinct conformers, including transiently populated ones, and transitions between them. This feature of ESI-MS is widely employed in the studies of protein dynamics (2–4) and protein–ligand (5) and protein–protein (6, 7) interactions.

The charge acquired by biomolecules during ESI is strongly influenced by their three-dimensional structure in the sprayed solution. Thus tightly folded protein conformation, with minimal

Vladimir N. Uversky and A. Keith Dunker (eds.), *Intrinsically Disordered Protein Analysis: Volume 2, Methods and Experimental Tools*, Methods in Molecular Biology, vol. 896, DOI 10.1007/978-1-4614-3704-8_24, © Springer Science+Business Media New York 2012

solvent exposure, gives rise to ESI-generated ions carrying small number of charges, whereas less compact conformers will accommodate larger number of charges upon ESI process depending on the extent of their unfolding. Accordingly, ESI-MS spectrum of multiple protein conformers represents a linear combination of ionic contributions from individual conformers (8). It is deduced (9) that the ionic signal from each individual protein conformer is characterized by its "own" charge-state distribution (CSD) or "pure" spectrum, which can effectively be approximated in practice by normal (or Gaussian) distribution (see Note 1). While such parameters as average charge and width of the "pure" spectra remain nearly constant under different solvent conditions (see Note 2), amplitudes of the signals may change dramatically due to changes in protein environment (1).

1.2. Chemometrics for CSD Analysis

CSD of individual conformers often overlap in ESI-MS owing to their heterogeneity. A chemometric approach based on factor analysis (FA) is devised (1) to deal with this problem. Recent work has demonstrated its great potential for protein structure studies (3, 10) (see Note 3). Basic aspects of the approach are briefly provided below, whereas practical guidelines for its use can be found in Subheading 3.1.

First, an array of ESI-MS spectra is acquired over a wide range of near-native and denaturing conditions to ensure adequate sampling of various protein states and significant variation of their fractional concentrations (1). For example, CSD of highly dynamic α-synuclein protein undergoes considerable changes within pH range from 2.5 to 9, and addition up to 60% of ethanol further contributes to the changes of CSD (see Fig. 1). Note that common protein denaturing reagents utilized in ESI-MS are pH modifiers and alcohol (see Subheading 2.2). Thus, the array of data on α-synuclein, needed for complete chemometric analysis, should include a range of ESI-MS spectra acquired under the given solution conditions. In average, approximately 8–20 solvent conditions are preselected with different pH factors and alcohol fractions for comprehensive investigation of CSD of proteins (see Note 4).

Next, mathematical approach based on singular value decomposition (SVD) algorithm (11) is applied to the ESI-MS spectra. SVD of the array of ESI-MS spectra yields a number of independent components or distinct conformations, which are responsible for the observed variations of CSDs. This number is equal to a number of protein conformers whose geometries are different enough to allow at least some distinction to be made as far as their individual contributions to the overall CSD. A set of Gaussian functions is then constructed, each representing a "pure signal" of a certain conformer and the entire data array is fitted using this set, yielding ionic profiles of individual protein states over the range of solution conditions (12).

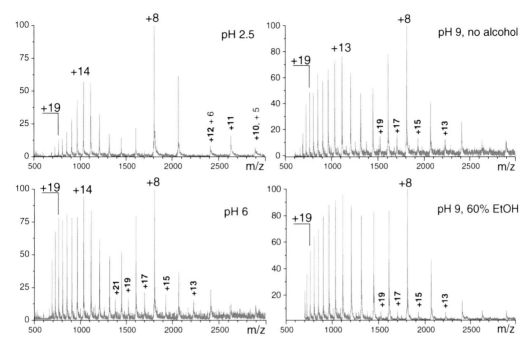

Fig. 1. ESI-MS spectra of human α-synuclein, which were used for CSD analysis. The measurements were carried out using two-sector JMS-700 MStation mass-spectrometer (JEOL, Tokio, Japan). *Adapted with permission from* (10).

Utilization of SVD algorithm is straightforward, since it is incorporated into many mathematical software packages, including MatLab, MathCAD, etc. The Gaussian fitting of the ESI-MS spectra is easiest to implement by using ORIGIN software (*OriginLab Corp., Northampton, MA*), as it incorporates versatile nonlinear fitting utilities based on Levenberg–Marquardt algorithm.

2. Materials

2.1. ESI Solutions and Reagents

All solutions for ESI-MS should be prepared using ultrapure water (prepared by purifying deionized water to attain 18 MΩ cm at 25°C). Common buffers (Tris–HCl, HEPES, phosphate buffers, etc.) used during isolation or storage of proteins contain nonvolatile electrolytes, which cause suppression of useful ESI signal and/or extensive adduct formation. Therefore, most of the protein solutions need to be buffer exchanged to ESI-friendly solvent systems prior to mass analysis (*refer to the list of ESI-friendly reagents below*).

Both standard dialysis by diffusion across cellulose tubing/resin-based column and centrifuge ultrafiltration (using centrifugal micro-concentrators) are reliable techniques with regard to MS standards for desalting, concentration, buffer exchange, and removal of low-molecular-weight impurities (see Note 5).

Common ESI-friendly reagents used in the approach are listed below:

1. Ammonium acetate (>98%, CH_3COONH_4).
2. Ammonium bicarbonate (>98%, NH_4HCO_3).
3. Ammonium trifluoroacetate (>99%, CF_3OONH_4).
4. Ammonium hydroxide solution (99.99%, NH_4OH).
5. Methanol (HPLC grade; CH_3OH).
6. Formic acid (98 + % pure, HCOOH).
7. Ethanol (absolute, ACS grade, C_2H_5OH).
8. Acetronitrile (HPLC grade; CH_3CN).
9. Acetic acid (99.99%, CH_3COOH).

2.2. Sample Preparation: Use of pH Modifiers and Alcohol as Denaturing Agents

1. Protein stock sample should at first be buffer exchanged/desalted by ultrafiltration (using *Vivacon, Amicon, Microcon* centrifugal concentrators) or dialysis into freshly prepared 10–20 mM ammonium acetate (*ca.* 0.8–8 mg/ml) solution with physiologically relevant pH factor. In average, 1×10^{-5}–2×10^{-5} mol of protein is usually required for the whole set of MS experiments.

2. Next, prepare a batch of protein samples of 250 μL volume each, at near-native and various (preselected) denaturing conditions, at fixed 2–10 μM protein concentration in a 10 mM (*ca.* 0.8 mg/ml) ammonium acetate solution. Adjust pH factor of these samples by using acetic acid, formic acid, or ammonium hydroxide. Allow the samples to equilibrate at room-temperature bath for an hour prior to MS.

2.3. Mass Spectrometry: Choice and Instrumental Settings

1. In principle, CSD analysis of the protein of interest could be done using any mass spectrometer with proper mass range. However, it is recommended only to run a mass spectrometer that (1) minimal bias across the entire m/z range (2) could be operated with no/minimal changes to the settings of the ion source and the mass analyzer ion optics within the entire range of conditions used in the data acquisition process (see Note 6). In general, targeted mass range will be lying between 500 and 6,000 m/z for proteins from 10 to 150 kDa in mass.

2. Optimize experimental settings (ion-source parameters, ion-transfer optics, detector voltage, etc.) and calibrate the instrument (see the instrument's *User Manual* for calibration procedure), using easy-to-handle and low-cost proteins, e.g., Ubiquitin (MW 8,565 Da), horse heart myoglobin (MW 16,951 Da), and any available protein, which is close in mass to the protein of interest. Optimization of instrumental parameters should be aimed at both improving signal intensity from a

protein and minimizing its fragmentation at all relevant solution conditions (see Note 7). All MS parameters (desolvating plate temperature, electrostatic potentials of ion-optics elements, etc.), once tuned, should be kept constant throughout the actual measurements to insure constancy in protein ion desorption and transmission conditions.

3. Methods

3.1. Data Acquisition and Chemometric Processing of CSD

1. Acquire each ESI-MS spectrum for all prepared samples (see Note 8) as long as needed to ensure high signal-to-noise (S/N) ratio.

2. Convert the array of acquired raw mass spectra into files with TXT or DAT format and upload them into ORIGIN-6 software.

3. Plot spectra, and normalize every spectrum by normalizing its most abundant peak to 100 (to facilitate application of Levenberg–Marquardt fitting algorithm, see below). Determine ionic intensities for each charge state of the spectra by integrating corresponding peak areas (or just their peak intensities for simplicity) in the ORIGIN.

4. Arrange calculated values as a matrix $K = M \times N$, where each column represents one of the N mass spectra measured under certain conditions. M is a number of charge states detected in all N experiments. For instance, a similar matrix 18×8 was constructed out of the set of experimental data obtained for α-synuclein protein (see Fig. 1).

5. Introduce this matrix with name K into MathCAD for SVD procedure. Execute MathCAD function svds(K), which returns a vector containing the singular values in decreasing order. For instance, it has resulted into (701, 75, 65, 22, 19, 14, 11, 6) vector for the above-mentioned matrix 18×8.

6. Transform the singular values into the percentage scale by normalizing their total sum to 100%; so, e.g., our vector transforms into (76.5%, 8.2%, 7.2%, 2.4%, 2.1%, 1.6%, 1.2%, 0.6%). Construct a total signal variance function $F = (F, i)$ from the resulted vector, where F represents cumulative variance accounted for the first i singular values and i—corresponding index numbers (see Table 1, total signal variance function from our example). The index number i, corresponding to approx $93\% < F \leq 95\%$, usually represents the number of principal components in the system, provided that all treated experimental data are accurate and have high S/N ratios (see Note 9). Hence, α-synuclein populates four distinct conformations

Table 1
Total signal variance function $F = (F, i)$ calculated from SVD of matrix K, constructed out of the set of experimental data obtained for α-synuclein (see Subheading 3 for more detail)

i	1	2	3	4	5	6	7	8
F (%)	76.5	84.7	91.9	94.3	96.4	98	99.2	99.8

under given solution conditions, together accounting for ≈94.3% of the total signal variance (see Table 1).

7. The deconvolution is done by Gaussian fitting using Levenberg–Marquardt algorithm, incorporated into ORIGIN-6 software. Plot all acquired mass spectra in the format of ionic intensities (Y-axis) against charge states (X-axis) using plot type "scatter" (a spectrum per graph). It is recommended to fit one spectrum at the time (see Note 10). Load one of the spectra (open its graph). Now, activate "Non-linear Curve Fitting" under "Analysis" tab. It will establish "Non-linear Curve Fitting" session. Choose "Gauss" in the "Origin Basic Functions" category. Specify the number of Gauss-functions by defining "Replicas," which should be equal to the number of principal components determined earlier by SVD. Set up "weighting method" as "Statistical" with χ^2. Finally, access the "Basic mode" and start fitting (see Note 11). The results of the deconvolution should meet both criteria: (a) both average charge and width values of basis Gaussians (principal components) should remain nearly the same throughout all spectra; (b) good χ^2 values. Both criteria are met in our example with α-synuclein protein (see Fig. 2).

4. Notes

1. In general, these "pure" spectra should be approximated by binomial distribution. Normal (or Gaussian) distribution, as an approximation to binomial distribution, is used to facilitate CSD analysis.

2. Sometimes changes in solvent composition may trigger certain gas phase processes, which also influence the extent of multiple charging of protein ions (12). However, in many cases it is possible to choose experimental conditions in such a way that

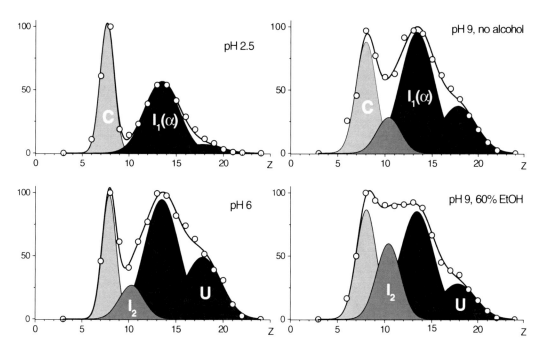

Fig. 2. CSD deconvolution results for the ESI-MS spectra of α-synuclein (shown in Fig. 1). Individual contributions from highly structured **C**, intermediate I_1, $I_2(\alpha)$, and "random coil" **U** states are determined using the chemometric approach (10). *Adapted with permission from* (10).

the influence of gas-phase ion-chemistry on protein ion CSDs is not altered over a wide range of solvent conditions (3).

3. For instance, in the report (10), correlating the abundance evolution of individual conformers of α-synuclein protein as a function of pH from ESI-MS with earlier spectroscopic measurements allowed conclusions to be made regarding the nature of the individual conformations.

4. Selection of solvent conditions could be based upon preliminary knowledge acquired from CD, fluorescence spectroscopy data, etc.

5. The simplest means of buffer exchange involves dialysis of a protein solution several times at 4°C against deionized water, following by its lyophilization and subsequent re-dissolution into a relevant solution for MS analysis. However, many proteins "crash out" of deionized water solution and precipitate. A loss of protein due to precipitation during solution exchange cannot be avoided in some cases, especially for proteins which are prone to aggregation, but can be decreased. Elimination/ minimization of protein-solution temperature changes at any time of buffer exchange (for example, between centrifuge runs) and use of lesser of protein concentration may reduce precipitation/loss of the protein. Thus, care should be taken during protein handling. Overall, centrifuge ultrafiltration is often

preferred over dialysis, because it is faster and better suited for low-amount protein samples. In our practice, we found that the buffer exchange of protein samples at high concentration of sucrose and/or glycerol with centrifuge ultrafiltration works much better than with standard dialysis for MS analysis.

6. For instance, mass spectrometers with trap-based analyzers may have these unwanted features due to effects like space-charge repulsion between ions.

7. It is easiest to start working on denatured protein samples. Most important parameters include flow rate, cone voltage (or de-clustering potential), de-solvating temperature, and electro-spray voltage.

8. It is recommended to thoroughly syringe-flush tubing and needle of electro-spray source between MS runs with 50%/ 50% of isopropyl alcohol/ultrapure water (≈ 1 ml) and then with 50%/50% of acetonitrile/ultrapure water (≈ 1 ml), following by 1–2 ml of ultrapure water.

9. The given range of F-values, obtained from empirical evidence, is below 100% due mostly to unavoidable presence of experimental noise/error.

10. Ideally, deconvolution should be performed in parallel for all ESI-MS spectra.

11. Deconvolution can be carried out in supervised, semi-supervised, and automatic modes. All parameters (in the ORIGIN format: x—average charge, w—width, A—area under Gaussian curve) could be "released" for variations or manually adjusted/fixed. Each parameter contains a "Vary" check box in "Basic mode" window. Select this check box to vary the parameter value during the iterative procedure. Otherwise, the parameter remains fixed at its current value.

Acknowledgements

This work was supported by a grant CHE-0750389 from the National Science Foundation.

References

1. Mohimen A, Dobo A, Hoerner JK, Kaltashov IA (2003) A chemometric approach to detection and characterization of multiple protein conformers in solution using electrospray ionization mass spectrometry. Anal Chem 75:4139–4147

2. Konermann L, Douglas DJ (1998) Equilibrium unfolding of proteins monitored by electrospray ionization mass spectrometry: distinguishing two-state from multi-state transitions. Rapid Commun Mass Spectrom 12:435–442

3. Frimpong AK, Abzalimov RR, Eyles SJ, Kaltashov IA (2007) Gas-phase interference-free analysis of protein ion charge-state distributions: detection of small-scale conformational transitions accompanying pepsin inactivation. Anal Chem 79:4154–4161

4. Invernizzi G, Grandori R (2007) Detection of the equilibrium folding intermediate of beta-lactoglobulin in the presence of trifluoroethanol by mass spectrometry. Rapid Commun Mass Spectrom 21:1049–1052

5. Zhang M, Gumerov DR, Kaltashov IA, Mason AB (2004) Indirect detection of protein-metal binding: interaction of serum transferrin with In^{3+} and Bi^{3+}. J Am Soc Mass Spectrom 15:1658–1664

6. Griffith WP, Kaltashov IA (2003) Highly asymmetric interactions between globin chains during hemoglobin assembly revealed by electrospray ionization mass spectrometry. Biochemistry 42:10024–10033

7. Simmons DA, Wilson DJ, Lajoie GA, Doherty-Kirby A, Konermann L (2004) Subunit disassembly and unfolding kinetics of hemoglobin studied by time-resolved electrospray mass spectrometry. Biochemistry 43:14792–14801

8. Gumerov DR, Dobo A, Kaltashov IA (2002) Protein-ion charge-state distributions in electrospray ionization mass spectrometry: distinguishing conformational contributions from masking effects. Eur J Mass Spectrom 8:123–129

9. Dobo A, Kaltashov IA (2001) Detection of multiple protein conformational ensembles in solution via deconvolution of charge state distributions in ESI MS. Anal Chem 73:4763–4773

10. Frimpong AK, Abzalimov RR, Uversky VN, Kaltashov IA (2010) Characterization of intrinsically disordered proteins with electrospray ionization mass spectrometry: conformational heterogeneity of α-synuclein. Proteins 78(3):714–722

11. Watkins DS (2005) Fundamentals of matrix computations, 2nd edn. http://onlinelibrary.wiley.com/book/10.1002/0471249718

12. Kaltashov IA, Eyles SJ, Mohimen A, Hoerner JK, Abzalimov RR, Griffith WP (2007) NMR and EPR Spectroscopies, Mass-Spectrometry and Protein Imaging, In: Uversky V, Eugene A. Permyakov (eds), Methods in Protein Structure and Stability Analysis. Nova science, U S A. ISBN 978-1-60021-705-0, 175–196

Chapter 25

Localizing Flexible Regions in Proteins Using Hydrogen–Deuterium Exchange Mass Spectrometry

Cedric E. Bobst and Igor A. Kaltashov

Abstract

Hydrogen–deuterium exchange (HDX) can provide invaluable structural information for proteins. The incorporation of deuterium into a protein's backbone amide is readily monitored by mass spectrometry (MS). Assuming that the molecular weight of the protein is not a limiting factor of the MS, HDXMS can be performed on intact proteins; however, digesting the protein prior to MS allows one to assign HDX information to specific peptides within the protein. Here, we describe HDXMS data collection and analysis to identify regions based on their degree of protection in the pharmaceutical protein glucocerebrosidase (GCase).

Key words: Hydrogen–deuterium exchange, Mass spectrometry, Protein conformation, Flexible loop, Backbone protection

1. Introduction

Monitoring the exchange of hydrogen for deuterium within a protein by mass spectrometry (HDXMS) coupled to high-performance liquid chromatography (HPLC) (1, 2) is becoming a widespread method for studying structural characteristics of proteins in solution. MS methods are well suited for measuring HDX along a protein's amide backbone due to the rate of exchange of this hydrogen. Other labile hydrogen atoms exchange much more rapidly by comparison and exchange back during sample processing and are therefore not observed by this technique. The rate of exchange of a backbone amide hydrogen is determined by its chemical environment and solvent accessibility. Labile hydrogen deeply buried within the protein core or involved in hydrogen bonding would require a substantial local or global conformational change before exchange is possible and therefore would be considered to have a high level of

Vladimir N. Uversky and A. Keith Dunker (eds.), *Intrinsically Disordered Protein Analysis: Volume 2, Methods and Experimental Tools*, Methods in Molecular Biology, vol. 896, DOI 10.1007/978-1-4614-3704-8_25, © Springer Science+Business Media New York 2012

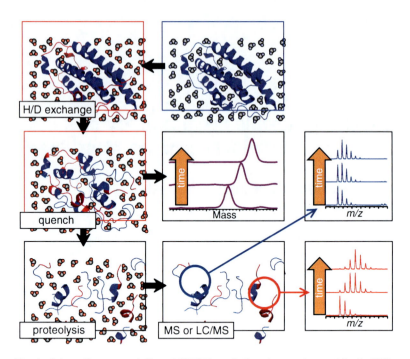

Fig. 1. Schematic representation of HDXMS work flow to examine protein flexibility. The exchange is initiated by placing the unlabeled protein into a D_2O-based solvent system (e.g., by a rapid dilution). Unstructured and highly dynamic protein segments undergo fast exchange (*blue* and *red* colors represent protons and deuterons, respectively). Following the quench step (rapid solution acidification and temperature drop), the protein loses its native conformation, but the spatial distribution of backbone amide protons and deuterons across the backbone is preserved (all labile hydrogen atoms at side chains undergo fast back-exchange at this step). Rapid cleanup followed by MS measurement of the protein mass reports the total number of backbone amide hydrogen atoms exchanged under native conditions (a global measure of the protein stability under native conditions), as long as the quench conditions are maintained during the sample workup and measurement. Alternatively, the protein can by digested under the quench conditions using acid-stable protease(s), and LC/MS analysis of masses of individual proteolytic fragments will provide information on the backbone protection of corresponding protein segments under the native conditions.

protection from HDX. Conversely, labile hydrogen located on the surface of the protein or not constrained by hydrogen bonding would be unprotected and readily undergo exchange. This ability to readily distinguish protein regions by their level of protection makes HDXMS an ideal technique for detecting intrinsic disorder.

In the typical workflow, as illustrated in Fig. 1, HDX is initiated by diluting the protein into a deuterated buffer of choice and allowed to incubate at a defined temperature. At various time points the rate of the HDX reaction is greatly reduced by rapidly lowering the temperature to ~0°C and adjusting the pH to ~2.6. These quench conditions slow down but do not completely halt exchange, therefore all steps following the quench must be performed in a rapid manner. The quenched samples can either be analyzed directly or first subjected to digestion by an acid stable protease. Identification of the proteolytic peptides allows localizing the HDX data to the protein's primary structure. The protein or produced peptides are then desalted and analyzed by LC-MS. Rates of HDX can be ascertained by monitoring the increase in mass of the protein or peptides over time.

Here, we demonstrate the utilization of HDX-MS by detecting flexible regions within the protein glucocerebrosidase (GCase) (3). To readily adapt this method, each step is described in a general manner and then specific details for the example data are given.

2. Materials

Prepare all solutions using ultrapure water and analytical grade reagents. Prepare all reagents at room temperature and store as appropriate.

2.1. HDX Components

1. Exchange Buffer: Prepare a 10–20× stock solution of preferred buffer ionic strength and pH in H_2O. Ensure that the desired pH is achieved upon dilution to 1× in H_2O. In the example data shown, a 20× stock of 1 M sodium citrate buffer was prepared that produced a pH of 4.5 when diluted to 1× in H_2O.

2. Quench Buffer: Using a pH meter, determine by titration the amount of formic acid (% v/v) required to adjust the pH of a known volume of 0.5× exchange buffer to a value of pH 2.4–2.8. A solution that has twice the concentration of formic acid used in the proceeding step will be the 2× quench solution. Additional components frequently added to the quench solution include varying concentrations of the denaturant guanidine hydrochloride (GndCl) and/or the reductant tris (2-carboxyethyl)phosphine (TCEP) (see Note 1). The 2× quench solution used in the sample data consisted of 2 % formic acid, 200 mM TCEP, and 2 M GdnCl.

3. Endpoint Buffer: Prepare a solution of formic acid such that can be added in the same volume as the exchange buffer to the exchange sample and yield the same final concentration as the quenched sample. In the sample data, 0.5 μL of an 80 % solution of formic acid was used.

4. Endpoint Quench Buffer: Prepare a solution to match the 2× sample quench solution except omit the formic acid and instead add exchange buffer to yield a final post-quench concentration that matches the exchange samples. A solution of

100 mM sodium citrate buffer, 200 mM TCEP, and 2 M GndCl was used for the sample data.

5. D$_2$O: Deuterium oxide can be used directly out of the bottle. In order to mitigate incorporation of atmospheric H$_2$O, it is recommended to aliquot large bottles of deuterium oxide upon first opening into smaller single use sized vials leaving a minimal air headspace.

6. Protein Sample: Prepare a 10–40× stock solution of the protein or proteins under study. Since the final volume is flexible (we routinely use volumes from 10 to 250 μL), the 1× final concentration will largely be a function of investigator's preference and instrument sensitivity and requires empirical optimization to determine. For the example data, we used 0.5 μL of a 65 μM protein stock concentration per time point.

2.2. HPLC Components

1. An HPLC system is used to provide the elution gradient using mobile phases A and B (see below and Fig. 2). A separate high pressure pump is also needed to provide flow using mobile phase C (see below Fig. 2) for the digestion/desalting step.

2. Mobile phases A: 0.1 % formic acid in water.

3. Mobile phase B: 0.2 % formic acid in acetonitrile.

4. Mobile phase C: 0.1 % formic acid in 2 % acetonitrile.

5. Separation column: An analytical sized (e.g., 2.1 × 50 mm) reverse phase (C$_8$ or C$_{18}$ are good choices for most peptides) HPLC column.

6. Protease: Either a stock solution prepared in mobile phase C or immobilized to a support matrix and packed in an empty column (see Note 2).

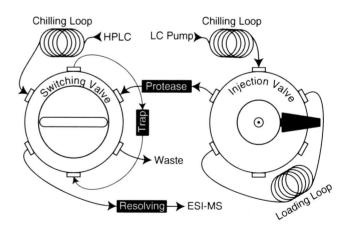

Fig. 2. Schematic representation of an HDX MS setup utilizing an immobilized protease column and a peptide trap prior to LCMS analysis. All lines, loops, and columns are immersed in an ice bath; the metal chilling loops introduce a dead volume to allow mobile phase sufficient time to equilibrate in the ice bath.

7. Peptide micro trap cartridge: An analytical sized peptide trap cartridge that is compatible with HPLC (e.g., Michrom Bioresources Inc, Auburn, CA, USA).

8. Six-port two-position switching valve, injection valve, and tubing as diagrammed in Fig. 2. Place the setup in a waterproof container (e.g., a styrofoam or plastic cooler) to allow immersion in an ice bath.

2.3. MS Component

An ESI-MS of adequate resolution (at least 5,000) and preferably capable of controlling an LC-MS system.

2.4. Software for Analysis and Display

In addition to MS software, spreadsheet software for analysis of data (i.e., Microcal Origin, Microsoft Excel or Open Office). If a structure has been released for the protein of interest then a pdb viewer (i.e., SPDB viewer (4) or Pymol (Schrödinger, LLC, New York, NY)) that allows for coloring of data is very helpful interpreting and comparing results. There are also various programs freely available to assist in analysis of HDXMS (HD Desktop (5), Hydra (6), HX-Express (7)) and LC MS/MS (Mascot (8), Protein Prospector (9)) data.

3. Methods

For the highest quality data, many of the steps below require optimization of various parameters for each protein or condition being studied. Collection of large HDX data sets should be delayed until optimal conditions have been determined. The parameters listed for the example data can be used as a starting point. Once conditions have been established it is recommended to collect data for nonexchanged protein utilizing LC-MS/MS data collection in combination with a long elution gradient. This will allow more high confidence peptide identifications. In addition to a nonexchanged control, it is recommended that an endpoint control is prepared (see Note 3).

3.1. HDX of the Protein

1. Prepare the deuterated exchange buffer (see Note 4) by diluting to an appropriate volume with D_2O. Equilibrate to the desired temperature. In the example data, a separate exchange reaction was prepared in triplicate for each time point. In a 1.5-ml Eppendorf tube, 0.5 μL of the 20× buffer was added to 19 μL of D_2O, mixed by aspirating, capped, and placed in a water bath set to 25°C and allowed to equilibrate for at least 5 min.

2. Initiate HDX by diluting the protein stock solution (typically 10- to 20-fold) into the deuterated exchange buffer and allow

to incubate at set temperature until ready for the quench step (see Note 5). For the example data, 0.5 μL of a 65 μM protein stock was added to the deuterated buffer.

3. Prepare endpoint control sample by diluting both the protein sample and the endpoint exchange buffer to the appropriate volume with D_2O. This is incubated at a similar temperature as the experimental time point samples for as long as the longest time point or longer.

3.2. Quenching HDX

1. Equilibrate quench solution on ice.

2. At the desired exchange time the sample is transferred to an ice bath and allowed to cool for 30 s. From this point forward carry out all procedures on ice in a timely fashion.

3. Add the quench solution and mix by aspirating several times (see Note 6). In the example data 20 μL of quench solution was added to the sample.

4. The quenched sample can either be injected for analysis or flash-frozen and stored at −80°C until analysis can be performed (see Note 7).

3.3. LC-MS Data Collection

1. Equilibrate the setup from Fig. 2 in ice and position as close to the ESI source as possible in order to minimize warming of the elution solvents and delay time. Equilibrate column and trap by running mobile phase at the optimized initial flow rates and composition. Ensure that the loading valve is in the load position to allow sample injected into the injection port to enter the loading loop. Ensure that the switching valve is set to allow flow from the digestion to enter the trap column.

2a. Digestion with protease in solution (*option 1*). For a solution digest, add the protease to the quenched sample, mix by aspirating, and then inject into the loading loop. The extent of digestion is controlled by incubation time in the loading loop (typically 1–3 min) and the protease to substrate ratio (typically 1:20 to 1:100). Both of these parameters should be optimized for each protein under study. Trigger the injection valve from the load to inject position to halt the digest and initiate peptide trapping and desalting.

2b. Digestion with immobilized protease (*option 2*). Carefully inject quenched sample into loading loop. If a preincubation of the quenched sample has been determined advantageous (increased reduction of disulfides by TCEP or enhanced digest by longer denaturing time), this can be performed in the sample tube on ice or in the loading loop. Protein digestion is initiated upon triggering the injection valve from the load to inject position. The extent of digest is controlled by the flow rate of mobile phase C, which should be optimized for each protein under study. After passing through the immobilized protease column, peptides flow into the peptide trap. In the example data, the mobile phase C flow rate was 0.1 mL/min and the 40 μL quenched sample volume was allowed 1 min to pass through the pepsin column.

3. Trapping and desalting of the peptides. Trapping of the peptides occurs automatically. The peptides should be washed with enough mobile phase C to remove most of the buffer and quench components present in the sample. Typically 1–2 min at a flow rate of 0.1 mL/min is adequate to wash a 1 mm × 8 mm microtrap. In the example data 2 min were allocated to this step.

4. Elution of peptides and detection by ESI MS. The switching valve is triggered, effectively placing the peptide trap in line with the resolving column.

5. Initiate the optimized HPLC elution gradient. For the sample data the following elution gradient was used: equilibrate with 2 % B, 0.25 mL/min; 0 min, 20 %B, 0.25 mL/min; 3.7 min, 0.25 mL/min; 3.8 min, 0.15 mL/min; 6 min, 35 %B; 8 min, 0.15 mL/min, 95 %B; 9 min, 95 %B; 10 min, 5 %B (see Note 8).

6. Initiate MS data collection. MS data collection can begin shortly prior to the first detected peptides begin to elute. In the sample data, MS collection was initiated 2 min after the switching valve was triggered.

3.4. LC-MS Data Analysis

The data analysis phase is the most time-consuming. We describe below the common methodology we employ in the analysis of all our data.

1. Peptide mapping. The first stage of data analysis should be to identify as many peptides as possible. This is typically performed as a preliminary stage to HDX data collection to determine whether the sequence coverage is adequate to provide useful structural information. Since commonly employed acid proteases are relatively nonspecific, making identification with high confidence often requires MS/MS data. Compile a list of peptides and their masses.

2. Data extraction. In this step, the average mass for each peptide of interest is extracted from the raw data file. The precise manner of calculating the average mass will vary by MS software. These values can be copied into a template worksheet formatted to aid in the next step.

3. Data reduction. For each peptide, we have an amino acid sequence, a nonexchanged average mass, a series of average masses at different exchange times, and an endpoint mass. A common way to condense the data is by calculating the increase in mass versus time for each peptide (see Fig. 3, lower plots).

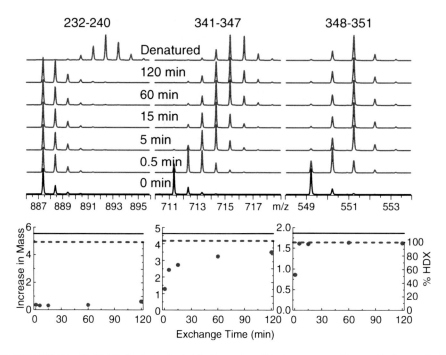

Fig. 3. Example HDX data displaying the isotopic distributions of peptide ions representing peptic fragments (232–240, 341–347 and 348–351) of GCase following 0.5, 5, 15, 60, and 120 min of HDX in solution as well as the denatured endpoint (*upper plots*). Kinetic plots of the same data (*lower plots*) expressing the exchange of backbone amide hydrogen atoms within these peptides as increase in mass or as %HDX. The *solid line* represents the theoretical maximum HDX for each peptide and the *dotted line* represents the endpoint measured HDX.

These values can be normalized using the mass increase measured for the endpoint sample to express their values as percentage exchange. Comparing the endpoint values with the calculated maximum exchange will allow a determination of back exchange (see Note 9). Back exchange values that are consistently large (greater than 40 %) indicate a considerable degree of back exchange is occurring during sample workup and steps should be taken to ensure the primary factors (time, pH, and temperature) that contribute to back exchange are controlled favorably.

4. Data presentation/summary. Data can be presented in tabular format as a list of values, although a more meaningful result can be obtained by aligning the data to the amino acid sequence of the protein (see Fig. 4). This graphical format readily conveys the degree of sequence coverage and allows for quick comparison of HDX at known key locations. This format can be used to present data for multiple time points and to compare HDX behavior for different samples. The results can also be applied to a three-dimensional structure, if one is available, using a color or gray scale to indicate HDX at a single time point.

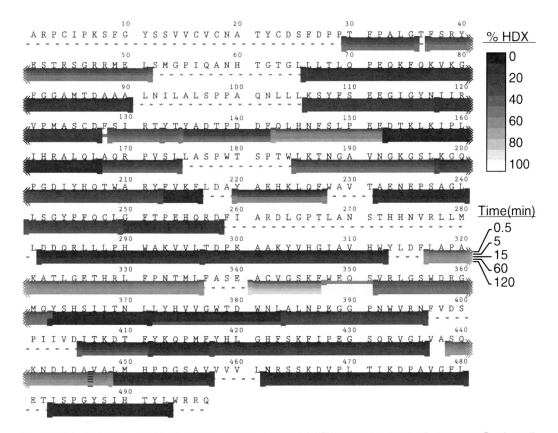

Fig. 4. HDX flexibility data, expressed as back exchange corrected %HDX, mapped to the protein sequence. *Bars* beneath the amino acid sequence represent peptides for which HDX data was extracted. Each *row of bars* represents a separate HDX time point (0.5, 5, 15, 60, or 120 min). Highly flexible regions are indicated by *lightly shaded bars*.

4. Notes

1. Reduction of disulfides under the low pH required for quenching HDX remains problematic. It can be achieved usually with limited success using a high concentration of TCEP (limited by its solubility) in the quench. When using a denaturant such as GdnCl, the concentration should be limited to avoid self-proteolysis by the protease. We begin to observe this with immobilized pepsin when the GdnCl concentration exceeds 2.5 M.

2. There is currently a limited selection of commercially available acid proteases with pepsin by far being the most popular choice. An immobilized protease column can be prepared in house (10) or purchased commercially. Though more costly, an immobilized protease adds a level of convenience and ensures reproducibility of digest from run to run.

3. In addition to a nonexchanged control it is recommended an endpoint control is prepared. Analyzing a fully exchanged endpoint sample allows the investigator to evaluate their method and setup by determining the extent of back exchange. While any method that destabilizes a protein structure could be used to produce near complete deuterium exchange within the protein, we have had considerable success with most proteins using acid to unfold the protein. A convenient way to perform this while maintaining identical conditions with the time point exchange samples prior to the digest is to trade the acid in the quench for the buffer in the exchange. This is what is described in the above method.

4. The shortest time point that can be collected reliably will depend on the speed of sample manipulation, typically ~10 s if the mixing is done by hand. The longest time point can vary greatly from 1 h to several days and will depend on the aim of the experiment and stability of the protein being studied. If collecting fewer time points it is recommended to collect replicates so a measure of standard error can be determined. If fitting HDX data to exponential functions for calculation of exchange rates, then a larger number of time points may be preferable and replicates of only one time point may be feasible.

5. Consistency of time and temperature during sample handling steps is a key factor in minimizing error.

6. An incubation of 1–5 min could be favorable at this point if reduction of disulfides is desired. This should be determined empirically prior to collecting a full data set.

7. If flash freezing the quenched samples for storage at −80°C, it is recommended to add glycerol to a final concentration of 10 %. This will speed up subsequent thawing of the sample on ice (to ~5 min) prior to analysis.

8. It is useful to know the dead volume within the HPLC. The gradient used in the example data was optimized for an Agilent 1100 system with a dead volume of approximately 0.5 ml. The initial higher flow rate of 0.25 mL/min decreases the amount of time for the acetonitrile portion of the gradient to reach the column. Then, decreasing the flow to 0.15 mL/min when the peptides begin to elute allows the MS more time to collect data.

9. When calculating back exchange for a peptide it is typically assumed that the first amide hydrogen is lost during sample workup due to this hydrogen's accelerated exchange rate (11).

References

1. Smith DL, Deng Y, Zhang Z (1997) Probing the non-covalent structure of proteins by amide hydrogen exchange and mass spectrometry. J Mass Spectrom 32:135–146

2. Zhang Z, Smith DL (1993) Determination of amide hydrogen exchange by mass spectrometry: a new tool for protein structure elucidation. Protein Sci 2:522–531

3. Bobst CE, Thomas JJ, Salinas PA, Savickas P, Kaltashov IA (2010) Impact of oxidation on protein therapeutics: conformational dynamics of intact and oxidized acid-beta-glucocerebrosidase at near-physiological pH. Protein Sci 19:2366–2378

4. Kaplan W, Littlejohn TG (2001) Swiss-PDB Viewer (Deep View). Brief Bioinform 2:195–197

5. Pascal BD, Chalmers MJ, Busby SA, Griffin PR (2009) HD desktop: an integrated platform for the analysis and visualization of H/D exchange data. J Am Soc Mass Spectrom 20:601–610

6. Slysz GW, Baker CA, Bozsa BM, Dang A, Percy AJ, Bennett M, Schriemer DC (2009) Hydra: software for tailored processing of H/D exchange data from MS or tandem MS analyses. BMC Bioinformatics 10:162

7. Weis DD, Engen JR, Kass IJ (2006) Semi-automated data processing of hydrogen exchange mass spectra using HX-Express. J Am Soc Mass Spectrom 17:1700–1703

8. Perkins DN, Pappin DJC, Creasy DM, Cottrell JS (1999) Probability-based protein identification by searching sequence databases using mass spectrometry data. Electrophoresis 20:3551–3567

9. Clauser KR, Baker P, Burlingame AL (1999) Role of accurate mass measurement (+/−10 ppm) in protein identification strategies employing MS or MS MS and database searching. Anal Chem 71:2871–2882

10. Wang L, Pan H, Smith DL (2002) Hydrogen exchange-mass spectrometry: optimization of digestion conditions. Mol Cell Proteomics 1:132–138

11. Bai Y, Milne JS, Mayne L, Englander SW (1993) Primary structure effects on peptide group hydrogen exchange. Proteins 17:75–86

Chapter 26

Mass Spectrometry Tools for Analysis of Intermolecular Interactions

Jared R. Auclair, Mohan Somasundaran, Karin M. Green, James E. Evans, Celia A. Schiffer, Dagmar Ringe, Gregory A. Petsko, and Jeffrey N. Agar

Abstract

The small quantities of protein required for mass spectrometry (MS) make it a powerful tool to detect binding (protein–protein, protein–small molecule, etc.) of proteins that are difficult to express in large quantities, as is the case for many intrinsically disordered proteins. Chemical cross-linking, proteolysis, and MS analysis, combined, are a powerful tool for the identification of binding domains. Here, we present a traditional approach to determine protein–protein interaction binding sites using heavy water (^{18}O) as a label. This technique is relatively inexpensive and can be performed on any mass spectrometer without specialized software.

Key words: Mass spectrometry, Cross-linking, In-gel digest, Heavy (^{18}O) water

1. Introduction

Intrinsically disordered regions and proteins by their disordered nature make it inherently difficult to crystallize and thus determine their three dimensional structures as well as identifying intermolecular interactions (1). Mass spectrometry (MS) is a powerful tool that has become indispensible for proteomic studies, affording the identification of thousands of proteins in a single experiment (2). In recent years, new mass spectrometry techniques have been developed to move beyond protein identification and analyze the binding domains formed during protein–protein interactions. One such approach has been to use chemical cross-linking in conjunction with proteolysis and mass spectrometric analysis (3–9). This approach takes advantage of several characteristics of MS measurements (10): (1) Cross-linkers with different spacer arms and chemistries have inherently different masses and can give insight into

Vladimir N. Uversky and A. Keith Dunker (eds.), *Intrinsically Disordered Protein Analysis:*
Volume 2, Methods and Experimental Tools, Methods in Molecular Biology, vol. 896,
DOI 10.1007/978-1-4614-3704-8_26, © Springer Science+Business Media New York 2012

distance constraints and thus highlight networks of nearby residues. (2) The mass of the protein complex is not limiting because proteolysis precedes mass measurement, resulting in peptides that are tractable for LC-MS/MS analysis. (3) Only nanomoles, and in some cases as little as femtomoles, of protein are needed. (4) Cross-linking is conducted in solution, therefore flexible regions of proteins can undergo any necessary disorder-to-order transitions in order to form intermolecular interactions (10, 11).

Although chemical cross-linking-MS is a powerful approach to identifying protein binding domains; it is also a difficult one. The major problem associated with cross-linking mass spectrometry is that it increases the complexity of the sample by orders of magnitude. Take for example, a 10 kDa protein, which would produce roughly ten 1 kDa peptides following trypsin proteolyis. After intramolecular cross-linking, however, the number of potential cross-linked peptides increases from 10 to 10! (~3.5 million). If intermolecular cross-linking with an entire proteome is considered, this number becomes astronomical. In either case, traditional methods for matching an experimental mass to database derived theoretical masses are no longer sufficient, and novel techniques are required to reduce the complexity (e.g., purification of proteins and the use of isotopically labeled cross-linkers). The use of isotopically labeled cross-linkers allows the experimenter to discern which of the "peaks" in a mass spectrum arise from cross-linked peptides, so that they may concentrate efforts on these species (11, 12).

In addition to cross-linking-MS other strategies exist, of which we mention a few, to identify protein intermolecular interactions such as traditional proteomic strategies. For example, coimmunoprecipitation or pull down assays can be used in combination with denaturing or 2D gel electrophoresis and mass spectrometry to identify protein–protein interactions (reviewed in (13)). These proteomic studies can also be performed on a large scale, or high throughput, using such things as tandem-affinity purification (TAP) mass spectrometry (14, 15). Alternatively, techniques such as stable isotopic amino acids in cell culture (SILAC) and immunoprecipitation (IP) can be used to identify protein binding partners where isotopic distributions are used to rule out background interactions (16).

Here, we present a traditional cross-linking and mass-spectrometry approach, in which we use heavy water (^{18}O) as an isotopic label (17, 18). In order to obtain detailed information about binding domains such as residues involved and their proximity to each other cross-linking is advantageous. Specifically, cross-links between two peptides within the same protein provide distance/structural constraints, whereas those between two peptides in different proteins (including respective monomers of oligomeric proteins) provide a region of intermolecular interaction or binding (11, 19).

2. Materials

Use HPLC grade reagents (water, acetonitrile, etc.) for all steps following the SDS-PAGE.

2.1. Cross-Linking

1. EDC (1-ethyl-3-[3-dimethylaminopropyl] carbodiimide) and sulfo-NHS (both obtained from Thermo Pierce) are prepared as 0.1 M stocks. To make a 0.1 M stock of EDC dissolve 1.6 g EDC in milliQ water and to make a 0.1 M stock of sulfo-NHS dissolve 2.2 g sulfo-NHS in DMSO.

2. Activation Buffer: 0.1 M MES, pH 6.0, 0.1 M sodium chloride (Dissolve 2.0 g MES and 0.6 g sodium chloride in 100 mL milliQ water).

2.2. Sample Preparation for Mass Spectrometry

1. Precast 16 % acrylamide SDS PAGE (Novex; Invitrogen). Lower percentages of acrylamide may be used for larger proteins, and user-casting of gels is optional.

2. SDS Sample Buffer: Add 2.5 mL 0.5 M Tris–HCl, pH 6.8; 2 mL glycerol; 4 mL 10 % SDS; 0.5 mL 0.1 % (w/v) Bromophenol blue; 0.5 mL 2-mercaptoethanol; and 0.5 mL milliQ water.

3. Coomassie Brilliant Blue Stain. Add 100 mL of glacial acetic acid to 450 mL of milliQ water, dissolve 3 g of Coomassie brilliant blue R250 in 450 mL of methanol, combine the two and filter-sterilize.

4. Destain: 40 % methanol, 10 % glacial acetic acid, 50 % milliQ water. To make 2 L combine 800 mL of methanol, 200 mL of glacial acetic acid, and 1,000 L of milliQ water. Mix.

5. Heavy Water ($^1H_2^{18}O$): Cambridge Isotopes (product number: OLM-240-97).

6. 8 ng/µL Trypsin (Princeton Separations; EN-151).

7. Wash buffer: 50 mM ammonium bicarbonate in 50 % ethanol. Dissolve 0.4 g of ammonium bicarbonate in 50 mL milliQ water and 50 mL ethanol.

8. 100 % HPLC grade ethanol (sigma).

9. 50 mM ammonium bicarbonate and 10 mM DTT: Add 101.7 mg of DTT to 10 mL 50 mM ammonium bicarbonate, pH 8.0. *Critical Step*: Make shortly before use.

10. 55 mM iodoacetamide in 50 mM ammonium bicarbonate, pH 8.0: Add 15.4 mg of iodoacetamide to 10 mL 50 mM ammonium bicarbonate. *Critical Step*: Make shortly before use.

11. Extraction buffer: 50 mM ammonium bicarbonate buffer, pH 8.0 and 50 % N, N-dimethyl formamide.

12. 2 % acetonitrile, 0.1 % TFA. To make a 50 mL stock, dilute 1 mL 100 % acetonitrile, 0.1 % TFA (sigma) into 49 mL of HPLC water, 0.1 % TFA (sigma).

13. Omix C18 Tip (Agilent).

14. 50 % acetonitrile. Dilute 50 mL HPLC grade 100 % acetonitrile (sigma) with 50 mL HPLC grade water (Sigma) and store in a glass vial.

15. 50 % acetonitrile, 0.1 % TFA. Dilute 50 mL HPLC grade 100 % acetonitrile, 0.1 % TFA (Sigma) with 50 mL HPLC grade water, 0.1 % TFA (Sigma) and store in a glass vial.

16. 0.1 % TFA.

2.3. Mass Spectrometry and Data Analysis

1. Any mass spectrometer. The experiments shown here employ MALDI-TOF Microflex (Bruker Daltonics) and a 9.4 T Apex MALDI-ESI-Q_e-FTICR (Bruker Daltonics).

2. 10 mg/mL α-cyano-4-hydroxycinnamic acid matrix (α-cyano) in 50 % acetonitrile, 0.1 % TFA. Weigh approximately 40 mg of α-cyano and resuspend in 40 μL 50 % acetonitrile, 0.1 % TFA.

3. Mobile phase: 100 % acetonitrile, 0.1 % FA and HPLC water, 0.1 % FA (Sigma).

4. Solid phase: NS-AC-10-C18 Biosphere C18 5 μm column (Nanoseparations).

2.4. Data Analysis

1. Flex Analysis (Bruker).

2. GPMAW (General Protein Mass Analysis for Windows) (Lighthouse Data).

3. Data Analysis (Bruker).

4. ProteinProspector (UCSF; http://prospector.ucsf.edu/prospector/mshome.htm).

3. Methods

Be sure to dispose of all hazardous chemicals appropriately. Be sure to wear the appropriate personal protective equipment as outlined in MSDS. Wear gloves to avoid keratin contamination in mass spectrometric analysis. A work flow of mass spectrometry analysis of protein binding is outlined in Fig. 1.

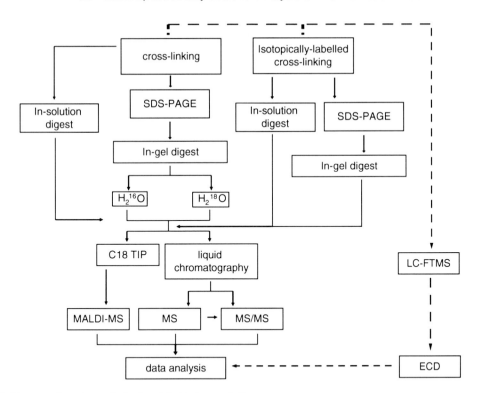

Fig. 1. Mass spectrometry analysis of protein binding work-flow.

3.1. Cross-Linking

1. Dilute protein(s) of interest to approximately 1 mg/mL in 50 μL activation buffer. If the stock protein concentration is approximately 1 mg/mL to start dialyze into activation buffer or buffer exchange using size exclusion chromatography.

2. Add 1 μL 0.1 M EDC stock, giving a final concentration of 2 mM, and 2.5 μL 0.1 M sulfo-NHS, giving a final concentration of 5 mM, to the protein solution (see Note 1). Vortex the samples slightly and centrifuge at 3,000 RPM (837 rcf) for 30 s.

3. Incubate the cross-linking reaction at room temperature for at least 15 min (see Note 2). The extent (or completeness) of cross-linking can be monitored using MALDI-TOF or SDS PAGE analysis.

3.2. Sample Preparation: SDS PAGE and In-Gel Digest

1. Incubate a non-cross-linked control and all cross-linked samples with an equal volume of 2× sample loading buffer containing 2-mercaptoethanol at 95°C for 5 min, which quenches the cross-linking reaction.

2. Load 10–100 μL sample (if using 100 μL load into ten different lanes; 10 μL per lane) onto a 16 % SDS PAGE gel and run at 200 V until the dye front is a few centimeters from the bottom of the gel (approximately 45 min for a mini-gel apparatus).

3. Following electrophoresis, pry the plates apart using a spatula, remove the gel taking caution not to tear it, and stain in a plastic container with Coomassie brilliant blue stain for a few hours on a rocker (see Note 3).

4. Decant the Coomassie stain (save for reuse) and destain the gel in 50 % water, 40 % methanol and 10 % acetic acid in the same plastic container while will rocking over night. Change destain as needed until the gel is completely destained.

5. Place the destained gel on a solid glass/plastic surface that has been wiped down with 100 % acetonitrile and excise each band of interest using a new razor blade for each unique complex isolated (alternatively wipe clean the razor blade with 100 % acetonitrile between each use).

6. Cut each excised band into smaller pieces and place in a 1.5-mL eppendorf tube (precleaned with acetonitrile and methanol) for in-gel trypsin digestion (20) (see Note 4).

7. Wash gel slices twice with 50 mM ammonium bicarbonate in 50 % ethanol at room temperature. Add enough 50 mM ammonium bicarbonate in 50 % ethanol to each respective eppendorf tube to completely cover the gel slices (typically 50–100 µL), pipet up and down a few times, and aspirate using a pipette. Alternatively, gel slices can be vortexed to wash and it is not necessary to remove all the Coomassie stain from the gel slices.

8. Shrink gel slices by adding enough 100 % ethanol to cover them (typically 50–100 µL), aspirate using a pipette, and then incubate for 1 h at 56°C with approximately 50 µL 50 mM ammonium bicarbonate containing 10 mM DTT.

9. Warm the gel slices to room temperature, aspirate the DTT solution, and incubate with 55 mM iodoacetamide for 30 min in the dark.

10. Aspirate the iodoacetamide, wash using 50 mM ammonium bicarbonate in 50 % ethanol, shrink using 100 % ethanol, and dry using a Vacufuge (speedvac).

11. Split each respective gel slice into two tubes and digest overnight at 37°C with 1 µL 8 ng/µL Trypsin in the presence of ^{16}O or ^{18}O water, respectively (see Note 5).

12. Extract peptides using 50 µL 50 mM ammonium bicarbonate and 50 % N, N-dimethyl formamide.

13. Evaporate peptides to dryness using a speedvac (Vacufuge) and resuspend in 10 µL of 2 % acetonitrile, 0.1 % trifluoroacetic acid.

14. Extract peptides from the ^{16}O digest and mix with peptides extracted from the ^{18}O digest.

3.3. C18 Purification and MALDI-MS

1. Purify the above peptide samples using a 10 μL Omix C18 Ziptip (Varian/Agiliant).

2. Prewash the C18 Omix tip with 100 % acetonitrile and equilibrate with 0.1 % TFA.

3. To bind the peptides to the C18 tip, aspirate peptides approximately 30 times and any unbound or weakly bound peptides are removed by three 10 μL 0.1 % TFA washes.

4. Elute peptides by aspirating 5 μL of 50 % acetonitrile, 0.1 % TFA, three times into an eppendorf tube.

5. Then wash the C18 tip with 100 % acetonitrile and reuse.

6. Make a 10 mg/mL α-cyano-4-hydroxycinnamic acid (HCCA, α-cyano) (Sigma) by measuring out 40 mg of HCCA and diluting it in 40 μL of 50 % acetonitrile, 0.1 % TFA.

7. Spot 1 μL of the C18 tip eluate on a MALDI-TOF target with 1 μL of HCCA and air dry (or alternatively dried using a bench top fan).

8. Load the MALDI-TOF target into a Bruker Daltonics Microflex operated in reflectron mode.

9. Calibrate the MALDI-TOF using a peptide calibration standard (Bruker) with each use.

10. Spectra are the average of 500–1,000 laser shots and are collected using 50–70 % laser power.

11. Collect data over a range of 400–6,000 m/z and analyze using Flex Analysis software (Bruker Daltonics). Data are collected to 6,000 m/z in order to be sure we isolate any potentially large cross-linked peptides.

3.4. LC-MS/MS

1. Aspirate the digested sample (1 μL), no C18 tip purification, into an Eksigent 2D UPLC or Proxeon 1D HPLC system onto a NS-AC-10-C18 Biosphere C18 5 μm column (see Note 6).

2. The following gradient is used to elute peptides: 3–50 % buffer B for 30 min, 50–95 % buffer B for 7 min, 95 % buffer B for 5 min, 95–3 % buffer B for 1 min, and 3 % buffer B for 15 min. Buffer A is HPLC water + 0.1 % FA and buffer B is acetonitrile + 0.1 % FA (both Sigma).

3. Eluted peptides from the above C18 column are introduced into a 9.4 T FTICR (Bruker Daltonics) using a nanospray source.

4. Collect mass spectra (MS data) during the elution using the Apex Control Expert software (Bruker Daltonics) in chromatography mode over the length of the gradient (approximately 70 min).

5. Data dependent MS/MS data are collected using collision-induced dissociation (CID). One full scan (MS) is followed by MS/MS product ion spectra of the four most intense ions

and sequenced; once an ion is selected for MS/MS analysis it is added to an exclusion list to prevent it from being selected numerous times.

3.5. Data Analysis

1. Use GPMAW (General Protein Mass Analysis for Windows) (Lighthouse Data) software for the prediction of theoretical m/z's for both cross-linked and non-cross-linked samples (see Note 7).

2. Analyze MALDI-TOF data using Flex Analysis software (Bruker Daltonics) and FTICR data using Data Analysis software (Bruker Daltonics).

3. Analyze each individual sample (data set) and generate a mass list of monoisotopic masses.

4. Identify peptides and cross-linked peptides using the following criteria:

   a. Trypsin-specific cleavages are present (arginine or lysine).

   b. Contain a potentially cross-linkable residue. For example, a lysine would be present in both peptides involved in the cross-link if an amine cross-linker is used.

   c. m/z specific to a cross-linked sample.

   d. The m/z is reproducibly observed in duplicate or triplicate experiments.

   e. Intramolecular cross-links are identified by the presence of a unique m/z in monomer alone cross-linked samples.

   f. Intermolecular cross-links are seen in only Protein A and B cross-linked samples.

   g. Two water molecules are added per peptide in trypsin digests. Therefore, for a non-cross-linked peptide the m/z will shift +4 Da in the presence of heavy water and for a cross-linked peptide the m/z will shift +8 Da in the presence of heavy water (see Fig. 2).

   h. When possible, MS/MS data provides sequence data for a cross-linked peptide. Cross-linking complicates MS analysis, and thus obtaining MS/MS sequence data of a cross-linked peptide is often not possible; however, the reproducibility and heavy water label is sufficient to confirm cross-linked peptides.

   Two important criteria above being the reproducibility of the data and the presence of a +4 Da shift for a peptide and +8 Da shift for a cross-linked peptide.

5. After identification of cross-links using the above criteria, map the data back onto the respective primary structure (sequence) of each respective protein, thus highlighting the binding domains.

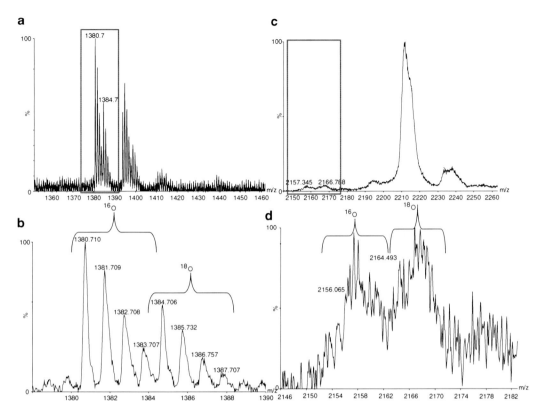

Fig. 2. Heavy water labeled peptides and cross-linked peptides. (**a**) Mass spectra of a peptide; box zoomed in (**b**). (**b**) An unlabelled m/z at 1,380.7 and its labeled counterpart at 1,384.7; the shift in mass of four indicates a non-cross-linked peptide. (**c**) Mass Spectra of a cross-linked peptide (box; zoomed in (**d**)). Analyzing the spectra by hand allows for the majority of cross-linked pairs to be identified. (**d**) An unlabelled m/z at 2,156.1 and its labeled counterpart at 2,164.5; the shift in mass of eight indicates a cross-linked peptide. The cross-linking complicates the spectral analysis; however, it is still possible to distinguish the pair of related m/z values based on the isotopic label.

4. Notes

1. Cross-linking: multiple cross-linkers with different spacer arms can be used to give an idea of the distance at which one residue is from another.

2. Cross-linking can be performed on ice or at 4°C; however, the cross-linking reaction is slowed and therefore incubation times should be at least an hour and preferably overnight. In addition, if an SDS-PAGE is not run, 2-mercaptoethanol can be used to quench the cross-linking reaction.

3. Silver staining is not as amenable to mass spectrometry analysis as Coomassie staining; therefore, if possible it should not be used.

4. If the samples being tested are pure enough, an in-solution trypsin digest can be preformed instead of and in-gel trypsin digest.

5. Isotopically labeled cross-linkers/biotinylated cross-linkers can be used instead of heavy water as a label to aide in cross-linked peptide identification (e.g., deuterated cross-linkers—Pierce).

6. Samples that have been C18 tip purified should be dried to completeness using a speed vac (Vacufuge) and resuspended in 10 μL of 2 % acetonitrile, 0.1 % trifluoroacetic acid prior to LC-MS/MS analysis. This is done because the ZipTip elution buffer is 50 % ACN, 0.1 % TFA which will prevent peptide binding to the C18 LC column.

7. Peptides and cross-links are identified using Data Analysis software (Bruker Daltonics) and exported to a generic mascot file. The MASCOT file is uploaded into the MASCOT search engine selecting none as the enzyme, NCBIr as the database, with a 1.2 Da (MS error tolerance) and a 0.6 Da (MS/MS error tolerance). Cross-linked and non-cross-linked sample data are compared and m/z's unique to the cross-linked sample are analyzed further using MS Bridge (ProteinProspector, UCSF, http://prospector.ucsf.edu/prospector/mshome.htm) with the specific cross-linker molecular weight and sequence inputed by the user. MS Bridge performs an in silico digest of a cross-linked protein generating m/z's for theoretical peptides and cross-linked peptides taking into account the molecular weight of the cross-linker. Potential cross-linked peptides can be further characterized using extracted ion chromatograms and MS/MS data.

8. One can also detect noncovalent binding of proteins and compounds using "soft ionization" techniques (electrospray ionization) and the Fourier transform mass spectrometer (FTMS) (reviewed in (21)). Purified protein is incubated with four unique compounds for 30 min on ice. After incubation the protein-compound complex is directly infused into a Bruker Daltonics. 4 T FTICR using a nanosource. The protein–compound complex is sprayed in 10 mM ammonium acetate, pH 6.8 with a skimmer one voltage between 20 and 40 V; 75 spectra are averaged together and the protein alone control is compared to the protein–compound spectra for the addition of any new peaks consistent with the molecular weight of a complex (see Fig. 3).

9. In addition to cross-linking and noncovalent binding one can also use a "footprinting" type of experiment (22). First, biotinylate protein A and B, separately, and detect sites of biotinylation using LC-MS/MS. Then incubate protein A with protein B for 30–60 min (or longer) and then biotinylate and identify sites of biotinylation using LC-MS/MS. Any sites of biotinylation in the complex that are lost indicate a binding surface.

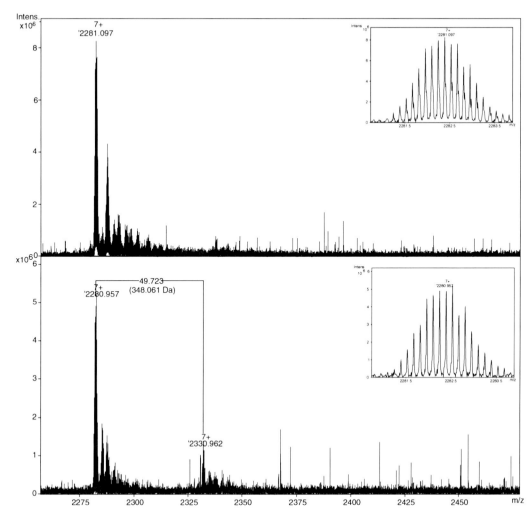

Fig. 3. Noncovalent Interactions using the FTICR. (*Top*) Spectra of protein alone. (*Bottom*) Protein incubated with a panel of small molecules. The addition of a new peak indicates protein–compound complex (noncovalent interaction).

Acknowledgement

This work was supported in part by a National Institutes of Health R21 (to J.N.A.) (1R21NS071256), R21 (to C.A.S.) (R21 A1067021), P01 (to C.A.S.) (P01 GM091743), GM 32415 and GM 26788 (to G.A.P. and D.R.), and Fidelity Biosciences Research Initiative (to G.A.P. and D.R.).

References

1. Dunker AK et al (2001) Intrinsically disordered protein. (Translated from eng). J Mol Graph Model 19(1):26–59 (in eng)

2. Yates JR, Ruse CI, Nakorchevsky A (2009) Proteomics by mass spectrometry: approaches, advances, and applications. (Translated from eng). Annu Rev Biomed Eng 11:49–79 (in eng)

3. Back JW, de Jong L, Muijsers AO, de Koster CG (2003) Chemical cross-linking and mass spectrometry for protein structural modeling. (Translated from eng). J Mol Biol 331(2): 303–313 (in eng)

4. Eyles SJ, Kaltashov IA (2004) Methods to study protein dynamics and folding by mass spectrometry. (Translated from eng). Methods 34(1):88–99 (in eng)

5. Farmer TB, Caprioli RM (1998) Determination of protein-protein interactions by matrix-assisted laser desorption/ionization mass spectrometry. (Translated from eng). J Mass Spectrom 33(8):697–704 (in eng)

6. Kalkhof S, Ihling C, Mechtler K, Sinz A (2005) Chemical cross-linking and high-performance Fourier transform ion cyclotron resonance mass spectrometry for protein interaction analysis: application to a calmodulin/target peptide complex. (Translated from eng). Anal Chem 77(2):495–503 (in eng)

7. Schulz DM, Ihling C, Clore GM, Sinz A (2004) Mapping the topology and determination of a low-resolution three-dimensional structure of the calmodulin-melittin complex by chemical cross-linking and high-resolution FTICRMS: direct demonstration of multiple binding modes. (Translated from eng). Biochemistry 43(16):4703–4715 (in eng)

8. Sinz A (2003) Chemical cross-linking and mass spectrometry for mapping three-dimensional structures of proteins and protein complexes. (Translated from eng). J Mass Spectrom 38 (12):1225–1237 (in eng)

9. Trester-Zedlitz M et al (2003) A modular cross-linking approach for exploring protein interactions. (Translated from eng). J Am Chem Soc 125(9):2416–2425 (in eng)

10. Sinz A (2006) Chemical cross-linking and mass spectrometry to map three-dimensional protein structures and protein-protein interactions. (Translated from eng). Mass Spectrom Rev 25(4):663–682 (in eng)

11. Auclair JR et al (2007) Mass spectrometry analysis of HIV-1 Vif reveals an increase in ordered structure upon oligomerization in regions necessary for viral infectivity. (Translated from eng). Proteins 69(2):270–284 (in eng)

12. Leitner A et al (2010) Probing native protein structures by chemical cross-linking, mass spectrometry, and bioinformatics. (Translated from eng). Mol Cell Proteomics 9(8):1634–1649 (in eng)

13. Pandey A, Mann M (2000) Proteomics to study genes and genomes. (Translated from eng). Nature 405(6788):837–846 (in eng)

14. Gavin AC et al (2002) Functional organization of the yeast proteome by systematic analysis of protein complexes. (Translated from eng). Nature 415(6868):141–147 (in eng)

15. Ho Y et al (2002) Systematic identification of protein complexes in Saccharomyces cerevisiae by mass spectrometry. (Translated from eng). Nature 415(6868):180–183 (in eng)

16. Blagoev B et al (2003) A proteomics strategy to elucidate functional protein-protein interactions applied to EGF signaling. (Translated from eng). Nat Biotechnol 21(3):315–318 (in eng)

17. Back JW et al (2002) Identification of cross-linked peptides for protein interaction studies using mass spectrometry and 18O labeling. (Translated from eng). Anal Chem 74(17): 4417–4422 (in eng)

18. Yao X, Freas A, Ramirez J, Demirev PA, Fenselau C (2001) Proteolytic 18O labeling for comparative proteomics: model studies with two serotypes of adenovirus. (Translated from eng). Anal Chem 73(13):2836–2842 (in eng)

19. Auclair JR, Boggio KJ, Petsko GA, Ringe D, Agar JN (2010) Strategies for stabilizing superoxide dismutase (SOD1), the protein destabilized in the most common form of familial amyotrophic lateral sclerosis. (Translated from Eng). Proc Natl Acad Sci U S A 107(50): 21394–21399 (in Eng)

20. Shevchenko A, Tomas H, Havlis J, Olsen JV, Mann M (2006) In-gel digestion for mass spectrometric characterization of proteins and proteomes. (Translated from eng). Nat Protoc 1 (6):2856–2860 (in eng)

21. McCullough, BJ and Gaskell, SJ (2009) Using electrospray ionisation mass spectrometry to study non-covalent interactions. Combo Chem High Throughput Screen 12(2): 203–211.

22. Kvaratskhella M, Miller JT, Budihas SR, Pannell LK, and Le Grice SFJ (2002) Identification of specific HIV-1 reverse transcriptase contacts to the viral RNA:tRNA complex by mass spectrometry and a primary amine selective reagent. Proc Natl Acad Sci U S A 99(25): 15988–15993

Chapter 27

Characterization of Oligomerization–Aggregation Products of Neurodegenerative Target Proteins by Ion Mobility Mass Spectrometry

Camelia Vlad, Marius Ionut Iurascu, Stefan Slamnoiu,
Bastian Hengerer, and Michael Przybylski

Abstract

Protein amyloidogenesis is generally considered to be a major cause of two most severe neurodegenerative disorders, Parkinson's disease (PD) and Alzheimer's disease (AD). Formation and accumulation of fibrillar aggregates and plaques derived from α-synuclein (α-Syn) and ß-amyloid (Aß) polypeptide in brain have been recognized as characteristics of Parkinson's disease and Alzheimer's disease. Oligomeric aggregates of α-Syn and Aß are considered as neurotoxic intermediate products leading to progressive neurodegeneration. However, molecular details of the oligomerization and aggregation pathway(s) and the molecular structure details are still unclear. We describe here the application of ion-mobility mass spectrometry (IMS-MS) to the identification of α-Syn and Aß oligomerization–aggregation products, and to the characterization of different conformational forms. IMS-MS is an analytical technique capable of separating gaseous ions based on their size, shape, and topography. IMS-MS studies of soluble α-Syn and Aß-aggregates prepared by in vitro incubation over several days were performed on a quadrupole time of flight mass spectrometer equipped with a "travelling wave" ion mobility cell, and revealed the presence of different conformational states and, remarkably, truncation and proteolytic products of high aggregating reactivity. These results suggest that different polypeptide sequences may contribute to the formation of oligomeric aggregates of heterogeneous composition and distinct biochemical properties.

Key words: Parkinson's disease, α-Synuclein, Alzheimer's disease, ß-Amyloid, Oligomerization, Ion mobility mass spectrometry

1. Introduction

Soluble amyloid oligomers have been generally thought to be formed as common intermediates in the amyloid fibrillization pathway, and have been implicated as neurotoxic species of amyloid-related neurodegenerative diseases, such as Alzheimer's disease (AD) and Parkinson's Disease (PD). In the last few years, increasing

Vladimir N. Uversky and A. Keith Dunker (eds.), *Intrinsically Disordered Protein Analysis:*
Volume 2, Methods and Experimental Tools, Methods in Molecular Biology, vol. 896,
DOI 10.1007/978-1-4614-3704-8_27, © Springer Science+Business Media New York 2012

evidence has been obtained that soluble oligomers, and not high molecular weight fibrillar end products, are highly toxic, even when they are formed from proteins that are not normally related to neurodegenerative processes (1, 2). Formation and accumulation of fibrillar plaques and "misfolding"- aggregation products of ß-amyloid peptide (Aß) and α-synuclein (α-Syn) in brain have been recognized as characteristics of Alzheimer's disease (AD) and Parkinson's disease (PD) (3–6). Moreover, in recent years an increasing number of proteins have been identified as being intrinsically disordered (Intrinsically Disordered Proteins; IDPs), suggesting that they lack stable, folded tertiary structures under physiological conditions (7). A number of intrinsically disordered proteins have been shown, or are suspected to be associated with human diseases as cancer, cardiovascular disease, amyloidoses, neurodegenerative diseases, and diabetes. IDPs, such as α-synuclein involved in Parkinson's disease (8, 9), protein-Tau in Alzheimer's disease (AD) and related tauopathies (10), and prion protein (PrP) in prion disorders (PrD) such as Creutzfeldt–Jakob Disease (CJD), Gerstmann–Straussler–Scheinker Syndrome and Fatal Familial Insomnia (FFI) (11, 12) represent potentially crucial targets for drugs that modulate their protein–protein interactions.

α-Synuclein, a protein of 140 amino acids, is mainly expressed in the presynaptic terminals of neurons and has been strongly correlated with Parkinson's disease, one of the most severe neurodegenerative motor disorders (8, 9). α-Syn is a natively unfolded protein with unknown function and unspecific conformational heterogeneity and properties. The main evidence for a causal role of α-Syn in PD came from the discovery of three rare point mutations of the amino acid sequence, A53T (13), A30P (14), and E46K (15) in families with autosomal dominant Parkinson's Disease that lead to α-Syn accumulation in Lewy bodies and other pathological inclusions at conditions inducing PD (16).

ß-Amyloid (Aß) is a neurotoxic polypeptide containing 39–43 amino acid residues and is derived from proteolytic cleavage of the transmembrane Aß precursor protein (APP). Recently, the formation of Aß-oligomers has become of particular interest, since oligomeric aggregates have been suggested to be key neurotoxic species for progressive neurodegeneration (3, 17, 18); however, molecular details of the pathophysiological degradation of APP, and of the Aß-aggregation pathways and structures are hitherto unclear (19).

In recent years, mass spectrometry (MS) has become a major analytical tool in structural biology (20). In particular, electrospray mass spectrometry (ESI-MS) has emerged as a powerful technique for producing intact gas- phase ions from large biomolecular ions, and ions due to supramolecular complexes (21). However, in contrast to the large number of ESI- mass spectrometric studies of protein structures, structure modifications and applications to proteomics (22–28), the molecular characterization of protein "misfolding" and aggregation species by mass spectrometry had little success; possible explanations are (1) the low intermediate concentrations and (2) and slow rates of aggregate formation in vitro (6, 7). Conventional "soft-ionization" mass spectrometry methods such as ESI-MS and HPLC-MS are not suitable to direct "in-situ" analysis of conformational states and intermediates at different concentrations. Recently, ion mobility mass spectrometry (IMS-MS) is emerging as a new tool to probe complex biomolecular structures from solution phase structures, due to the potential of IMS-MS for separation of mixtures of protein complexes by conformational state, and spatial shape and topology (29–36). The IMS-MS instrument employed in this study consists of two parts: (1) an ion mobility drift cell where ions are separated within an electric field according to their collisional cross sections and (2) a quadrupole time-of-flight mass spectrometer (Synapt-QTOF-MS) (33, 34, 36–38). Thus, IMS-MS implements a new mode of separation that allows the differentiation of protein conformational states. In the present study, ESI-MS (21) and IMS-MS have been applied to the direct analysis of mixtures of α-Syn and Aß (38) aggregation products in vitro, formed during prolonged incubation times.

2. Materials

2.1. Preparation and Characterization of α-Synuclein and Aß(1–40) Oligomers

Alpha-Syn and Aß (1–40) oligomers were prepared by an in vitro incubation procedure for up to 7 days at 37°C and at 20°C (38, 39) (see Note 2). α-Syn oligomers type A1 were prepared by dissolving the protein in 50 mM sodium phosphate buffer (pH 7.0) containing 20 % ethanol, to a final concentration of 7 μM (39). In the case of oligomers type A2, 10 μM $FeCl_3$ was added, while oligomers type A1 were prepared without adding $FeCl_3$. After 4 h of shaking, both types of oligomers were lyophilized and resuspended in 50 mM sodium phosphate buffer (pH 7.0) containing 10 % ethanol with a half of the starting volume. This was followed by incubation for 7 days at room temperature For Aß-oligomerization, Aß(1–40) was solubilized at a concentration of 1 μg/μL in a buffer containing 50 mM Na_3PO_4, 150 mM NaCl, 0.02 % NaN_3 at pH 7.5 (38). α-Syn and Aß(1–40) oligomers were separated by gel electrophoresis and subjected to IMS-MS measurements.

2.2. Gel Electrophoresis of α-Synuclein and Aß(1–40) Oligomers and Aggregates

Separation of α-Syn and Aß (1–40) oligomers was performed by one-dimensional Tris–tricine polyacrylamide gel electrophoresis (Tris–tricine PAGE) on a Mini-Protean II or Mini-Protean 3 electrophoresis cell (BioRad, München, Germany) using 12 % and 15 % polyacrylamide gel electrophoresis, and protein bands visualized by Coomassie Blue staining. The dimensions of the gel were

90 × 60 × 1 mm. The proteins were solubilized and denatured using a stock solution of the twofold concentrated sample buffer containing 4 % SDS, 25 % Glycerin, 50 mM Tris–HCl, 0.02 % Coomassie, 6 M urea, pH 6.8. Gels were run at 60 V until the tracking dye entered the separating gel and at 120 V until the tracking dye reached the bottom of the gel.

2.3. Buffers and Stock Solutions

The following solutions were prepared in ultrapure water (prepared by purifying deionized water to attain a conductivity of 18 MΩ):

1. α-Syn oligomers type A1 buffer: 50 mM sodium phosphate buffer, pH 7.0. Weigh 1.779 g Na_2HPO_4 and transfer to a 200-ml graduated glass beaker. Add ultrapure water to a volume of 200 ml. Mix and adjust pH with HCl (see Note 3). Add 40 ml ethanol. Store at 4°C.

2. α-Syn oligomers type A2 buffer: 50 mM sodium phosphate buffer, pH 7.0. Weigh 1.779 g Na_2HPO_4 and transfer to a glass beaker. Add ultrapure water to a volume of 200 ml. Mix and adjust pH with HCl (see Note 3). Add 40 ml ethanol and 0.3244 mg $FeCl_3$. Store at 4°C.

3. Aß(1–40) oligomers buffer: 50 mM Na_3PO_4, 150 mM NaCl, 0.02 % NaN_3 at pH 7.5. Weigh 1.9 g Na_3PO_4 and 0.88 g NaCl and transfer to a 100-ml graduated glass beaker containing about 80 ml ultrapure water. Mix and adjust pH with HCl (see Note 3). Make up to 100 ml with ultrapure water. Store at 4°C.

For Tris–tricine PAGE, the following solutions were prepared in ultrapure water.

4. Tris–HCl/SDS buffer: 3 M Tris, 0.3 % SDS, pH 8.45. Weight 36.4 g Tris and transfer to a 100 ml graduated glass beaker containing about 60 ml ultrapure water. Mix and adjust pH with HCl to 8.45 (see Note 3). Add 0.3 g SDS and make up to 100 ml with ultrapure water.

5. 2× Tricine sample buffer: 100 mM Tris pH 6.8, 24 % Glycerol, 8 % SDS, 0.02 % Coomassie brilliant blue G-250. Weigh 1.21 g Tris and transfer to a 10-ml graduated glass beaker. Adjust pH with HCl to 6.8 (see Note 3). Add 2.4 ml Glycerol, 0.8 g SDS (see Note 4), and 2 mg Coomassie brilliant blue G-250. Store at 4°C.

6. 1× Cathode buffer (upper chamber): 100 mM Tris, 100 mM Tricine, 0.1 % SDS. Weigh 6.055 g Tris, 8.96 g Tricine, 0.5 g SDS and transfer to a 500-ml graduated glass beaker and make up with ultrapure water.

7. 1× Anode buffer (lower chamber): 20 mM Tris, pH 8.96. Weigh 12.11 g Tris and transfer to a 500-ml graduated glass beaker containing about 400 ml ultrapure water. Mix and

adjust pH with HCl to 8.96 (see Note 3). Make up with ultrapure water to 500 ml.

8. Ammonium persulfate (APS): 10 % solution in water (see Note 5).

9. N', N', N', N'-tetramethyl-ethylenediamine (TEMED).

10. 15 % separating gel: Mix 7.5 ml Acrylamide, 5 ml Tris–HCl/SDS and 0.9 ml ultrapure water in a glass beaker. Add 1.58 ml Glycerol and sonicate until total solubilization. Add 75 µl APS, 10 µl TEMED and pour the gel into casting plates leaving room at the top for the stacking gel and comb. Add ultrapure water to cover top of gel and make flat interface. Polymerize for about 30 min. Check that remaining gel in glass beaker is polymerized before proceeding.

11. Stacking gel: Mix in a glass beaker 972 µl Acrylamide, 1.86 ml Tris–HCl/SDS, 4.67 ml ultrapure water, 40 µl APS and 7.5 µl TEMED and pour gel on top of already polymerized separating gel. Insert comb and allow to polymerize (about 30 min).

3. Methods

3.1. Gel Electrophoretic Isolation and Primary Structure Characterization of α-Synuclein Oligomerization–Aggregation Products

The in vitro oligomerization products of α-Syn were first characterized by Tris–tricine PAGE, as described in Subheadings 2.1 and 2.2, and the isolated protein bands analyzed by mass spectrometry. The electrophoretic separation revealed bands with molecular weights corresponding to monomeric $(\alpha\text{-Syn})_1$, dimeric $(\alpha\text{-Syn})_2$, oligomeric $(\alpha\text{-Syn})_n$ proteins, and three additional bands below the α-Syn monomer that were assigned to truncated forms of α-Syn (see Fig. 1). The band in the region of approximately 37 kDa was excised from the gel, digested with trypsin, and analyzed by HPLC-MS with an ESI-ion trap mass spectrometer (see Note 6). The mass spectrometric data ascertained the presence of α-Syn peptides by identification of the corresponding peptide fragments, and identified an α-Syn dimer composed of two intact α-Syn polypeptides. This result was ascertained by N-terminal Edman sequence determination (see Fig. 1c) (see Note 7). Following electrophoretic separation, the bands were transferred onto a PVDF membrane and subjected to Edman sequencing which provided the identification of N-terminal truncation of the band at 13.9 kDa, α-Syn(5–140). Direct mass spectrometric analysis of monomer $(\alpha\text{-Syn})_1$ was performed with an Esquire 3000+ ion trap mass spectrometer (Bruker Daltonics, Bremen, Germany). The protein band was solubilized in 1 % formic acid as a volatile solvent, to a final concentration of 30 µM, pH 2.5.

a

¹MDVFMKGLSK AKEGVVAAAE KTKQGVAEAA GKTKEGVLYV GSKTKEGVVH GVATVAEKTK EQVTNVGGAV

⁷¹
VTGVTAVAQK TVEGAGSIAA ATGFVKKDQL GKNEEGAPQE GILEDMPVDP DNEAYEMPSE EGYQDYEPEA   ¹⁴⁰

b   **c**

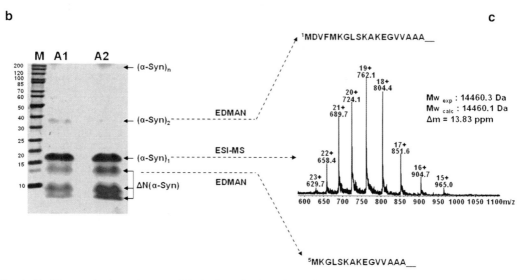

Fig. 1. (**a**), Primary structure of α-synuclein (α-Syn). (**b**), Tris–Tricine PAGE of α-Syn stained with Coomassie Blue: molecular weight marker (*lane 1*); (A1) α-Syn incubated in PBS buffer at room temperature for 7 days (*lane 2*); (A2) α-Syn incubated in PBS buffer with 10 μM FeCl₃ at 20°C for 7 days (*lane 3*). (**c**), Edman sequence determination of (α-Syn)₂; ESI-MS of α-Syn in 1 % aqueous formic acid at a final concentration of 34.5 μM, pH 2.5 showing the 15+ to 23+ charged molecular ions; Edman sequencing of N-terminal truncated polypeptide sequence of 13.9 kDa, α-Syn(5–140).

The spectrum was obtained in the positive ion mode (see Fig. 1b). In the ESI mass spectrum, a series of multiply charged ions were obtained with the most abundant ion at charge state +19.

3.2. Ion-Mobility Mass Spectrometry

Ion-mobility mass spectrometry was performed with a SYNAPT-QTOF-MS (Waters, Manchester, UK) equipped with an nano-electrospray ionization source (22–25). The ions were passed through a quadrupole and either set to transmit a substantial mass range or to select a particular m/z before entering the Triwave ion mobility cell. The Triwave system (22) is composed of three T-Wave devices (Waters SYNAPT). The first device, the Trap T-Wave, accumulates ions which are stored and gated (500 μsec) into the second device, the ion mobility T-Wave cell, where they are separated according to their cross section-dependent mobilities (22, 23). The final transfer T-Wave is used to transport the separated ions into the orthogonal acceleration oa-TOF for MS analysis. The injection volume was 5 μl, and a prior desalting step was performed on a cartridge for 10 min at 20 μl/min with gradient of 10–90 % acetonitrile. The mass spectrometric acquisition characteristics were: 350–4,000 m/z, cone voltage 25 V, IMS pressure 0.45 bar, wave height 5–15 V. IMS-MS data are inherently three-dimensional, consisting of mass, drift time and intensity (relative abundance) for all ions observed.

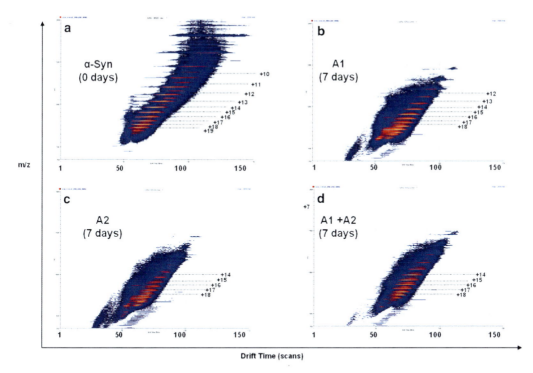

Fig. 2. Ion-mobility drift time of α-Syn after 7 days of incubation in 50 mM sodium phosphate buffer, containing 20 % ethanol to a final concentration of 7 μM (A1); Syn after 7 days of incubation in 50 mM sodium phosphate buffer, containing 20 % ethanol, 10 μM $FeCl_3$ (A2); freshly prepared α-Syn. Signals corresponding to monomers, dimers, and truncated peptide sequences are observed.

3.3. Ion Mobility Mass Spectrometry of α-Syn- Oligomers

Figure 2 summarizes the ion-mobility- MS drift time profiles of a freshly prepared α-Syn solution, and α-Syn- oligomers in incubation mixtures type A1, A2 and A1 + A2 after 7 days of incubation in phosphate-buffered saline with 20 % ethanol, 10 μM $FeCl_3$ (drift scope data view). In the different charge state series of ion signals, α-Syn monomer, dimer as well as proteolytic and truncated peptide sequences were identified by molecular weight determinations and partial tandem-MS sequence determinations (see Figs. 3 and 4a, b). The comparison of the extracted drift time "mobilograms" of the ion at m/z 804.1 (18+) between α-Syn oligomers type A1, A2, and A1 + A2 is shown in Fig. 3. In the α-Syn oligomers type A1, two drift time (conformational) states are present, while in type A2 and A1 + A2 even three drift time states are observed indicating three different structural types. The extracted, deconvoluted mass spectra of the drift scope profile due to the α-Syn oligomers, type A1 revealed the α-Syn monomer and the intact full-length dimer (see Fig. 4a; m.w. 14,459; 29.919 Da), while the oligomer peak, type A1 + A2 showed an additional truncated polypeptide, α-Syn (40–140) (see Fig. 4b; m.w. 10.437 Da); further proteolytic peptide fragments of α-Syn with high aggregating reactivity, corresponding to the proteolytic bands observed by gel electrophoresis (see Fig. 1) have been recently identified (40).

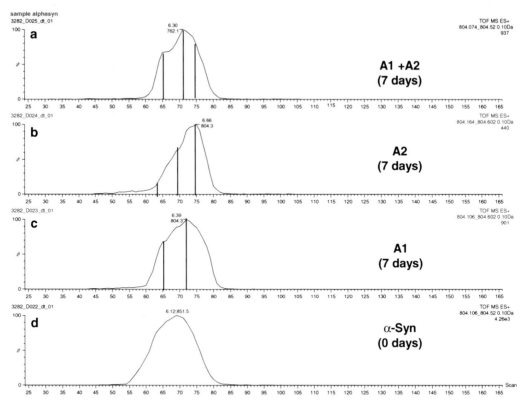

Fig. 3. Comparison of the extracted mobilograms of the ion m/z 804.1 between α-Syn oligomers type A1, A2, A1 + A2. In α-Syn oligomers type A1 two structures are observed; in type A2 and A1 + A2 three main forms are present.

3.4. Ion Mobility Mass Spectrometry of Aß-Oligomers

The soluble fraction of the Aß(1–40) fibril preparation obtained by incubation over 5 days at 37°C, and a freshly prepared Aß(1–40) peptide solution (220 µM) were subjected to comparative analysis by ion mobility- MS (see Fig. 5a, b). In the freshly prepared solution of Aß(1–40), the $[M+5H]^{5+}$ ion was predominant, while in soluble fibril preparations ions with increasingly higher charge states ($[M+6H]^{6+}$, $[M+7H]^{7+}$) were most abundant, indicative of oligomer formation. In the gel electrophoretic analysis, the formation of Aß- oligomers can be directly observed (38). The signal-to-noise ratios of the $[M+5H]^{5+}$ and $[M+6H]^{6+}$ ions was found to be significantly lower in the fibril preparation sample compared to the sample of freshly prepared Aß(1–40), suggesting a decreased amount of Aß(1–40) monomer due to the formation of oligomers. The extracted ion mobility profiles for the $[M+5H]^{5+}$ ion of the freshly prepared Aß(1–40) (see Fig. 5c) compared to that of the fibril preparation (see Fig. 5d) indicate the presence of different conformational states (A and B) with different ion mobilities (cross sections), indicative of the oligomerization process. Furthermore, partial oxidation of Aß at the Met-35 residue has been observed in samples upon formation of oligomers (38).

a

¹
MDVFMKGLSK AKEGVVAAAE KTKQGVAEAA GKTKEGVLYV GSKTKEGVVH GVATVAEKTK EQVTNVGGAV
⁷¹   ¹⁴⁰
VTGVTAVAQK TVEGAGSIAA ATGFVKKDQL GKNEEGAPQE GILEDMPVDP DNEAYEMPSE EGYQDYEPEA

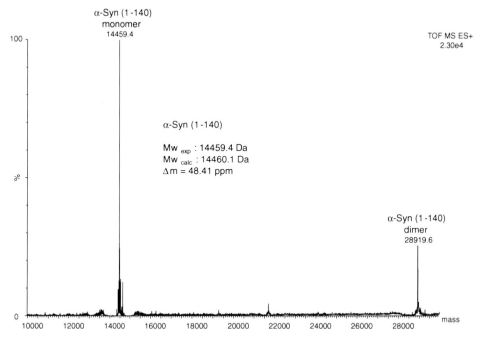

b

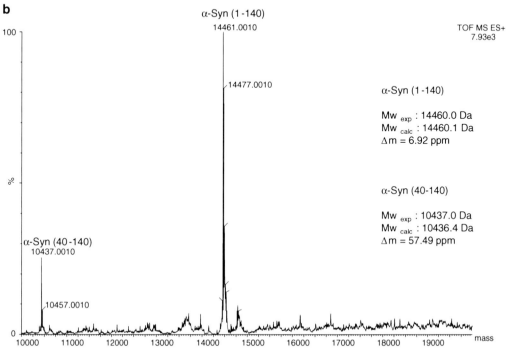

Fig. 4. Extracted deconvoluted spectra of α-Syn oligomers: (**a**), Type A1 oligomers are showing α-Syn monomer plus dimer; (**b**), type A1 + A2 oligomers are containing truncated polypeptide sequence at 10.5 kDa (α-Syn(40–140)).

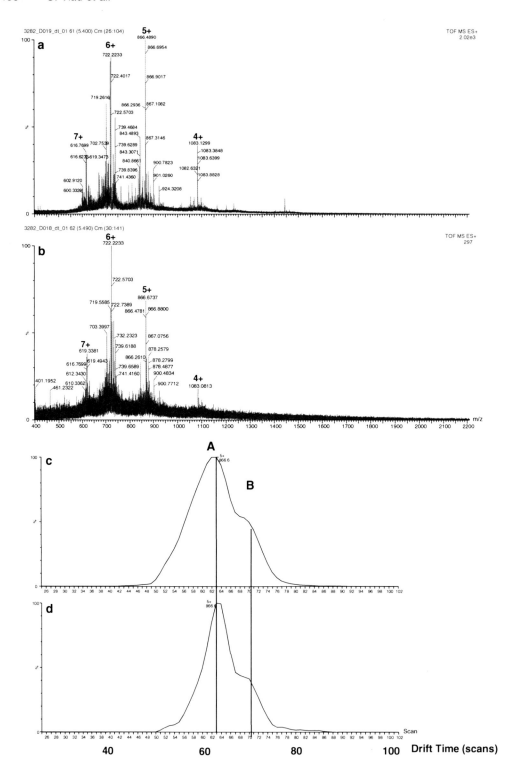

Fig. 5. Ion mobility MS analysis of (**a**) freshly prepared Aß(1–40); (**b**) supernatant after 5 days incubation; (**c**) and (**d**) extracted ion mobility profiles for m/z 866.6, [M+5H]$^{5+}$ from freshly prepared Aß(1–40) and supernatant, respectively.

4. Notes

Ion mobility mass spectrometry has been shown to be an efficient tool for the characterization of in vitro α-Syn and Aß oligomerization–aggregation products, and provides evidence of (1), the presence of distinct conformational forms of α-Syn and (2), the presence of at least two different conformational forms involved in Aß- aggregation. These results suggest IMS-MS as a powerful tool for the analysis of reactive intermediates in the aggregation of intrinsically disordered proteins due to its "affinity"-like gas phase separation capability. In the case of α-Syn, the IMS-MS data identified proteolytic fragments as key molecular species involved in "misfolding" and aggregation pathways, suggesting that the IMS-MS method can also be successfully applied to the identification of hitherto undetected proteolytic truncation products of neurodegenerative target proteins. Thus, gel electrophoresis protein Tau, another key protein in Alzheimer's disease shows the formation of truncation/proteolytic fragments in addition to oligomerization products (see Fig. 6). It will be of interest to systematically explore IMS-MS in the study of aggregation pathways of intrinsically disordered proteins.

1. α-Syn was purified prior to use by analytical RP-HPLC on a UltiMate 3000 system (Dionex, Germering, Germany),

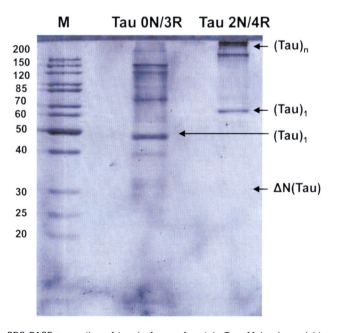

Fig. 6. SDS-PAGE separation of two isoforms of protein-Tau: Molecular weight marker (*lane 1*); 2 μg Tau2N/4R protein (*lane 2*) and 2 μg Tau0N/3R protein (*lane 3*) showing bands corresponding to monomeric (Tau)$_1$, oligomeric (Tau)$_{nv}$ and truncated forms of Tau, isoforms ΔN(Tau).

equipped with a Vydac C4 column (250 × 4.6 mm I.D.) with 5 μm silica (300 Å pore size) (Hesperia CA). Linear gradient elution (0 min—0 % B; 5 min—0 % B; 50 min—90 % B) with eluent A: (0.1 % trifluoroacetic acid in water) and eluent B: (0.1 % trifluoroacetic acid in acetonitrile–water 80:20 v/v) was employed at a flow rate of 1 ml/min.

2. Aß(1–40) was synthesized by solid phase peptide synthesis on a NovaSyn TGR resin using Fmoc chemistry by a semiautomated synthesizer EPS-221 (Abimed, Germany). Fmoc amino acids, NovaSyn TGR resin, PyBop, and other reagents were obtained from Novabiochem (Laufelfingen, Switzerland). The crude products were purified by analytical RP-HPLC (see Note 1).

3. Concentrated HCl (12N) can be used at first to narrow the gap from the starting pH to the necessary pH and further use HCl (1N) until the pH 7 is obtained.

4. SDS precipitates at 4°C. Therefore, the 2× Tricine sample buffer needs to be warmed prior to use.

5. The best is to prepare fresh solution each time.

6. Protein spots were excised manually from the gel, cutting as close to the spots as possible, in order to minimize the contamination with SDS. After distaining, the dried gel pieces were swollen in digestion buffer (12 ng/μl TPCK-trypsin in 50 mM NH_4HCO_3) at 4°C (on ice) for 45 min, and then they were incubated at 37°C overnight (12 h) and at last the peptides were extracted twice with acetonitrile–0.1 % trifluoroacetic acid 3:2.

7. The spots of interest were excised, destained with 100 % methanol, and applied into the sequencing cartridge. Sequence determinations were performed on an Applied Biosystems Model 494 Procise Sequencer attached to a Model 140 C Microgradient System, a 785A Programmable Absorbance Detector, and a 610A Data Analysis System.

Acknowledgments

We thank Adrian Moise for expert assistance with Edman sequence and Nick Tomczyk and Emmanuelle Claude, Waters MS- Technology Centre Manchester, UK, for assistance with the ion mobility-MS. We are grateful to David Clemmer and Nicolas Pearson for helpful discussion and comments on IMS-MS. This work has been supported by the Graduate School Chemical Biology, the Research Centre Proteostasis, University of Konstanz, and the Landesstiftung für Wissenschaft und Forschung Baden- Württemberg.

References

1. Bucciantini M, Giannoni E, Chiti F, Baroni F, Formigli L, Zurdo J, Taddei N, Ramponi G, Dobson CM, Stefani M (2002) Inherent toxicity of aggregates implies a common mechanism for protein misfolding diseases. Nature 416:507–511

2. Kayed R, Sokolov Y, Edmonds B, McIntire TM, Milton SC, Hall JE, Glabe CG (2004) Permeabilization of lipid bilayers is a common conformation-dependent activity of soluble amyloid oligomers in protein misfolding diseases. J Biol Chem 279:46363–46366

3. Morgan D (2006) Immunotherapy for Alzheimer's disease. J Alzheimers Dis 9:425–432

4. Shults CW (2006) Lewy bodies. Proc Natl Acad Sci U S A 103:1661–1668

5. Goedert M (2001) Alpha-synuclein and neurodegenerative diseases. Nat Rev Neurosci 2:492–501

6. Uversky VN (2007) Neuropathology, biochemistry, and biophysics of alpha-synuclein aggregation. J Neurochem 103:17–37

7. Dyson HJ, Wright PE (2005) Intrinsically unstructured proteins and their functions. Nat Rev Mol Cell Biol 6:197–208

8. Maroteaux L, Campanelli JT, Scheller RH (1988) Synuclein: a neuron-specific protein localized to the nucleus and presynaptic nerve terminal. J Neurosci 8:2804–2815

9. Dawson TM, Dawson VL (2003) Molecular pathways of neurodegeneration in Parkinson's disease. Science 302:819–822

10. Tolnay M, Probst A (1999) Review: tau protein pathology in Alzheimer's disease and related disorders. Neuropathol Appl Neurobiol 25:171–187

11. Johnson RT, Gibbs CJ (1998) Creutzfeldt-Jakob disease and related transmissible spongiform encephalopathies. N Engl J Med 339:1994–2004

12. Horiuchi M, Caughey B (1999) Prion protein interconversions and the transmissible spongiform encephalopathies. Structure (London, England) 7:231–240

13. Polymeropoulos MH, Lavedan C, Leroy E, Ide SE, Dehejia A, Dutra A, Pike B, Root H, Rubenstein J, Boyer R et al (1997) Mutation in the alpha-synuclein gene identified in families with Parkinson's disease. Science 276:2045–2047

14. Kruger R, Kuhn W, Muller T, Woitalla D, Graeber M, Kosel S, Przuntek H, Epplen JT, Schols L, Riess O (1998) Ala30Pro mutation in the gene encoding alpha-synuclein in Parkinson's disease. Nat Genet 18:106–108

15. Zarranz JJ, Alegre J, Gomez-Esteban JC, Lezcano E, Ros R, Ampuero I, Vidal L, Hoenicka J, Rodriguez O, Atares B et al (2004) The new mutation, E46K, of alpha-synuclein causes Parkinson and Lewy body dementia. Ann Neurol 55:164–173

16. Saha AR, Hill J, Utton MA, Asuni AA, Ackerley S, Grierson AJ, Miller CC, Davies AM, Buchman VL, Anderton BH et al (2004) Parkinson's disease alpha-synuclein mutations exhibit defective axonal transport in cultured neurons. J Cell Sci 117:1017–1024

17. Salminen A, Ojala J, Kauppinen A, Kaarniranta K, Suuronen T (2009) Inflammation in Alzheimer's disease: amyloid-beta oligomers trigger innate immunity defence via pattern recognition receptors. Prog Neurobiol 87:181–194

18. Crews L, Tsigelny I, Hashimoto M, Masliah E (2009) Role of synucleins in Alzheimer's disease. Neurotox Res 16:306–317

19. Hull M, Berger M, Heneka M (2006) Disease-modifying therapies in Alzheimer's disease: how far have we come? Drugs 66:2075–2093

20. Heck AJ, Van Den Heuvel RH (2004) Investigation of intact protein complexes by mass spectrometry. Mass Spectrom Rev 23:368–389

21. Fenn JB, Mann M, Meng CK, Wong SF, Whitehouse CM (1989) Electrospray ionization for mass spectrometry of large biomolecules. Science 246:64–71

22. Schulenborg T, Schmidt O, van Hall A, Meyer HE, Hamacher M, Marcus K (2006) Proteomics in neurodegeneration–disease driven approaches. J Neural Transm 113:1055–1073

23. Mann M, Hendrickson RC, Pandey A (2001) Analysis of proteins and proteomes by mass spectrometry. Annu Rev Biochem 70:437–473

24. Perdivara I, Petrovich R, Allinquant B, Deterding LJ, Tomer KB, Przybylski M (2009) Elucidation of O-glycosylation structures of the beta-amyloid precursor protein by liquid chromatography-mass spectrometry using electron transfer dissociation and collision induced dissociation. J Proteome Res 8:631–642

25. Perdivara I, Deterding L, Moise A, Tomer KB, Przybylski M (2008) Determination of primary structure and microheterogeneity of a beta-amyloid plaque-specific antibody using high-

performance LC-tandem mass spectrometry. Anal Bioanal Chem 391:325–336

26. Przybylski M, Glocker MO (1996) Electrospray mass spectrometry of biomacromolecular complexes with noncovalent interactions—new analytical perspectives for supramolecular chemistry and molecular recognition processes. Angew Chem Int Ed Engl 35:806–826

27. Kebarle P, Verkerk UH (2009) Electrospray: from ions in solution to ions in the gas phase, what we know now. Mass Spectrom Rev 28:898–917

28. Nemes P, Goyal S, Vertes A (2007) Conformational and noncovalent complexation changes in proteins during electrospray ionization. Anal Chem 80:387–395

29. Ruotolo BT, Benesch JL, Sandercock AM, Hyung SJ, Robinson CV (2008) Ion mobility-mass spectrometry analysis of large protein complexes. Nat Protoc 3:1139–1152

30. Bernstein SL, Dupuis NF, Lazo ND, Wyttenbach T, Condron MM, Bitan G, Teplow DB, Shea JE, Ruotolo BT, Robinson CV, Bowers MT (2009) Amyloid-ß protein oligomerization and the importance of tetramers and dodecamers in the aetiology of Alzheimer's Disease. Nat Chem 1:326–331

31. Thalassinos K, Slade SE, Jennings KR, Scrivens JH, Giles K, Wildgoose J, Hoyes J, Bateman RH, Bowers MT (2004) Ion mobility mass spectrometry of proteins in a modified commercial mass spectrometer. Int J Mass Spectrom 236:55–63

32. Bleiholder C, Dupuis NF, Wyttenbach T, Bowers MT (2011) Ion mobility-mass spectrometry reveals a conformational conversion from random assembly to ß-sheet in amyloid fibril formation. Nat Chem 3:172–177

33. Ruotolo BT, Hyung SJ, Robinson PM, Giles K, Bateman RH, Robinson CV (2007) Ion mobility-mass spectrometry reveals long-lived, unfolded intermediates in the dissociation of protein complexes. Angew Chem Int Ed Engl 46:8001–8004

34. Trimpin S, Clemmer DE (2008) Ion mobility spectrometry/mass spectrometry snapshots for assessing the molecular compositions of complex polymeric systems. Anal Chem 80:9073–9083

35. Kanu AB, Dwivedi P, Tam M, Matz L, Hill HH Jr (2008) Ion mobility-mass spectrometry. J Mass Spectrom 43:1–22

36. Henderson SC, Valentine SJ, Counterman AE, Clemmer DE (1999) ESI/ion trap/ion mobility/time-of-flight mass spectrometry for rapid and sensitive analysis of biomolecular mixtures. Anal Chem 71:291–301

37. Zhou M, Sandercock AM, Fraser CS, Ridlova G, Stephens E, Schenauer MR, Yokoi-Fong T, Barsky D, Leary JA, Hershey JW et al (2008) Mass spectrometry reveals modularity and a complete subunit interaction map of the eukaryotic translation factor eIF3. Proc Natl Acad Sci U S A 105:18139–18144

38. Iurascu MI, Cozma C, Tomczyk N, Rontree J, Desor M, Drescher M, Przybylski M (2009) Structural characterization of beta-amyloid oligomer-aggregates by ion mobility mass spectrometry and electron spin resonance spectroscopy. Anal Bioanal Chem 395:2509–2519

39. Danzer KM, Haasen D, Karow AR, Moussaud S, Habeck M, Giese A, Kretzschmar H, Hengerer B, Kostka M (2007) Different species of alpha-synuclein oligomers induce calcium influx and seeding. J Neurosci 27:9220–9232

40. Vlad C, Lindner K, Karreman C, Schildknecht S, Leist M, Tomczyk N, Rontree J, Langridge J, Danzer K, Ciossek T, Petre A, Gross ML, Hengerer B, Przybylski M (2011) Autoproteolytic fragments are intermediates in the oligomerization/aggregation of the Parkinson's disease protein alpha-synuclein as revealed by ion mobility mass spectrometry. Chem Bio Chem 12:2740–2744

Part V

Expression and Purification of IDPs

Chapter 28

Identifying Solubility-Promoting Buffers for Intrinsically Disordered Proteins Prior to Purification

Kelly A. Churion and Sarah E. Bondos

Abstract

Intrinsically disordered proteins are anticipated to be more prone to aggregation than folded, stable proteins. Chemical additives included in the buffer can help maintain proteins in a soluble, monomeric state. However, the array of chemicals that impact protein solubility is staggering, precluding iterative testing of chemical conditions during purification. Herein, we describe a filter-based aggregation assay to rapidly identify chemical additives that maintain solubility for a protein of interest. A hierarchical approach to buffer selection is provided, in which the type of chemical which best improves solubility is first determined, followed by identifying the optimal chemical and its most effective concentration. Finally, combinations of chemical additives can be assessed if necessary. Although this assay can be applied to purified protein, partially purified protein, or aggregated protein, this protocol specifically details the use of this assay for crude cell lysate. This approach allows identification of solubility-promoting buffers prior to the initial protein purification.

Key words: Filter-based aggregation assay, Osmolyte, Chaotrope, Kosmotrope, Chaperone, Protein solubility, Precipitation, Purification

1. Introduction

Intrinsically disordered regions of a protein often mediate important functions and implement key regulatory mechanisms. For instance, intrinsically disordered regions permit specific, yet reversible protein interactions with numerous protein partners to regulate protein function (1–6). In addition, intrinsically disordered regions may also directly regulate the function of structured domains within the same protein (7, 8). In both cases, the activity of intrinsically disordered regions may be modulated by posttranslational modifications (9). Consequently, intrinsically disordered regions can have a major impact on the cellular function of a protein.

Vladimir N. Uversky and A. Keith Dunker (eds.), *Intrinsically Disordered Protein Analysis: Volume 2, Methods and Experimental Tools*, Methods in Molecular Biology, vol. 896, DOI 10.1007/978-1-4614-3704-8_28, © Springer Science+Business Media New York 2012

However, intrinsically disordered proteins can be prone to aggregation, and thus are often challenging to handle in vitro. Intrinsically disordered proteins form amyloids more easily (10–12). Because intrinsically disordered regions are often designed to mediate protein–protein interactions (13, 14), they may also readily form amorphous aggregates. For example, p53, a protein that participates in many protein interactions, is extremely aggregation prone (15, 16).

There are several approaches used to mitigate such aggregation, including truncation to remove the intrinsically disordered regions, or fusion to small soluble proteins. However, these methods may perturb or even remove the interesting regulatory features of a protein. A useful alternative is to identify buffer conditions that will maintain the protein as a soluble monomer. Iteratively probing buffer conditions during protein purification can require weeks and waste precious resources (17). Furthermore, protein purification protocols often require specific buffers or buffer changes that promote insolubility in aggregation-prone proteins. Herein, we describe a simple aggregation assay that can be used on crude cell extract to rapidly identify buffers that maintain protein solubility for purification and subsequent assays (17). This assay can be applied to purified or partially purified protein, aggregated protein, or crude cell lysate.

We use an iterative approach to optimizing buffers for protein solubility. The first assay identifies the type of chemical that improves protein stability, whereas subsequent assays optimize chemical identity and concentration (see Fig. 1). Chemicals added to buffers can impact protein solubility by a wide variety of mechanisms (see Table 1). The impact of any one chemical will vary for different proteins. For instance, chaotropes weaken protein–protein interactions. Consequently, while chaotropes may destabilize aggregates formed by proteins in or near the native state, they may also enhance aggregation in proteins that aggregate from partially denatured states (18). Conversely, kosmotropes enhance protein–protein associations and thus may increase aggregation in proteins that associate in the native state, but stabilize structure and thus decrease aggregation for proteins that associate only in partially unfolded states (18). Arginine and its derivatives have both chaotropic and kosmotropic moieties and thus are often successful in improving protein solubility (19, 20). Inclusion of osmolytes favors protein structures with the least surface area (21), and addition of surfactants can mask exposed hydrophobic regions (22). Importantly, each category of additive will improve solubility of some proteins while decreasing the solubility of others, highlighting the need for experimental trails to identify the most solubility-promoting buffer components for specific proteins.

Finally, it is important to note that even chemicals which improve protein solubility at low concentrations can enhance aggregation or denaturation of that same protein at high concentrations. For instance, high concentrations of chaotropes such as

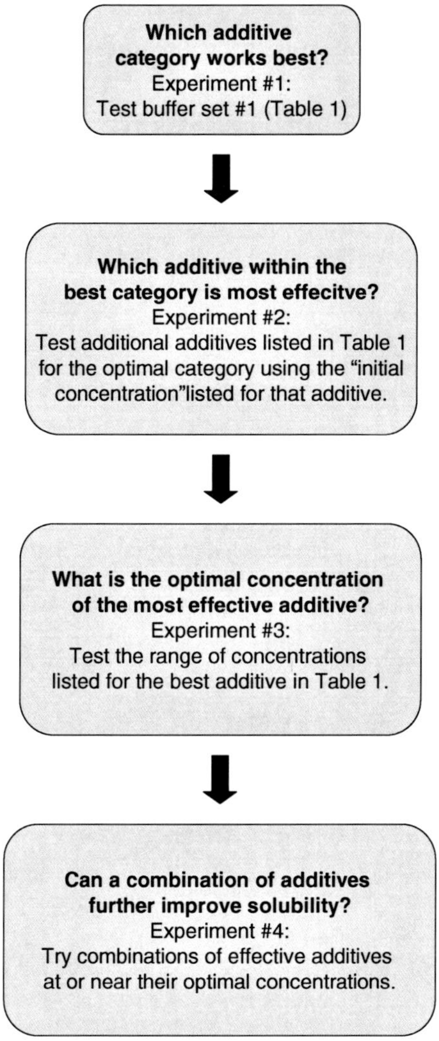

Fig. 1. Flowchart for selecting chemical additives in the filter-based aggregation assay.

urea or guanidinium hydrochloride are often used to denature proteins. In this chapter, we provide recommended concentration ranges for the additives listed that are generally useful in our experience, although results for specific proteins vary.

2. Materials

2.1. Selection of Assay Buffers

In this approach, the optimal buffer is progressively defined in a series of experiments summarized in Fig. 1 and Table 1. First, a standard set of additives, each representing a class of chemicals with characteristic effects on protein solubility, is examined to identify which category of buffer is most effective. Second, different chemicals within that

Table 1
Selection of additives for the filter assay

Additives for Assay 1: Determining the correct category of buffer additive

1. 300 mM NaCl (500 mM NaCl total concentration)
2. 10 % glycerol
3. 0.5 M Urea
4. 500 mM Arginine
5. 0.5 M TMAO
6. 1 % Nonidet P40

Additives for Assays 2 and 3: Optimizing additive identity and concentration within a category

Additive	Recommended initial concentration	Recommended concentration range[1]
If increasing NaCl works best, try other weak kosmotropes (1, 2) or strong kosmotropes (4–6):		
1. NaCl	200 mM	0–1 M
2. KCl	200 mM	0–1 M
3. $MgSO_4$	100 mM	0–0.4 M
4. $(NH_4)2SO_4$	50 mM	0–0.2 M
5. Na_2SO_4	50 mM	0–0.2 M
6. Cs_2SO_4	50 mM	0–0.2 M
If adding glycerol or TAMO works best, try other osmolytes:		
1. Glycerol	10 %	0–40 % v/v
2. TMAO	20 mM	0–1 M
2. Glucose	1 M	0–2 M
4. Sucrose	500 mM	0–1 M
5. Ethylene glycol	10 %	0–60 %
6. Trehalose	500 mM	0–1 M
If adding urea works best, try mild chaotropes:		
1. Urea	0.5 M	0–1.5 M
2. NaI	0.2 M	0–0.4 M
3. $CaCl_2$	0.1 M	0–0.2 M
4. $MgCl_2$	0.1 M	0–0.2 M
If adding arginine works best, try other amino acids, amino acid derivatives, as well as osmolytes listed above:		
1. Glycine		0.5–2 %
2. Arginine	200 mM	0–2 M
3. Arginine ethylester	200 mM	0–500 mM
4. Proline	200 mM	0–1 M
5. Potassium glutamate	200 mM	0–500 mM
If adding Nonidet P40 works best, try other detergents:		
1. Nonidet P40	0.2 %	0–1 %
2. Tween 80	0.01 %	0.02 % w/v

All buffers should contain 50 mM NaH_2PO_4, 200 mM NaCl pH 8.0 or a different standard buffer. Additional additive suggestions and more detailed information about their mechanism of action can be found elsewhere (1, 17, 23–27)
[1]Although these concentrations have been used with other proteins, they may induce aggregation in specific proteins

category are examined to determine the optimal chemical additive within that category. Third, the concentration of the best additive is optimized. Finally, combinations of additives can be assessed to further improve solubility (see Note 1).

2.2. Preparation of Assay Buffers

1. Assay buffers should be selected and prepared prior to lysing cells. This assay is iteratively applied to rapidly identify buffers that improve solubility. All assay buffers include the standard buffer (50 mM NaH_2PO_4, 200 mM NaCl pH 8.0) (see Note 2). In addition to the standard buffer, the chemical additives listed in Table 1 under "Buffer Additives for Assay 1" should be included to create the first six different test solutions, one testing each additive.

2.3. Preparing Crude Cell Lysate

1. Cell lysis buffer: 50 mM NaH_2PO_4, 200 mM NaCl, pH 8.0 with 2 mg/mL hen egg white lysozyme and one Complete Mini Protease Inhibitor Cocktail Tablet per liter (Roche Diagnostics, catalogue number 11836153001) (see Notes 2–4).

2. DNase stock: Dissolve a bottle of DNase I (Roche Diagnostics catalogue number D4263) in 5 mL cold 0.15 M NaCl to yield a 2,000 Kunitz units/mL solution. Store DNase in 40 μL aliquots in a −80°C freezer.

2.4. Filter-Based Aggregation Assay Components

1. Protein concentrators in which the molecular weight cutoff is larger than the native protein. For proteins less than 100 kDa, use a 100 kDa MW cutoff Amicon Ultra centrifugal filters (0.5 mL, 100 K membrane, Amicon catalogue number UFC510096). For larger proteins, use the Ultrafree-MC filter unit centrifugal filter (0.1 mm pore size, Millipore catalogue number UFC30VV25).

2. Test buffers: Table 1. For the first experiment, test the buffer Additives for Assay 1 in a standard buffer, such as 50 mM NaH_2PO_4, 200 mM NaCl, pH 8.0.

2.5. SDS Polyacrylamide Gel Reagents

1. Running gel buffer: 1.5 M Tris/SDS Buffer, pH 8.8. Mix ~100 mL of water in a glass beaker or graduated cylinder with 181.7 g of Tris Base and 4 g of 0.4 % SDS and transfer to the beaker. Add water to a volume of 1 L. Mix and adjust pH with HCl to pH 8.8.

2. Stacking gel buffer: 0.5 M Tris/SDS buffer, pH 6.8. Mix ~100 mL of water in a glass beaker or graduated cylinder with 60.57 g Tris base and 4 g SDS. Add water to a volume of 1 L. Mix and adjust pH with HCl to pH 6.8.

3. 40 % 29:1 acrylamide solution (Fisher Scientific catalogue number BP1408-1) (see Note 5).

4. Ammonium persulfate (APS): 10 % solution in water (see Note 6).

5. *N, N, N, N'* -tetramethyl-ethylenediamine (TEMED) (Fisher Scientific Inc).

6. 2× Sample Buffer: Add about 40 mL of water in a glass beaker or graduated cylinder. Weigh 1.52 g of Tris Base, 20 mL glycerol, 2 g SDS (wear mask), 2 mL 2-mercaptoethanol (BME) and 1 mg Bromophenol Blue. Add water to a volume of 100 mL. Mix and adjust pH with HCl to pH 6.8.

7. Electrophoresis Running Buffer: Add about 100 mL of water in a glass beaker or graduated cylinder. Weigh 3.02 g Tris base, 14.4 g glycine, 1.0 g SDS. Add water to a volume of 1 L.

8. Mini-Protean 3 System glass plates (catalog number 1653310) (Bio-Rad) with 0.75 mm spacers and a compatible gel electrophoresis unit.

9. Medium binder clips (1¼ in.).

10. A pre-mixed, pre-stained protein marker such as Prestained Protein Marker, Broad Range (New England Biolabs, catalogue number P7708S).

11. 1 % agarose solution. Add 1 g agarose to 10 mL ddH$_2$O and microwave until the agarose is dissolved. This solution can be stored at room temperature.

12. Water-saturated isobutanol.

2.6. Western Blot Reagents

1. Nitrocellulose membranes (Whatman Protran Nitrocellulose Membranes, VWR catalogue number 74330-034).

2. Western Blot Transfer Buffer: 0.025 M Tris Base, 0.19 M glycine, and 20 % methanol (see Note 7).

3. Phosphate-buffered saline Tween (PBST): 0.08 M sodium phosphate, dibasic (Na$_2$HPO$_4$), 0.02 M sodium phosphate, monobasic (NaH$_2$PO$_4$), 0.1 M NaCl, and 0.1 % Tween 20 (see Note 8).

4. Blocking solution: 10 % milk solids in PBST (see Note 9).

5. Western blot cassette and transfer system.

6. Container to wash and incubate the nitrocellulose membrane, ideally as small as possible yet larger than 7 cm × 10 cm.

7. Filter paper (BioRad extra thick blot paper, catalogue number 170-3958).

3. Methods

3.1. Preparation of Crude Cell Lysate (see Note 10)

1. Grow *E. coli* in Erlenmeyer or Fernbach flasks and induce expression of your protein of interest according to the standard protocols for the selected cell line and expression vector (see Note 11).

2. Remove seven 100 μL aliquots from the cell culture and place each aliquot in a 1.5-mL Eppendorf tube (see Note 12).

3. Centrifuge the Eppendorf tubes for 5–10 min at $18,000 \times g$.

4. Remove all of the supernatant from each tube (see Note 13). Before disposal of the supernatant, ensure any *E. coli* are dead by treating the supernatant with 10 % bleach or autoclaving.

5. At this point, the *E. coli* cell pellets may be frozen at −80° for later use or you may proceed directly to step 6.

6. Add 200 μL of lysis buffer to the cell pellets and resuspend the cells to begin lysis. Incubate the lysis on ice first with pipetting for 5 min, and subsequently without pipetting for an additional 5 min. Lysis should become very viscous (see Note 14).

7. Add 5 μL of (20 mg/mL) DNase stock to each Eppendorf tube and mix by pipetting. Incubate for 5 min on ice. The solution should return to being fluid. This solution is the crude cell lysate, and should be used immediately in the filter-based aggregation assay to prevent proteolysis.

3.2. Filter-Based Aggregation Assay

1. Collect 10 μL of the crude cell lysate (see Note 11) and dilute with 100 μL of each test buffer (one at a time).

2. Incubate at 4°C for approximately 2 h. During this time you can pour the gels as directed in Subheading 3.4. Add each total 110 μL test solution to a separate filter unit.

3. Spin the filter unit at $15,000 \times g$ for 15–20 min.

4. After spinning, the soluble protein should be in the filtrate and the aggregated protein should be retained by the filter.

5. Resuspend the aggregated protein in 30 μL of ddH$_2$O (pipette repeatedly across membrane surface).

6. Take 30 μL samples of the soluble protein (filtrate) and ~30 μL of the aggregated protein and process these samples for SDS-PAGE as described in Subheading 3.5.

3.3. SDS-PAGE

1. Clean two plates and spacers with water (ddH$_2$O) and ethanol (EtOH); use Kimwipes to dry.

2. Assemble gel cassette using four medium binder clips using the clips to both hold the gel cassette together and allow it to stand vertically.

3. Melt the 1 % agarose solution in a microwave and pipette an 11 cm line of melted agarose on to a clean surface. Quickly set gel sandwich down into the agarose to seal the bottom of the cassette. Leave the cassette standing in the agarose while the gel is cast.

4. Mix 1.25 mL running gel buffer (pH 8.8), 1.25 mL 40 % 29:1 acrylamide solution, and 2.5 mL ddH$_2$O in a 15 mL falcon tube. Mix thoroughly by gently pipetting the mixture. When

ready to pour the gel, add 25 μL 10 % APS and 5 μL TEMED (see Note 15).

5. Quickly mix well and pour the gel solution into preassembled gel plate sandwich measuring 7.25 cm × 10 cm × 0.75 mm gel cassette to desired height, ~ 1 cm below comb. Overlay running gel with isobutanol solution (water-saturated isobutanol). Allow gel to solidify for about 30 min (see Note 16).

6. After gel is solidified, decant the isobutanol layer (in appropriate disposal location) and rinse the remaining interior of the gel cassette with ddH₂O. Dry gel cassette in between the plates with a paper towel.

7. Mix 0.5 mL stacking buffer (pH 6.8), 0.2 mL 40 % acrylamide solution (29:1 mixture), and 1.3 mL ddH₂O in a 15 mL falcon tube. Mix thoroughly by gently pipetting the mixture. When ready to pour stacking gel, add 8 μL 10 % APS and 4 μL TEMED.

8. Quickly mix well and pour stacking gel then place desired comb into the gel sandwich based on the number of samples you need to run (make sure that there are no bubbles in between the teeth of the comb). Allow the gel to solidify for about 30 min.

3.4. Sample Preparation

1. Add approximately 20 μL of protein sample and an equal volume (20 μL) of the 2× sample buffer. Dilute molecular weight markers as instructed by the supplier.

2. Add 5 μL of 2-mercaptoethanol to each sample.

3. Spin down briefly in a desktop centrifuge to collect all of the sample at the bottom of the tube.

4. Heat aliquots at 95°C for 5 min.

5. After gel has solidified carefully remove the comb and fill gel box with electrophoresis running buffer to appropriate level according to the number of gels being set up for electrophoresis.

6. Load the gel carefully with 10 μL sample per well.

7. Electrophorese at 15 mA until the dye front reaches the bottom of the gel (see Note 17).

3.5. Western Blot (see Note 18)

1. Immediately following SDS-PAGE, turn off the power supply. Separate the gel plates with the help of a spatula or similar tool. Remove the stacking gel.

2. Cut out a rectangular piece of the nitrocellulose membrane held only with forceps at least 0.5 cm larger than the gel on all sides. Do not touch membrane with bare hands or gloves.

3. Prepare Western transfer cassette as directed by the supplier. Insert the cassette in the Western transfer unit, taking care that the gel side of the cassette faces the anode, whereas the membrane side of the cassette faces the cathode (see Note 19).

4. Pour western transfer buffer into western transfer box to the appropriate level.

5. Run western transfer: 250 mA for 35 min (see Note 20).

6. Once western transfer is complete, turn off the power supply and remove the sandwich.

7. Open the sandwich and lift one corner of the membrane to determine whether the prestained marker proteins similar in size to your protein have sufficiently transferred (see Note 21). Be careful that the gel and membrane do not move relative to one another in case transfer is incomplete. If transfer is sufficient, remove the membrane from the Western transfer cassette.

8. Using forceps and scissors remove the portion of the membrane that was not in contact with the gel and cut a notch in the top left corner of membrane to mark the original orientation relative to the gel. Wash the membrane in PBST buffer in a small container that can hold enough volume to cover the nitrocellulose membrane.

9. Wash your membrane in PBST by placing the container on a rocker (or provide gentle agitation) at a low speed for approximately 7 min.

10. Incubate membrane in 15–30 mL of 10 % milk solids for 60 min on rocker (see Note 22).

11. Wash the membrane with PBST three times for 7 min each on rocker.

12. Incubate the membrane in primary antibody for 60 min on rocker (see Note 23).

13. Wash the membrane with PBST three times for 7 min each on rocker.

14. After washing the membrane incubate the membrane in secondary antibody for 60 min on rocker (see Note 24).

15. Wash the membrane with PBST three times for 7 min each on rocker.

16. To develop the western, place the nitrocellulose membrane on Saran Wrap.

17. Apply the developing solution ECL (Enhanced Chemiluminescence, Invitrogen, catalogue number WP20005) onto the membrane using the transfer pipette. Make sure that you cover the entire membrane. Incubate the membrane in the liquid for 1 min.

18. Drain the excess solution off the membrane, wrap the membrane in Saran Wrap with only one covering the side of the membrane previously in contact with the gel during transfer, and place inside a cassette.

19. In a dark room with the lights off, place the film over the membrane (Be sure to never expose the film to any light.).

20. Allow the film to be exposed for a desired amount of time (see Note 25).

21. Remove the film after the desired exposure time and develop it accordingly.

22. After developing your film you should see bands that correspond with the molecular weight of the protein being tested. Depending on your results after you see your film, you can re-expose the film for a shorter or longer period of time.

3.6. Analysis and Planning the next Experiment

1. For each test solution, both the soluble (filtrate) and aggregated (retained on the filter) fractions are assessed on the gel or western. For the ideal chemical additive, all of the protein will be in the soluble lane. However, initial experiments may only show that some conditions have more soluble protein or less aggregates than other conditions. The condition that yields the most soluble protein is the best candidate, although it is best to avoid urea or moderate chaotropes if other additives also improve solubility. To select assay buffers for the second experiment, refer to Table 1. Find the category that matches the best candidate additive, and test other additives in that category at the initial test concentration listed. Once the best chemical has been identified, vary the concentration of this additive within the concentration range listed in Table 1. Finally, if multiple additives work, these can be tested in combination to further improve solubility (see Note 26).

4. Notes

1. The test buffers can be most rapidly generated by making stocks of each component at ten times the final concentration, and subsequently diluting to create each individual buffer. The pH of the final diluted solutions may need readjustment.

2. Our basic lysis buffer is designed to be compatible with his-tag chromatography. Other lysis buffers are perfectly acceptable. However, be aware that buffer components such as osmolytes or large quantities of salt may harm protein solubility. In these cases, we recommend using a simple version of the lysis buffers for the purposes of this aggregation assay. The pH of the standard buffer should ideally be more than 1 pH unit above or below the isoelectric point of the protein, since proteins are generally least soluble at their isoelectric point. To predict the isoelectric point for your favorite protein, use the program at the following web link: http://www.scripps.edu/~cdputnam/protcalc.html.

3. *E. coli* cell lines that express lysozyme (include the pLysS or pLysE plasmids) to control protein expression prior to induction do not need to include additional lysozyme in the lysis buffer.

4. The buffer can be made ahead without the lysozyme and protease inhibitors and stored at 4°C. The lysozyme and protease inhibitors must be added just prior to use.

5. Always store acrylamide in the dark if possible, or at least wrapped in aluminum foil. Acrylamide will last longer if stored at 4°C.

6. Store at 4°C for no more than 1 month.

7. 4 L of western transfer buffer are usually made at once and can be reused until the solution begins to yellow or transfer efficiency decreases.

8. 4 L of PBST buffer are usually made at once and stored at room temperature.

9. Add 100 mL of PBST to a glass beaker and transfer 20 g milk solids powder. Stir until dissolved then add PBST to a volume of 200 mL. Allow it to stir for a couple of minutes.

10. These steps can be omitted if purified, partially purified, or aggregated sample is being tested.

11. If you already have cell pellets frozen, you can scrape from the cell pellet the equivalent of 100 μL of cells and proceed directly to step 6. You may also start with partially purified or aggregated protein in lieu of crude cell lysis (17).

12. The remaining cell culture may be set aside for later centrifugation and protein purification if desired.

13. Failure to remove all of the supernatant will result in a more acid pH during lysis which will impede or prevent lysozyme function.

14. If cells fail to lyse, test the pH, which must be greater than 7.6 for hen egg white lysozyme to function efficiently.

15. This protocol produces a 10 % 29:1 acrylamide gel. Small proteins may require a 12.5 % or 15 % gel, and large proteins may require a 7.5 % gel for adequate separation.

16. This overlay prevents contact with atmospheric oxygen (which inhibits acrylamide polymerization) in addition to leveling the resolving gel solution.

17. If you are running one gel, this number should be 15 mA. If you are running two gels it should be 30 mA, for three gels it should be 45 mA, and for four gels it should be 60 mA).

18. If the protein is sufficiently pure or sufficiently concentrated that it can be located on a gel without Western blotting, than Coomassie staining or silver staining can be used in place of the Western blot.

19. Everything should be wetted well with the transfer buffer. Make sure that there are no bubbles in your sandwich and everything is smoothed out. If the nitrocellulose membrane appears to have dry or white spots, then replace it with new membrane. Protein won't transfer to these areas.

20. The time and milliamps required will vary depending on the protein size. Smaller proteins transfer faster.

21. You should be able to see your molecular weight markers on the nitrocellulose membrane and your gel should have little to no markers on it. High molecular weight markers may still be present on the gel due to their slower migration.

22. You can incubate your membrane overnight at 4°C in milk solids if necessary.

23. Antibodies specific to your protein are vital to western blotting as they are able to specifically bind your protein. Dilute the antibody in PBST as specified by the supplier. Make sure that the nitrocellulose membrane is covered with the PBST and primary antibody dilution at all times while incubating.

24. Dilute the antibody in PBST as specified by the supplier. Anti-mouse, anti-rabbit, or other antibody conjugated with HRP enzyme for detection are commonly used. Make sure that the nitrocellulose membrane is covered with the PBST and secondary antibody dilution at all times while incubating.

25. Usually, we do 30 s first followed by 1 min exposure then a 5 min exposure to ensure data with visible, but not overexposed bands.

26. Keep in mind that soluble proteins are not necessarily native. For instance, arginine can solubilize a Zn-finger protein in the absence of Zn (17).

Acknowledgments

This work was supported by grants from the American Heart Association 422351 and the Texas A&M Health Science Center to SEB.

References

1. Bondos SE, Tan XX, Matthews KS (2006) Physical and genetic interactions link Hox function with diverse transcription factors and cell signaling proteins. Mol Cell Proteomics 5:824–834

2. Hazy E, Tompa P (2009) Limitations of induced folding in molecular recognition by intrinsically disordered proteins. Chemphyschem 10:1415–1419

3. Singh GP, Ganapathi M, Dash D (2007) Role of intrinsic disorder in transient interactions of hub proteins. Proteins 66:761–765

4. Dunker AK, Cortese MS, Romero P et al (2005) The roles of intrinsic disorder in protein interaction networks. FEBS J 272:5129–5148

5. Higurachi M, Ishida T, Kinoshita K (2007) Identification of transient hub proteins and the possible structural basis for their multiple interactions. Protein Sci 17:72–78

6. Patil A, Kinoshita K, Nakamura H (2010) Hub promiscuity in protein-protein interaction networks. Int J Mol Sci 11:1930–1943

7. Liu Y, Matthews KS, Bondos SE (2008) Multiple intrinsically disordered sequences alter DNA binding by the homeodomain of the *Drosophila* Hox protein Ultrabithorax. J Biol Chem 283:20874–20887

8. Liu Y, Matthews KS, Bondos SE (2009) Internal regulatory interactions determine DNA binding by a Hox transcription factor. J Mol Biol 390:760–774

9. Iakoucheva LM, Radivojac P, Brown CJ et al (2004) The importance of intrinsic disorder for protein phosphorylation. Nucleic Acids Res 32:1037–1049

10. Chiti F, Dobson CM (2006) Protein misfolding, functional amyloid, and human disease. Annu Rev Biochem 75:333–366

11. Linding R, Schymkowitz J, Rousseau F et al (2004) A comparative study of the relationship between protein structure and β-aggregation in globular and intrinsically disordered proteins. J Mol Biol 342:345–353

12. Tompa P (2009) Structural disorder in amyloid fibrils: its implication in dynamic interactions of proteins. FEBS J 276:5406–5415

13. Oldfield JC, Cheng Y, Cortese MS et al (2005) Coupled folding and binding with α-helix forming molecular recognition elements. Biochemistry 44:12454–12470

14. Wright PE, Dyson HJ (2009) Linking folding and binding. Curr Opin Struct Biol 19:31–38

15. Uversky VN, Oldfield CJ, Midic U et al (2009) Unfoldomics of human diseases: linking protein intrinsic disorder with diseases. BMC Genomics 10:S7

16. Sharma AK, Ali A, Gogna R et al (2009) p53 amino-terminus region (1–125) stabilizes and restores heat denatured p53 wild phenotype. PLoS One 4(10):e7159

17. Bondos SE, Bicknell A (2003) Detection and prevention of protein aggregation before, during, and after purification. Anal Biochem 316:223–231

18. Russo D (2007) The impact of kosmotropes and chaotropes on bulk and hydration shell water dynamics in a model peptide solution. Chem Phys 345:200–211

19. Tsumoto K, Umetsu M, Kumagai I et al (2004) Role of arginine in protein refolding, solubilization, and purification. Biotechnol Prog 20:1301–1308

20. Shukla D, Trout BL (2010) Interaction of arginine with proteins and the mechanism by which it inhibits aggregation. J Phys Chem B 114:13426–13438

21. Vagenende V, Yap MGS, Trout BL (2009) Mechanisms of protein stabilization and prevention of protein aggregation by glycerol. Biochemistry 48:11084–11096

22. Vidanovic D, Askrabic JM, Stankovic M et al (2003) Effects of nonionic surfactants on the physical stability of immunoglobulin G in aqueous solution during mechanical agitation. Pharmazie 58:399–404

23. Zou Q, Bennion BJ, Daggett V et al (2002) The molecular mechanism of stabilization of proteins by TMAO and its ability to counteract the effects of urea. J Am Chem Soc 124:1192–1202

24. Shiraki K, Kudou M, Nishikori S et al (2004) Arginine ethylester prevents thermal inactivation and aggregation of lysozyme. Eur J Biochem 271:3242–3247

25. Bondos SE (2006) Methods for measuring protein aggregation. Curr Anal Chem 2:157–170

26. Nielsen L, Khurana R, Coats A et al (2001) Effect of environmental factors on the kinetics of insulin fibril formation: elucidation of the molecular mechanism. Biochemistry 40:6036–6046

27. Han HY, Yao ZG, Gong CL et al (2010) The protective effects of osmolytes on yeast alcohol dehydrogenase conformational stability and aggregation. Protein Pept Lett 17:1058–1066

Chapter 29

Proteomic Methods for the Identification of Intrinsically Disordered Proteins

Agnes Tantos and Peter Tompa

Abstract

Intrinsically disordered proteins (IDPs) lack fixed 3D structure under physiological conditions, yet they often carry out critically important physiological functions. The first few disordered proteins have been discovered one-by-one from clues that suggested that a protein lacks structure. Since bioinformatic predictions suggest that a large portion of eukaryotic proteomes contains significant levels of protein disorder, a reliable method for the large-scale separation and identification of these proteins is needed. IDPs do not undergo large-scale structural changes and aggregation at low pH or elevated temperatures. Thus, such proteins are likely to remain soluble under these extreme conditions, making acid treatment and/or heat treatment suitable for substantial enrichment of intrinsically unfolded proteins in the soluble fraction.

Key words: Intrinsically disordered proteins, Large-scale separation, Proteomics, 2D electrophoresis, Heat-stable proteins

1. Introduction

Atom positions and backbone Ramachandran angles of many proteins under physiological conditions vary dynamically with no specific equilibrium values, leading to ensembles of interconverting forms (1). These proteins have been named intrinsically disordered. In this chapter we focus on the separation and identification of proteins that are mostly, if not totally, disordered, with few or no segments that fold into well-structured regions.

IDPs of extended conformations and high percentages of charged residues do not undergo large-scale structural changes at low pH. Thus, such proteins are likely to remain soluble under these extreme conditions (2). For structured proteins, the protonation of negatively charged side chains leads to charge imbalances

Vladimir N. Uversky and A. Keith Dunker (eds.), *Intrinsically Disordered Protein Analysis:*
Volume 2, Methods and Experimental Tools, Methods in Molecular Biology, vol. 896,
DOI 10.1007/978-1-4614-3704-8_29, © Springer Science+Business Media New York 2012

98 °C

MW
(kDa)

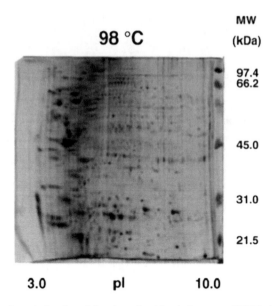

97.4
66.2

45.0

31.0

21.5

3.0   pI   10.0

Fig. 1. 2D gel analysis of proteins from heat-treated mouse NIH3T3 fibroblast cell extracts. Picture is reproduced from (12).

that disrupt salt bridges and causes the disassociation of subunits and cofactors, leading to the random coil state (3). Unlike in the case of IDPs, the induced random coils of structured proteins typically contain large numbers of hydrophobic groups, and the exposure of these normally buried hydrophobic residues usually leads to aggregation and precipitation (4). Some structured proteins adopt a less open conformation in acid pH, often referred to as the molten globular "A" state which may be "sticky" (5). For example, proteins that rely on buried salt bridges are typically among those that aggregate and precipitate upon transition to the molten globular state (6). Since the aggregation and precipitation of the A state can also depend on the anions present in the solution, not all acids precipitate such proteins quantitatively. Some structured proteins adopt non-sticky A states and thus remain in solution (7), others might be exceptionally stable and remain folded and soluble even at extremely low pH (8). Overall, the expectation is that acid treatment should lead to substantial enrichment of IDPs in the soluble fraction. Their intrinsic properties can be exploited to develop standard protocols to study them on a proteomic scale. That is, totally disordered proteins are separable from structured proteins by their intrinsic indifference to denaturing conditions that originates from their lack of tertiary structure (1) (Fig. 1).

2. Materials

Prepare all solutions using deionized water and analytical grade reagents. Prepare and store all reagents at room temperature (unless indicated otherwise).

2.1. Lysis Buffer for Bacterial and Yeast Cells

Prepare a 50 mM Tris–HCl and 150 mM NaCl containing buffer (see Note 1) by mixing 3.5 g of Tris–HCl and 4.4 g NaCl and dissolving them in about 150 mL water. Adjust the pH to 7.5 using 5 M HCl solution and adjust the volume to 500 mL. Add 2 mM dithioeritritol and one tablet of Complete Protein Inhibitor by Roche before use (see Note 2).

2.2. Acid Treatment

60 % Perchloric acid (PCA) or trichloroacetic acid (TCA): pour 860 of 70 % PCA or 600 mL of 99.9 % or TCA into a graded cylinder and add deionized water up to 1 L (see Note 3).

2.3. 2D Electrophoresis

2.3.1. Isoelectric Focusing

Rehydration

Rehydration solution: prepare a 8 M urea, 0.5 % CHAPS, 15 mM DTT, and 0.2 % ampholyte solution. To make 50 mL of the solution, dissolve 25.0 g of urea in deionized water and complete to 50 mL. To 48 mL of this solution add 0.25 g of CHAPS, 100 mg of DTT, and 0.25 mL IPG buffer (bought commercially in the appropriate pH range) and complete to 50 mL with deionized water. Rehydration solution should be prepared freshly on the day of use (see Note 4).

Sample solubilization buffer (9 M urea, 2 % CHAPS, 0.8 % ampholyte solution, 15 mM DTT): To make 50 mL of the solution, dissolve 28.0 g of urea in deionized water and complete to 50 mL. To 48 mL of this solution add 1 g of CHAPS, 100 mg of DTT, and 0.25 mL IPG buffer and complete to 50 mL with deionized water.

Strip Equilibration

Resolving gel buffer: 1.5 M Tris–HCl, pH 8.8 and 0.4 % (w/v) SDS. To make 100 mL, dissolve 18.2 g of Trizma base and 0.4 g of SDS in about 80 mL of deionized water. Adjust to pH 8.8 with 5 M HCl and fill up to 100 mL with deionized water and filter. The buffer can be stored at 4 °C up to 2 weeks.

Equilibration buffer: 6 M urea, 30 % (w/v) glycerol and 2 % (w/v) SDS in 0.05 M Tris–HCl buffer, pH 8.8. To make 500 mL add: 180 g of urea, 150 g of glycerol, 10 g of SDS and 16.7 mL of resolving gel buffer. Dissolve in deionized water and fill up to 500 mL. The buffer can be stored at room temperature up to 2 weeks.

Bromophenol Blue solution: 0.25 % (w/v) of Bromophenol blue in resolving gel buffer. To make 10 mL: dissolve 25 mg of Bromophenol blue in 10 mL of resolving gel buffer. Store at 4 °C.

2.3.2. Vertical SDS Gel

Acrylamide/Bisacrylamide solution (30.8 % T, 2.6 % C): 30 % (w/v) acrylamide and 0.8 % (w/v) methylenebisacrylamide in deionized water. To make 1,000 mL, dissolve 300.0 g of acrylamide and 8.0 g of methylenebisacrylamide in deionized water and fill up to 1,000 mL and filter.

Resolving gel buffer: 1.5 M Tris–HCl, pH 8.6 and 0.4 % (w/v) SDS. To make 500 mL, dissolve 90.85 g of Trizma base and 2 g of SDS in about 400 mL of deionized water. Adjust to pH 8.6 with 5 M HCl and fill up to 500 mL with deionized water and filter.

Ammonium persulfate solution: 10 % (w/v) of ammonium persulfate in deionized water. To prepare 10 mL of the solution, dissolve 1.0 g of ammonium persulfate in 10 mL of deionized water. This solution should be prepared freshly just before use.

Overlay buffer: buffer-saturated 2-butanol. To make 30 mL, mix 20 mL of resolving gel buffer with 30 mL of 2-butanol, wait for a few minutes and pipette off the butanol layer.

Displacing-solution: 50 % (v/v) glycerol in deionized water and 0.01 % (w/v) Bromophenol blue. To make 500 mL, mix 250 mL of glycerol (100 %) with 250 mL of deionized water, add 50 mg of Bromophenol blue and stir for a few minutes.

3. Methods

All procedures should be carried out at room temperature unless otherwise specified.

3.1. Protein Extracts

3.1.1. Bacterial Extract

Grow 500 mL culture of cells up to an O.D. of 0.8 then harvest the cells by centrifugation at 4,000 × g for 20 min at 4 °C. Resuspend cells in 10 mL of lysis buffer and disrupt cells by sonication (see Note 5). Centrifuge cells at 100,000 × g for 40 min at 4 °C. Use the supernatant for further experiments.

3.1.2. Yeast Cell Extract

Harvest and resuspend the cells as with bacterial cells. Disrupt cells with three freeze–thaw cycles by immersing the cells in liquid nitrogen (see Note 6). Homogenize cells with a Potter–Elvehjem tissue grinder and sonicate cells the same way as with bacterial cells. Centrifuge cells at 100,000 × g for 40 min at 4 °C. Use the supernatant for further experiments.

3.2. Acid Treatment

Add 60 % PCA or TCA to the soluble protein extract to reach a final concentration between 1.5 and 15 % (see Note 7). After incubation on ice for 15 min, stop the reaction by centrifuging at 4,500 × g for 20 min at 4 °C. Precipitate soluble proteins by adding 8 volumes of acetone to your sample and leave it at −20°C overnight. Collect the pellet by centrifuging the sample at 14,000 × g for 10 min at 4 °C.

3.3. Heat Treatment

Heat treatment can be applied alone or in combination with acid treatment. Place your sample in a heat-resistant container and boil it up to 15 min. Cool the sample on ice and remove precipitated proteins by centrifuging at $100,000 \times g$ for 30 min at 4 °C.

3.4. 2D Electrophoresis

3.4.1. First Dimension: Isoelectric Focusing

Solubilize proteins with sample solubilization buffer and dilute the extract (protein conc.: $\approx$10 mg/mL) with sample solubilization buffer to a final volume of 400–500 µl for 180 mm long IPG (immobilized pH gradient) strips.

In-Gel Rehydration of Sample (see Note 8)

Pipette 350–400 µl of sample-containing rehydration solution into the grooves of the reswelling tray. Peel off the protective cover sheets from the IPG strips and insert the IPG strips (gel side-down) into the grooves. Avoid trapping air bubbles. Cover the IPG strips with 1 mL of silicone oil, close the lid and let the strips rehydrate overnight (see Note 9).

Running IPG Strips After In-Gel Rehydration

Cover the flat-bed cooling block with 2–3 mL of kerosene and place the IPG gel strips, side by side and 1–2 mm apart, on it. The acidic end of the IPG gel strips must face towards the anode!

Cut two isoelectric focusing (IEF) electrode strips to a length corresponding to the width of all IPG gel strips (and spaces between them) lying on the cooling plate. Soak the electrode strips with deionized water and remove excessive moisture by blotting with filter paper. Place the IEF electrode strips on top of the aligned IPG gel strips at the cathodic and anodic gel ends (see Note 10). Position the electrodes and press them gently down on top of the IEF electrode strips. Place the lid on the electrofocusing chamber, start IEF. Running conditions depend on the pH gradient and the length of the IPG gel strips (see Note 11). For improved sample entry, voltage is limited to 150, 300, and 600 V for 1 h each, followed by 3,500 V to the steady state. Current is limited to 0.05 mA/IPG strip. Optimum focusing temperature is 20°C (9). After IEF, those IPG gel strips which are not used immediately for second dimension run or are kept for further reference are stored between two sheets of plastic film at −78°C up to several months.

3.4.2. Strip Equilibration (see Note 12)

Dissolve 100 mg of DTT in 10 mL of equilibration buffer. Place the focused IPG gel strips into individual test tubes. Add 10 mL of equilibration buffer and 50 µl of the Bromophenol Blue solution. Seal the test tubes and rock them 15 min on a shaker, then pour off the buffer.

Dissolve 400 mg of iodoacetamide in 10 mL of equilibration buffer (= equilibration buffer II). Add equilibration buffer II and 50 µL of Bromophenol Blue solution to the test tube and equilibrate for another 15 min on a rocker.

After equilibration, rinse the IPG gel strip with deionized water for a second and place it on a piece of filter paper at one edge for a few minutes to drain off excess buffer.

Table 1
Recipes for casting vertical SDS gels (10%T, 12.5 % T, or 15 % T)

	10 %	12.5 %	15 %
Acrylamide/Bisacrylamide (30.8%T, 2.6%C)	20 mL	25 mL	30 mL
Gel Buffer	15 mL	15 mL	15 mL
Glycerol (100 %)	3 g	3 g	3 g
TEMED (100 %)	3 μL	3 μL	3 μL
Ammonium persulfate (10 %)	0.3 mL	0.3 mL	0.3 mL
Water	up to 60 mL	up to 60 mL	up to 60 mL

Table 2
Running conditions for ten vertical SDS gels (13 % T homogeneous, 2.6 % C, size 250 × 200 × 1 mm)

Time (h)	Voltage (V)	Current (mA)	Power (W)	Temp. (°C)
18	150	150	50	15

3.4.3. Second Dimension: SDS PAGE Run

Stack a polymerization cassette vertically into the gel casting box.

Immediately before gel casting add TEMED and ammonium persulfate solutions to the gel solution (Table 1). To cast the gels, the gel solution (about 60 mL) is poured between the two glass plates. Avoid introduction of any air bubbles into solution.

Very carefully pipette about 1 mL of overlay buffer onto the top of the gel in order to obtain a smooth, flat gel top surface. Allow the gels to polymerize for at least 3 h at room temperature. Once the SDS gel is polymerized, drain the excess water from the surface and the second dimension is ready to be run. Fill the electrophoresis chamber with electrode buffer and turn on cooling (15 °C).

Equilibrate the IPG gel strips as described in the strip equilibration step and immerse them in electrode buffer for a few seconds. Place the IPG gel strip on top of the SDS gel and overlay it with 2 mL of hot agarose solution (75°C). Carefully press the IPG strip with a spatula onto the surface of the SDS gel to achieve complete contact. Allow the agarose to solidify for at least 5 min. Run the SDS-PAGE gels overnight (Table 2). Terminate the run when the Bromophenol Blue tracking dye has migrated off the lower end of the gel. Open the cassettes carefully and remove the agarose overlay from the polyacrylamide gel. Peel the gel off the glass plate carefully, lifting it by the lower edge and place it in a tray, then continue with fixing, protein staining, or blotting.

3.5. Staining

The gels can be stained either with Coomassie blue, silver, or fluorescent staining. The method of choice depends on the sensitivity needed. With Coomassie staining 36–47 ng protein can be detected, with silver staining this number is 0.5–1.2 ng, and with fluorescent staining it is 1–2 ng (see Note 13).

4. Notes

Here are a few practical notes, which help avoid most of the problems when using this method.

1. High concentrations of salt can pose problems to the integrity of the IEF gel. The salts present in the buffer can cause the amperage to soar resulting in deformation of the IPG strips which can be so severe that the strips fuse completely, halting the current.

2. The buffer can be stored at 4 °C for a couple of months, but DTE and the proteinase inhibitor mix must be added freshly since they are unstable in solution.

3. Always wear protective clothing (eye mask, gloves, and lab coat) when handling strong acids.

4. IPG strips are plastic backed, they can be bought in various lengths and in a variety of pH ranges. They are usually 3.0 mm wide and about 0.5 mm thick when rehydrated. Since they come dehydrated, they must be rehydrated before use.

5. Sonicate cells in a container immersed in ice and leave 10 s spaces between 10-s long sonication periods.

6. Be very careful when handling liquid nitrogen. Wear protective eye mask and freeze-resistant gloves.

7. In some cases 3 % PCA is sufficient to denature and precipitate all nonresistant proteins. TCA is less efficient than PCA and treatment with higher concentrations of TCA results in different sets of isolated proteins, so the combination of the two treatments is the best to cover all the potentially disordered proteins in a sample.

8. In-gel rehydration is beneficial since it allows you to load large quantities of your sample. Other methods have been suggested for applying larger sample volumes (10) that seem fairly simple to use.

9. After the IPG gel strips have been rehydrated, rinse them with deionized water for a second and place them, gel side up, on a sheet of water saturated filter paper. Wet a second sheet of filter paper with deionized water, place it onto the surface of the IPG gel strips and blot them gently for a few seconds. This is necessary since the remaining urea can crystallize on the surface of the gel during IEF.

10. When running basic IPGs (e.g. IPG 6-10 or 8-12), put an extra paper strip soaked with 20 mM DTT on the surface of the IPG strips next to the cathode.

11. To have a rough idea whether the IEF is running correctly, watch the bromophenol blue front carefully. It should slowly migrate toward the anode. If it does not migrate discretely or breaks down into a number of bands, be very suspicious, you may have a problem. Keep in mind though that the bromophenol blue migration cannot be used to measure the progress of the IEF run, it simply gives you a point of reference. Once the run is complete the strips can either be used immediately or frozen down at −78°C for use later.

12. Strip equilibration before using for SDS PAGE is necessary for a number of reasons (11) which are reflected in the components of the equilibration buffer. As the second dimension is SDS-PAGE, it is important that the proteins are treated with SDS to give them a negative charge. The equilibration is normally carried out in two steps. In the first step DTT is added to ensure that any reformed disulphide bridges are reduced. The second step is the addition of iodoacetamide. This alkylates the proteins and reacts with any unreduced DTT.

13. Be careful with fluorescent stains, read the product literature before choosing a product because many stains cannot be used with plastic backed gel systems as the plastic backing or the gelbond is fluorescent at the excitation wavelength of the stain and so produces unacceptable background levels.

References

1. Cortese MS, Baird JP, Uversky VN, Dunker AK (2005) Uncovering the unfoldome: enriching cell extracts for unstructured proteins by acid treatment. J Proteome Res 4:1610–1618
2. Uversky VN, Gillespie JR, Millett IS, Khodyakova AV, Vasiliev AM, Chernovskaya TV, Vasilenko RN, Kozlovskaya GD, Dolgikh DA, Fink AL, Doniach S, Abramov VM (1999) Natively unfolded human prothymosin alpha adopts partially folded collapsed conformation at acidic pH. Biochemistry 38:15009–15016
3. Dill KA, Shortle D (1991) Denatured states of proteins. Annu Rev Biochem 60:795–825
4. Fink AL, Calciano LJ, Goto Y, Kurotsu T, Palleros DR (1994) Classification of acid denaturation of proteins: intermediates and unfolded states. Biochemistry 33:12504–12511
5. Sivaraman T, Kumar TK, Jayaraman G, Yu C (1997) The mechanism of 2,2,2-trichloroacetic acid-induced protein precipitation. J Protein Chem 16:291–297
6. Schafer K, Magnusson U, Scheffel F, Schiefner A, Sandgren MO, Diederichs K, Welte W, Hulsmann A, Schneider E, Mowbray SL (2004) X-ray structures of the maltose-maltodextrin-binding protein of the thermoacidophilic bacterium Alicyclobacillus acidocaldarius provide insight into acid stability of proteins. J Mol Biol 335:261–274
7. Roychaudhuri R, Sarath G, Zeece M, Markwell J (2004) Stability of the allergenic soybean Kunitz trypsin inhibitor. Biochim Biophys Acta 1699:207–212
8. Tucker DL, Tucker N, Conway T (2002) Gene expression profiling of the pH response in Escherichia coli. J Bacteriol 184:6551–6558
9. Gorg A, Postel W, Friedrich C, Kuick R, Strahler JR, Hanash SM (1991) Temperature-dependent spot positional variability in two-dimensional polypeptide patterns. Electrophoresis 12:653–658

10. Sabounchi-Schutt F, Astrom J, Olsson I, Eklund A, Grunewald J, Bjellqvist B (2000) An immobiline DryStrip application method enabling high-capacity two-dimensional gel electrophoresis. Electrophoresis 21:3649–3656

11. Gorg A, Postel W, Domscheit A, Gunther S (1988) Two-dimensional electrophoresis with immobilized pH gradients of leaf proteins from barley (Hordeum vulgare): method, reproducibility and genetic aspects. Electrophoresis 9:681–692

12. Galea CA, Pagala VR, Obenauer JC, Park CG, Slaughter CA, Kriwacki RW (2006) Proteomic studies of the intrinsically unstructured mammalian proteome. J Proteome Res 5:2839–2848

Chapter 30

Selective Isotope Labeling of Recombinant Proteins in *Escherichia coli*

Kit I. Tong, Masayuki Yamamoto, and Toshiyuki Tanaka

Abstract

Selective stable-isotope labeling is a useful technique to study structures of proteins, especially intrinsically disordered proteins, by nuclear magnetic resonance spectroscopy. Here, we describe a simple method for amino acid selective isotope labeling of recombinant proteins in *E. coli*. This method only requires addition of an excess of unlabeled amino acids and, if necessary, enzyme inhibitors to the culture medium. Its efficiency has been demonstrated even in labeling with glutamine or glutamate that is easily converted to other amino acid types by the metabolic pathways of *E. coli*.

Key words: Amino acid selective isotope labeling, Intrinsically disordered protein, ^{15}N-Glutamine labeling, ^{15}N-Glutamate labeling

1. Introduction

There are increasing numbers of intrinsically disordered proteins that lack well-formed globular structures under physiological conditions, but can acquire rigid tertiary folds by binding with ligands or other proteins (1). Nuclear magnetic resonance (NMR) spectroscopy is one of the most suitable tools to investigate the structures of these proteins. However, it is usually challenging to study their structures in the free state due to extensive overlap of signals and weak intensities of sequential nuclear Overhauser effects. The problem of resonance degeneracy is also common when analyzing large proteins of molecular weight greater than 50 kDa. One of the solutions is to employ selective stable-isotope labeling to reduce ambiguities in resonance assignments. This labeling technique has already been widely used (2–5) before multidimensional heteronuclear NMR methods were developed (6–10) for more efficient completion of three-dimensional structural analysis using

Vladimir N. Uversky and A. Keith Dunker (eds.), *Intrinsically Disordered Protein Analysis: Volume 2, Methods and Experimental Tools*, Methods in Molecular Biology, vol. 896, DOI 10.1007/978-1-4614-3704-8_30, © Springer Science+Business Media New York 2012

uniformly [15]N- and/or ^{13}C-labeled proteins of interest, which are primarily prepared by expressing in *E. coli*.

To date, recombinant proteins have been selectively labeled with stable isotopes by expressing proteins directly in *E. coli* or insect cells (11) and by in vitro protein synthesis using bacterial or eukaryotic cell extracts. Cell-free protein synthesis, a technique described long time ago (12), has been improved enormously in the last 20 years (13–18). Several cell-free protein expression kits are now commercially available for mid- to large-scale protein production for NMR studies. Although proven successful in a number of ways such as low isotopic scrambling, possible expression of toxic or membranous proteins, lower consumption of expensive isotope-labeled amino acids, and ease of use, many users still experience low protein yield at the expense of high budget on the in vitro protein synthesis kit itself (17, 19). Furthermore, proteins cannot be specifically labeled with amino acids such as glutamine (Gln) and glutamate (Glu) by a popular commercial prokaryote-based cell-free protein expression system since large amount of glutamate salt is present in the reaction buffer (17, 20). Therefore, there is still a good demand for the use of selective stable-isotope labeling by expressing proteins directly in *E. coli*. However, in vivo interconversion of amino acids due to the metabolic pathways still poses a potential problem in labeling proteins with certain amino acids specifically. Here is the description of a selective stable-isotope labeling method to overcome the endogenous metabolic activities in bacteria by manipulating the product feedback inhibitory loops of amino acid metabolic pathways without using transaminase-deficient or auxotrophic mutant strains. This method has proved useful even for labeling with amino acids that are easy to be converted, such as Gln and Glu (21, 22). In addition to intrinsically disordered proteins, this method is also a good way to label well-folded proteins and, together with segmental isotope labeling techniques (23–25), proteins of high molecular weight in *E. coli*.

2. Materials

All reagents described in this protocol are prepared at room temperature with analytical grade chemicals and ultrapure water with a resistivity of 18 MΩ cm at 25°C prepared by a Millipore Milli-Q water purification system.

2.1. Protein Expression

1. LB medium: Combine 10 g of tryptone, 5 g of yeast extract, 5 g of NaCl, and 5 mL of 1 M Tris–HCl solution, pH 8.0. Add ultrapure water up to 1 L and autoclave on liquid cycle at 121°C for 20 min.

Table 1
Basic components of M9 medium[a]

Anhydrous sodium phosphate dibasic (Na_2HPO_4)	6 g
Anhydrous potassium phosphate monobasic (KH_2PO_4)	3 g
Sodium chloride (NaCl)	0.5 g
Ammonium chloride (NH_4Cl)	1 g
D-glucose	5 g

[a]Per liter of medium

2. 10-cm LB agar plates supplemented with appropriate antibiotics to select bacterial cells that contain the expression vector in use (see Note 1).

3. M9 medium base (see Table 1).

4. Anhydrous magnesium sulphate ($MgSO_4$), calcium chloride dihydrate ($CaCl_2 \cdot 2H_2O$), and sodium hydroxide (NaOH).

5. Antibiotics (for example, ampicillin) (see Note 1).

6. Unlabeled and selected isotopically labeled L-amino acids (see Tables 2 and 3).

7. Chemical enzyme inhibitors (see Table 4).

8. T7 promoter-driven prokaryotic expression vector with the gene of interest (see Note 2).

9. BL21-CodonPlus(DE3)-RIPL chemical competent cells (Stratagene) (see Note 2).

10. Isopropyl β-D-1-thiogalactopyranoside (IPTG) (see Note 3).

11. 0.22-μm filtration unit (for example, Corning vacuum filter/ bottle storage systems) to filter-sterilize medium.

12. Autoclave sterilizer.

13. Autoclaved 50-mL and 4-L Erlenmeyer flasks.

14. Incubator and incubator shaker.

15. Visible light spectrophotometer.

16. Disposable semimicrocuvettes.

17. Refrigerated centrifuge.

18. Centrifuge bottles.

2.2. Protein Purification and NMR Sample Preparation

1. Purification columns.

2. Protein concentrators (for example, Millipore Amicon Ultra-4 and -15 centrifugal filter units).

Table 2
Unlabeled amino acid supplement at the beginning of culture[a]

	Type of labeling					
	$^{15}N_2$-Lysine (g)	$^{15}N_1$-Leucine (g)	$^{15}N_4$-Arginine (g)	$^{15}N_1$-Glutamine (g)	^{15}N-Glutamate (g)	$^{15}N_2/^{13}C_5$-Glutamine and α-^{15}N-Lysine (g)
L-Alanine (A)	0.1	0.1	0.1	0.1	0.1	0.1
L-Cysteine (C)	0.1	0.1	0.1	0.1	0.1	0.1
L-Aspartate (D)	0.1	0.1	0.1	0.1	0.1	0.1
L-Glutamate (E)	0.1	0.1	0.1	–	–	–
L-Phenylalanine (F)	0.1	0.1	0.1	0.1	0.1	0.1
L-Glycine (G)	0.1	0.1	0.1	0.1	0.1	0.1
L-Histidine (H)	0.1	0.1	0.1	0.1	0.1	0.1
L-Isoleucine (I)	0.1	0.1	0.1	0.1	0.1	0.1
L-Lysine (K)	–	0.1	0.1	0.1	0.1	–
L-Leucine (L)	0.1	–	0.1	0.1	0.1	0.1
L-Methionine (M)	0.1	0.1	0.1	0.1	0.1	0.1
L-Asparagine (N)	0.1	0.1	0.1	0.1	0.1	0.1
L-Proline (P)	0.1	0.1	0.1	0.1	0.1	0.1
L-Glutamine (Q)	0.1	0.1	0.1	–	–	–
L-Arginine (R)	0.1	0.1	–	0.1	0.1	0.1
L-Serine (S)	0.1	0.1	0.1	0.1	0.1	0.1
L-Threonine (T)	0.1	0.1	0.1	0.1	0.1	0.1
L-Valine (V)	0.1	0.1	0.1	0.1	0.1	0.1
L-Tryptophan (W)	0.1	0.1	0.1	0.1	0.1	0.1
L-Tyrosine (Y)	0.1	0.1	0.1	0.1	0.1	0.1

[a]Per liter of medium

3. Methods

3.1. Medium Preparation

1. For 1 L of medium, combine the M9 medium base shown in Table 1 and the rest of the components shown in Table 2 (see Note 4) according to the type of labeling intended to make.
2. Add ultrapure water up to 1 L.

Table 3
Unlabeled and labeled amino acid supplements shortly before protein induction[a]

	Type of labeling					
	$^{15}N_2$-Lysine	$^{15}N_1$-Leucine	$^{15}N_4$-Arginine	$^{15}N_1$-Glutamine	^{15}N-Glutamate	$^{15}N_2/^{13}C_5$-Glutamine and α-^{15}N-Lysine
L-Alanine (A)	1 g	1 g	1 g	1 g	3 g	1 g
L-Cysteine (C)	1 g	1 g	1 g	1 g	1 g	1 g
L-Aspartate (D)	1 g	1 g	1 g	1 g	4 g	1 g
L-Glutamate (E)	1 g	1 g	1 g	2 g	–	2 g
L-Phenylalanine (F)	1 g	1 g	1 g	1 g	3 g	1 g
L-Glycine (G)	1 g	1 g	1 g	1 g	1 g	1 g
L-Histidine (H)	1 g	1 g	1 g	1 g	1 g	1 g
L-Isoleucine (I)	1 g	1 g	1 g	1 g	4 g	1 g
L-Lysine (K)	–	1 g	1 g	1 g	1 g	–
L-Leucine (L)	1 g	–	1 g	1 g	3 g	1 g
L-Methionine (M)	1 g	1 g	1 g	1 g	1 g	1 g
L-Asparagine (N)	1 g	1 g	1 g	1 g	3 g	1 g
L-Proline (P)	1 g	1 g	1 g	1 g	1 g	1 g
L-Glutamine (Q)	1 g	1 g	1 g	–	4 g	–
L-Arginine (R)	1 g	1 g	–	1 g	1 g	1 g
L-Serine (S)	1 g	1 g	1 g	1 g	1 g	1 g
L-Threonine (T)	1 g	1 g	1 g	1 g	1 g	1 g
L-Valine (V)	1 g	1 g	1 g	1 g	3 g	1 g
L-Tryptophan (W)	1 g	1 g	1 g	1 g	1 g	1 g
L-Tyrosine (Y)	1 g	1 g	1 g	1 g	4 g	1 g
$^{15}N_2$-Lysine	100 mg	–	–	–	–	–
$^{15}N_1$-Leucine	–	100 mg	–	–	–	–
$^{15}N_4$-Arginine	–	–	100 mg	–	–	–
$^{15}N_1$-Glutamine	–	–	–	100 mg	–	–
^{15}N-Glutamate	–	–	–	–	100 mg	–
$^{15}N_2/^{13}C_5$-Glutamine	–	–	–	–	–	100 mg
α-^{15}N-Lysine	–	–	–	–	–	100 mg

[a]Per liter of medium

Table 4
Enzyme inhibitor supplement shortly before protein induction[a]

	Type of labeling		
	$^{15}N_1$-Glutamine	^{15}N-Glutamate	$^{15}N_2/^{13}C_5$-Glutamine and α-^{15}N-Lysine
6-Diazo-5-oxo-L-norleucine	75 mg	–	75 mg
L-Methionine sulfoximine	180 mg	1 g	180 mg
L-Methionine sulfone	180 mg	1 g	180 mg
Disodium succinate	–	250 mg	–
Disodium maleate	–	250 mg	–
Aminooxyacetate	–	250 mg	–

[a]Per liter of medium

3. Add 1 mL of 1 M $MgSO_4$ and 1 mL of 100 mM $CaCl_2$ with gently rotating the container by hand (see Note 5).

4. Supplement the medium with appropriate antibiotics according to the expression vector in use (see Note 1).

5. Adjust pH to 7.4 with 1 M NaOH and filter-sterilize by a 0.22-μm filtration unit.

6. Transfer the sterilized medium into an autoclaved 4-L flask.

3.2. Transformation of E. coli

1. Add about 100 ng of the prokaryotic expression vector containing the gene of interest to an aliquot of the $CaCl_2$-treated competent cells (see Note 6) and mix gently.

2. After 15 min incubation on ice, heat-shock the cells at 42°C for 2 min and immediately place back on ice for at least 2 min (see Notes 7 and 8).

3. Spread the cells on LB agar plates containing antibiotics. The plate is then incubated at 37°C overnight (see Notes 9 and 10).

3.3. Small-Scale Preculture

1. On the second day, store the plate at 4°C until use (see Note 11).

2. Pick a single colony and transfer it into 15 mL of LB medium containing antibiotics in an autoclaved 50-mL flask. Incubate the liquid culture at 37°C in an incubator shaker at 150–200 rpm for overnight (usually about 16 h until the next day).

3.4. Large-Scale Protein Production

1. On the third day, start the large-scale culture by adding 10 mL of the overnight culture to the 4-L flask containing 1 L of the medium prepared (see Subheading 3.1) and incubate at 37°C with shaking at 200 rpm.

2. Monitor the growth of the cells until the optical density at 600 nm (OD_{600}) reaches approximately 0.80 (see Note 12).

3. Add the additional unlabeled amino acids shown in Table 3 and the enzyme inhibitors shown in Table 4, as solid or liquid (see Notes 13 and 14), to the culture according to the type of amino acid selective labeling (see Note 15).

4. Gently rotate the flask by hand until most of the solid chemicals are dissolved in the culture medium. Then add the isotopically labeled amino acid (see Table 3) in a solution form and continue to incubate at 37°C in the incubator shaker at 200 rpm (see Note 16).

5. Fifteen minutes after the addition of the labeled amino acid, add 1 mL of 0.5 M IPTG for protein induction (see Note 17).

6. After 90 min further incubation at 37°C in the incubator shaker at 200 rpm, transfer the culture to centrifuge bottles and centrifuge at 6,000 × *g* in a refrigerated centrifuge for 10 min. Decant and discard the used medium. Scoop up the bacterial cell pellet and place it into a clean 50-mL polypropylene conical centrifuge tube (see Note 18).

7. Proceed with protein purification or store the cells at −20°C until use.

3.5. Protein Purification and NMR Sample Preparation

1. Purify the protein of interest according to the procedure that is used to purify the unlabeled protein.

2. Exchange the buffer and concentrate the purified protein for NMR data collection.

4. Notes

1. Choose an antibiotic based on antibiotic-resistance genes of the expression vector in use.

2. For high-level protein expression, we usually use the pET vectors (Novagen) and the *E. coli* strain BL21-CodonPlus(DE3)-RIPL, which contains extra copies of tRNA genes that are rare in *E. coli* (*argU*, *ileY*, *proL*, and *leuW*). Other combination of prokaryotic expression vector and *E. coli* strain that show optimal expression of the protein of interest is also applicable.

3. IPTG is used for the induction of T7 RNA polymerase in an *E. coli* strain carrying the lambda DE3 lysogen, resulting in the expression of the T7 promoter-driven recombinant protein. Other induction methods are also applicable. Use an appropriate induction method for the expression system in use.

4. Table 2 includes the demonstrated examples of amino acid selective isotope labeling that were published in refs. (21) and (22). In general, 0.1 g of each unlabeled amino acid except the ones that are used to label proteins will be added to 1 L of medium. However, when you want to label proteins with either Gln or Glu, you should not add both unlabeled Gln and Glu, since they are easily converted to each other.

5. Add $MgSO_4$ and $CaCl_2$ solutions after filling up to 1 L, since magnesium and calcium ions are easy to precipitate in the presence of high concentration of phosphate ions.

6. Homemade or commercially available. For the procedure of preparing homemade competent cells by $CaCl_2$ treatment, please refer to the published literature such as Molecular cloning: a laboratory manual (26) or Current protocols in molecular biology (27).

7. Although we usually use this setting for heat shock transformation, other combination of incubation temperature and time is also successful: for example, 37°C for 5 min.

8. Alternatively, introduction of the plasmid DNA into the bacterial cells can be done by electroporation. In this case, an electroporator will be needed for this step and bacterial cells have to be prepared in a buffer of low ionic strength. Please refer to the published literature (26, 27) or the instrument's manual for details.

9. Usually 12–16 h incubation or until the colonies reach a size of about 1 mm in diameter. Longer incubation times may cause the formation of satellite colonies, as tiny dots around each single colony of interest, due to the degradation of antibiotics.

10. Other incubation temperature may be used depending on the expression vector in use.

11. The use of freshly transformed cells is recommended for optimal expression of protein.

12. Optimal OD_{600} for protein induction has to be determined empirically for each vector-host combination to obtain optimal protein yield before producing the selectively amino acid labeled protein. Note that a shorter induction time of 1.5 h is required to avoid metabolic scrambling or the appearance of undesired cross labeling (see step 6 under Subheading 3.4).

13. It is preferable to add additional unlabeled amino acids as solid in order not to change the culture volume. Protein expression is not affected by adding amino acid powder.

14. Enzyme inhibitors can be added as solid with the amino acid powder mixture or as a 0.22-μm filtered solution.

15. Tables 3 and 4 contain the demonstrated examples of amino acid selective isotope labeling described in refs. (21) and (22). In most cases, the addition of 1 g of each unlabeled amino acid except the ones that are used to label proteins to 1 L of medium will be sufficient to suppress undesired cross labeling. However, there are several amino acids that are easily converted to other amino acid types, such as Gln and Glu. For these kinds of amino acids, one may need to increase the amounts of unlabeled amino acids and/or add inhibitors of the enzymes responsible for metabolic conversion. In Tables 3 and 4, you can see such fine-tuning of the labeling conditions in the case of the selective labeling with Gln or Glu. The followings are useful online references: EcoCyc (http://EcoCyc.org) (28) for the metabolic network of *E. coli* and BRENDA (http://www.brenda-enzymes.org) (29) for enzyme inhibitors.

16. Isotopically labeled amino acid, due to its small amount, should be added as a solution form in order to minimize its loss during transfer.

17. Optimal concentration of IPTG for protein induction also has to be determined empirically for each vector–host combination to obtain optimal protein yield.

18. Sometimes you may notice some residual insoluble unlabeled amino acids at the bottom of the centrifuge bottles beneath the bacterial cell pellet. Care should be taken not to bring out the insoluble amino acids to a clean tube.

Acknowledgments

This work was supported in part by grants-in-aid from JST-ERATO and the Ministry of Education, Culture, Sports, Science, and Technology to MY. TT was supported by the 21st Century COE Program and the Protein 3000 project.

References

1. Dyson HJ, Wright PE (2005) Intrinsically unstructured proteins and their functions. Nat Rev Mol Cell Biol 6:197–208

2. Senn H, Eugster A, Otting G et al (1987) ^{15}N-labeled P22 c2 repressor for nuclear magnetic resonance studies of protein-DNA interactions. Eur Biophys J 14:301–306

3. Muchmore DC, McIntosh LP, Russell CB et al (1989) Expression and nitrogen-15 labeling of proteins for proton and nitrogen-15 nuclear magnetic resonance. Methods Enzymol 177:44–73

4. McIntosh LP, Dahlquist FW (1990) Biosynthetic incorporation of ^{15}N and ^{13}C for assignment and interpretation of nuclear magnetic resonance spectra of proteins. Q Rev Biophys 23:1–38

5. Lee KM, Androphy EJ, Baleja JD (1995) A novel method for selective isotope labeling of bacterially expressed proteins. J Biomol NMR 5:93–96

6. Clore GM, Gronenborn AM (1991) Structures of larger proteins in solution: three- and four-

dimensional heteronuclear NMR spectroscopy. Science 252:1390–1399

7. Bax A (1994) Multidimensional nuclear magnetic resonance methods for protein studies. Curr Opin Struct Biol 4:738–744

8. Kay LE (1995) Field gradient techniques in NMR spectroscopy. Curr Opin Struct Biol 5:674–681

9. Sattler M, Schleucher J, Griesinger C (1999) Heteronuclear multidimensional NMR experiments for the structure determination of proteins in solution employing pulsed field gradients. Prog Nucl Magn Reson Spectrosc 34:93–158

10. Wider G, Wüthrich K (1999) NMR spectroscopy of large molecules and multimolecular assemblies in solution. Curr Opin Struct Biol 9:594–601

11. Strauss A, Bitsch F, Cutting B et al (2003) Amino-acid-type selective isotope labeling of proteins expressed in Baculovirus-infected insect cells useful for NMR studies. J Biomol NMR 26:367–372

12. Zubay G (1973) In vitro synthesis of protein in microbial systems. Annu Rev Genet 7:267–287

13. Spirin AS, Baranov VI, Ryabova LA et al (1988) A continuous cell-free translation system capable of producing polypeptides in high yield. Science 242:1162–1164

14. Kigawa T, Muto Y, Yokoyama S (1995) Cell-free synthesis and amino acid-selective stable isotope labeling of proteins for NMR analysis. J Biomol NMR 6:129–134

15. Morita EH, Shimizu M, Ogasawara T et al (2004) A novel way of amino acid-specific assignment in ^{1}H-^{15}N HSQC spectra with a wheat germ cell-free protein synthesis system. J Biomol NMR 30:37–45

16. Kainosho M, Torizawa T, Iwashita Y et al (2006) Optimal isotope labelling for NMR protein structure determinations. Nature 440:52–57

17. Staunton D, Schlinkert R, Zanetti G et al (2006) Cell-free expression and selective isotope labelling in protein NMR. Magn Reson Chem 44:S2–S9

18. Schwarz D, Dötsch V, Bernhard F (2008) Production of membrane proteins using cell-free expression systems. Proteomics 8:3933–3946

19. Tyler RC, Aceti DJ, Bingman CA et al (2005) Comparison of cell-based and cell-free protocols for producing target proteins from the *Arabidopsis thaliana* genome for structural studies. Proteins 59:633–643

20. 5 PRIME RTS application manual for cell-free protein expression. http://www.5prime.com/media/430498/5%20prime%20rts%20application%20manual%20for%20cell-free%20protein%20expression.pdf. Accessed 30 Jan 2010

21. Tong KI, Katoh Y, Kusunoki H et al (2006) Keap1 recruits Neh2 through binding to ETGE and DLG motifs: characterization of the two-site molecular recognition model. Mol Cell Biol 26:2887–2900

22. Tong KI, Yamamoto M, Tanaka T (2008) A simple method for amino acid selective isotope labeling of recombinant proteins in *E. coli*. J Biomol NMR 42:59–67

23. Yamazaki T, Otomo T, Oda N et al (1998) Segmental isotope labeling for protein NMR using peptide splicing. J Am Chem Soc 120:5591–5592

24. Otomo T, Ito N, Kyogoku Y et al (1999) NMR observation of selected segments in a larger protein: central-segment isotope labeling through intein-mediated ligation. Biochemistry 38:16040–16044

25. Xu R, Ayers B, Cowburn D et al (1999) Chemical ligation of folded recombinant proteins: segmental isotopic labeling of domains for NMR studies. Proc Natl Acad Sci USA 96:388–393

26. Sambrook J, Russell D (2001) Molecular cloning: a laboratory manual, 3rd edn. Cold Spring Harbor Laboratory Press, New York

27. Seidman CE, Struhl K, Sheen J et al (1997) Introduction of plasmid DNA into cells. In: Ausubel FM, Brent R, Kingston RE et al (eds) Current protocols in molecular biology. Wiley, New York

28. Keseler IM, Bonavides-Martinez C, Collado-Vides J et al (2009) EcoCyc: a comprehensive view of *Escherichia coli* biology. Nucleic Acids Res 37:D464–D470

29. Scheer M, Grote A, Chang A et al (2011) BRENDA, the enzyme information system in 2011. Nucleic Acids Res 39:D670–D676

INDEX

Vladimir N. Uversky and A. Keith Dunker (eds.), *Intrinsically Disordered Protein Analysis:*
Volume 2, Methods and Experimental Tools, Methods in Molecular Biology, vol. 896,
DOI 10.1007/978-1-4614-3704-8, © Springer Science+Business Media New York 2012

Printed by Publishers' Graphics LLC
MO20120903-088